Joint/Interage SMARTBOOK

thelightnin

Third Edition (JIA1-3)

joint strategic & operational PLANNING

Planning for Planners (Third Edition)

The Lightning Press
Michael A. Santacroce

The Lightning Press

2227 Arrowhead Blvd.
Lakeland, FL 33813
24-hour Voicemail/Fax/Order: 1-800-997-8827
E-mail: SMARTbooks@TheLightningPress.com

www.TheLightningPress.com

(JIA1-3) Joint/Interagency SMARTbook 1: Joint Strategic & Operational Planning
Planning for Planners (3rd Edition)

Copyright © 2023 The Lightning Press

ISBN: 978-1-935886-93-8

JIA1-3 is the third edition of Joint/Interagency SMARTbook 1: Joint Strategic & Operational Planning (Planning for Planners), completely reorganized and updated with the latest joint publications for 2023.

Printed and bound in the United States of America.

View, download FREE samples and purchase online:
www.TheLightningPress.com

(JIA1-3)
Note to Readers

The dawn of the 21st Century presents an increasingly complex global security environment. Within this environment United States national interest, citizens, and territories are threatened by regional instability, failed states, increased weapons proliferation, global terrorism, unconventional threats and challenges from adversaries in every operating domain. If we are to be successful as a nation, we must embrace the realities of this environment and operate with clarity from within. It is this setting that mandates a flexible, adaptive approach to planning and an ever-greater cooperation between all the elements of national power, supported by and coordinated with that of our allies and various intergovernmental, nongovernmental and regional security organizations. It is within this chaotic environment that planners will craft their trade.

Joint/Interagency SMARTbook 1: Joint Strategic & Operational Planning (Planning for Planners), 3rd Edition (JIA1-3), was developed to assist planners at all levels in understanding how to plan within this environment utilizing the Joint Planning Process; an orderly, logical, analytical progression enabling planners to sequentially follow it to a rational conclusion. By utilizing this planning process, which is conceptually easy to understand and applicable in all environments, any plan can come to life. Paramount to planning is flexibility. The ultimate aspiration of this book is to help develop flexible planners who can cope with the inevitable changes that occur during the planning process in any environment.

Planning for Planners has been utilized since 2007 by war colleges, joint staffs, Services, combatant commands and allies as a step-by-step guide to understanding the complex world of global planning and force management.

JIA1-3 is the third edition of Joint/Interagency SMARTbook 1: Joint Strategic & Operational Planning (Planning for Planners), completely reorganized and updated with the latest joint publications for 2023. At 420-pgs, JIA1-3 is designed to give the reader a thorough understanding of the Joint Planning and Execution Process, where the Joint Planning Process resides. Topics and chapters include planning fundamentals, planning functions, global force management, JIPOE and IPIE, joint planning process, plan/order development, execution functions, and annexes.

SMARTbooks - DIME is our DOMAIN!

SMARTbooks: Reference Essentials for the Instruments of National Power (D-I-M-E: Diplomatic, Informational, Military, Economic)! Recognized as a "whole of government" doctrinal reference standard by military, national security and government professionals around the world, SMARTbooks comprise a comprehensive professional library.

SMARTbooks can be used as quick reference guides during actual operations, as study guides for educational and professional development courses, and as lesson plans and checklists in support of training. Visit **www.TheLightningPress.com**.

Introduction:
Planning for Planners

The criteria for deciding to employ United States (U.S.) military forces exemplifies the dynamic link among the people, the government, and the military. The responsibility for the conduct and use of U.S. military forces is derived from the people and loaned to the government. The people of the U.S. do not take the commitment of their armed forces lightly. They charge the government to commit forces, our fathers, mothers, sons and daughters, only after due consideration of the range of options and likely outcomes. Moreover, the people expect the military to accomplish its missions in compliance with national values. The American people expect decisive victory and abhor unnecessary casualties. They prefer quick resolution of conflicts and reserve the right to reconsider their support should any of these conditions not be met. They demand timely and accurate information on the conduct of military operations.

> *"True genius resides in the capacity for evaluation of uncertain, hazardous, and conflicting information."*
>
> Winston Churchill

The Department of Defense (DOD) commits forces only after appropriate direction from the President and in support of national strategy. The national strategy of the U.S. dictates where, when, and with what means the armed forces will conduct military campaigns and operations. The necessity to plan and conduct joint and combined operations across the operational continuum dictates a comprehensive understanding of the military strategy of the U.S., and proficiency in current Service and joint *doctrine*.

> *"One should know one's enemies, their alliances, their resources and nature of their country, in order to plan a campaign. One should know what to expect of one's friends, what resources one has, and foresee the future effects to determine what one has to fear or hope from political maneuvers."*
>
> Frederick the Great
> *Instructions for His Generals, 1747*

Never static, always dynamic, *"doctrine"* is firmly rooted in the realities of current capabilities. At the same time, it reaches out with a measure of confidence to the future. Doctrine captures the lessons of past wars, reflects the nature of war, conflict and crisis in its own time, and anticipates the intellectual and technological developments that will ensure victory now and in the future.

Doctrine derives from a variety of sources that profoundly affect its development: strategy, history, technology, the nature of the threats the nation and its armed forces face, inter-service relationships, and political decisions that allocate resources and designate roles and missions. Doctrine seeks to meet the challenges facing the armed forces by providing the guidance to deal with the range of threats to which its elements may be exposed. It reflects the strategic context in which armed forces will operate, sets a marker for the incorporation of developing technologies, and optimizes the use of all available resources. It also incorporates the lessons learned from the many missions, operations and campaigns of the U.S.

Scenario

It's a holiday weekend and you're new on a Joint Planning staff for a Combatant Command and your boss, the PlansO, is enjoying leave. It's a typical 1800 on a Friday and the Chief of Staff walks in as your headed out the door. The Chief relays with "some urgency" the following Warning Order to you:

[A magnitude 9.1 - 9.3 earthquake with its epicenter off the west coast of northern Sumatra at Coordinates: 3.316°N 95.854°E, occurred at 00:58 UTC. News reports it is the third largest earthquake ever recorded. It's reported that an extensive series of tsunamis up to 100 feet high were created by the earthquake and have flooded communities along the Indian Ocean. At least 15 independent countries to include Indonesia, Sri Lanka, India, Thailand, Bangladesh, Maldives, Malaysia, Myanmar, Madagascar, Somalia, Kenya, Tanzania, Seychelles, South Africa, and Yemen are affected. Sources indicate the possibility of over 250,000 casualties with millions displaced and homeless. The President wants to assist and needs options. Have an initial mission analysis brief for the CCDR by 0700. The CCDR will brief the Chairman, Secretary and President following your brief so make it clear, succinct, and have several separate courses of action with differing degrees of assistance. Bring the State Department into your planning. They'll be a tremendous resource of country, state and local populations. They may have people on the ground already. Get your team together and lets get to work!

What do you do?
Where do you begin?

You suddenly feel the full affect of the proverbial "planning fire hose."

#1- Take a breath.
#2- Pick up your well-worn and dog-eared *Planner's SMARTbook* and get to work.
#3- Delegate!

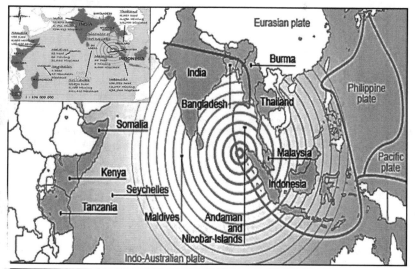

Continued on next page

Introduction (Cont.)

Continued from previous page

Baron von Steuben's 1779 "Regulations for the Order and Discipline of the Troops of the United States" was not penned in a setting of well-ordered formations and well-disciplined troops but, rather, at a time of turmoil during a winter at Valley Forge. Baron von Steuben's doctrine, maybe our first written doctrine, set forth principles and created a discipline that went on to defeat the greatest army on the face of the earth. This doctrine, written over 240 years ago, and followed by others, has led to a highly professional armed force that generations later stands foremost in the world. Doctrine reflects the collective wisdom of our armed forces against the background of history and it reflects the lessons learned from recent experiences and the setting of today's strategic and technological realities. It considers the nature of today's threats and tomorrow's challenges.[1]

Doctrinal principles set forth in planning are developed and written as the starting point for any variation or deviation from the planning process. One must understand doctrine prior to digressing from it. Doctrine should set forth principles and precious little more.[2] With that thought in mind Planning for Planners was designed to promulgate information from source documents, best practices, lessons learned and common sense from where the "principles" of Joint doctrine depart. This book explains and simplifies Joint planning and the often misunderstood and complex world of global force management.

Joint/Interagency SMARTbook 1 will assist planners at all levels with these challenges. It will furnish the planner with an understanding of doctrine and the intricate world of global planning. The ultimate aspiration is to develop planners who can cope with the inevitable change that occurs during the planning process.

Joint Planning and Variables

Joint planning is the overarching process that guides us in the development of plans for the employment of forces and capabilities within the context of national strategic objectives and national defense/military strategy to shape events, meet contingencies, and respond to unforeseen crises.

The Joint Planning Process consists of a set of logical steps to analyze a mission, develop and compare potential courses of action, select the best course of action, and produce a plan or order.

This planning process underpins planning at all levels and for missions across the full range of contingencies. It applies to all planners and helps them organize their planning activities, share a common understanding of the method, purpose and end state and to develop effective and executable plans and orders.

Continued from previous page

Planning provides an awareness and opportunity to study potential future events among multiple alternatives in a controlled environment. By planning we can evaluate complex systems and environments allowing us to break these down into small, manageable segments for analysis, assisting directly in the increased probability of success. In this way, deliberately planning for campaigns and contingencies allows us to manage identified risks and influence the operational environment in which we have chosen to interact, in a deliberate way. The plans generated in this process represent actions to be taken if an identified risk occurs or a trigger event has presented itself.

[1] *FM 100-5 Operations, Headquarters, Department of the Army*

[2] *Dr. Douglas V. Johnson II, Strategic Studies Institute, Doctrine That Works, www. StrategicStudiesInstitute.army.mil/pdffiles/pub724.pdf*

The variance in any plan is the constant change in the operational environment (system). Whether a contingency or crisis scenario, we plan in a chaotic environment. In the time it takes us to plan, the likelihood that the operational environment has changed is a certain, whether by action or inaction, affecting the plan (i.e., assumptions change or are not validated, leaders change, the operational environment fluctuates, apportionment tables are poor assumptions, disputed borders fluctuate, weather changes the rules, plans change at contact, enemy gets a vote, etc.).

Variables are hard to predict because each environment and situation has their own unique challenges which can certainly affect an orderly plan. Given the size and scope of an operational environment, a plan can only anticipate, or forecast, for a short duration without being updated. This is known as the plan's horizon. In a fluid crisis situation, the plan's horizon may be very short and contain greater risk, causing the planner to constantly re-evaluate and update the plan. Inversely, for a campaign or contingency plan, the plan's horizon may be relatively static with less risk allowing time for greater analysis. The number of variables within the operational environment and the interactions between those variables and known components of the operational environment increases exponentially with the number of variables, thus potentially allowing for many new and sometimes subtle planning changes to emerge.

As an example of a plan's horizon, or stability, let's look at an environment that constantly influences us, the weather:

> *A forecaster endeavors to anticipate the path of a tropical cyclone and utilizes historical models and probabilities to predict the tropical cyclone's path and warn residents. When a low-pressure area first forms and the storm begins to take shape along the equator, forecasters are working within a complex environment with constant and multiple variables (i.e., winds, temperatures, currents, pressures, etc.) and few facts (i.e., exact location at this moment, jet stream location, ocean temperatures, surface winds, etc.). As variables amplify and the storm begins to move, the storm's horizon shifts yet again, and the forecaster updates the assessment. Over days of surveillance, gathering information, updating, and studying the variables, the actual track of the storm begins to emerge and the storms horizon becomes more durable and predictable. The forecaster continuously narrows the storm's estimated track, eventually forecasting with some certainty the tropical cyclone's landfall.*

Planners employ the same technique by utilizing current knowledge of the operational environment to anticipate events, calculate what those may be by means of an in-depth analysis, update, and plan accordingly. But always remember, plans are orderly; probabilities and variables are not. Just as a tropical storm has a self-organizing phase within its environment, so must the planner.

So, the challenge is how to plan within an environment with continuously changing and emerging variables. The planner must understand that every plan is unique and never as perfect as you want it. There are too many variables. But with constant awareness each iteration of the plan will improve the prospect of success as the variables become known and are planned for.

Simplicity should be the aspiration for every plan. Prepare clear, uncomplicated plans and concise orders to ensure a thorough understanding. A plan need not be more complicated than the underlying principles which generate it.

(JIA1-3)
About the Author

Colonel (Ret) Michael A. Santacroce has 35 years of joint and interagency experience working within the Department of Defense as a Joint Staff, Combatant Command and Service Planner. As Faculty and Chair for the Joint Advanced Warfighting School, Campaign Planning and Operational Art, Mike taught advanced planning to leaders from all branches of the DoD, government agencies and our allies. His current SMARTbook, Planning for Planners, walks the prospective or advanced planner through joint strategic and operational planning as well as the complex world of global force management.

During his Marine Corps career Mike served in a multiple of demanding leadership, senior staff, strategic and operational planning positions. As a Marine aviator he flew the AV-8B Harrier Jump Jet and participated in operations globally. Mike served as the Operations Officer of Marine Air Weapons and Tactics Squadron One (MAWTS-1), commanded a Marine Harrier Squadron (VMA-214 Blacksheep) and later led a Marine Air Group (Forward) for combat operations in Iraq. A seasoned military professional and teacher, Mike has a unique understanding of operations and planning at all levels. Mike retired with more than 30 years of military service.

> *"The inspiration of a noble cause involving human interests wide and far, enables men to do things they did not dream themselves capable of before, and which they were not capable of alone."*
> *Joshua L. Chamberlain, October 3, 1889. Monument dedication ceremony, Gettysburg, Pa.*

Today's preparation determines tomorrow's achievements. Dedicated to all planners; may this work assist you in your planning endeavors'.

> *Joint/Interagency SMARTbook 1: Joint Strategic & Operational Planning (Planning for Planners) is reviewed continually and updated as required. Point of contact is the author, Col (Ret) Mike Santacroce, USMC, at mike.santa@yahoo.com.*

(JIA1-3)
Table of Contents

Chap 1

Planning
Fundamentals

Planning Activities & Functions

Global Force Management (GFM)

JIPOE & IPIE

Joint Planning Process (JPP)

Plan/Order Development

Execution

Annexes

Index

I. Strategic Organization

> "In pursuit of that future, we will look at the world with clear eyes and fresh thinking. We will promote a balance of power that favors the United States, our allies, and our partners. We will never lose sight of our values and their capacity to inspire, uplift, and renew. Most of all, we will serve the American people and uphold their right to a government that prioritizes their security, their prosperity, and their interests."
>
> *President Donald Trump, 2017 NSS*

1. Background

a. <u>Civilian Control of the Military</u>. Since the founding of the nation, civilian control of the military has been an absolute and unquestioned principle. The Constitution incorporates this principle by giving both the President and Congress the power to ensure civilian supremacy. The Constitution establishes the President as the Commander-in-Chief, but gives Congress the power "to declare war," to "raise and support Armies – provide and maintain a Navy – (and) to make rules for the government and regulation of the land and Naval forces."

b. <u>Joint Organization before 1900</u>. As established by the Constitution, coordination between the War Department and Navy Department was effected by the President as the Commander-in-Chief. Army and Naval forces functioned autonomously with the President as their only common superior. Despite Service autonomy, early American history reflects the importance of joint operations. Admiral MacDonough's Naval operations on Lake Champlain were a vital factor in the ground campaigns of the War of 1812. The joint teamwork displayed by General Grant and Admiral Porter in the Vicksburg Campaign of 1863 stands as a fine early example of joint military planning and execution. However, instances of confusion, poor inter-Service cooperation and lack of coordinated, joint military action had a negative impact on operations in the Cuban campaign of the Spanish-American War (1898). By the turn of the century, advances in technology and the growing international involvement of the United States (U.S.) required greater cooperation between the military departments.

c. <u>Joint History through World War I</u>. As a result of the unimpressive joint military operations in the Spanish-American War, in 1903 the Secretary of War and the Secretary of the Navy created the Joint Army and Navy Board charged to address "all matters calling for cooperation of the two Services." The Joint Army and Navy Board was to be a continuing body that could plan for joint operations and resolve problems of common concern to the two Services. Unfortunately, the Joint Board accomplished little, because it could not direct implementation of concepts or enforce decisions, being limited to commenting on problems submitted to it by the secretaries of the two military departments. It was described as "a planning and deliberative body rather than a center of executive authority." As a result, it had little or no impact on the conduct of joint operations during the First World War. Even as late as World War I, questions of seniority and command relationships between the Chief of Staff of the Army and American Expeditionary Forces in Europe were just being resolved.

d. <u>Joint History through World War II</u>. After World War I, the two Service secretaries agreed to reestablish and revitalize the Joint Board. Membership was expanded to six: the chiefs of the two Services, their deputies, the Chief of War Plans Division for the Army and Director of Plans Division for the Navy. More importantly, a working staff (named the Joint Planning Committee) made up of members of the Plans Divisions of both Service staffs was authorized. The new Joint Board could initiate recommendations on its own. Unfortunately, the 1919 board was given no more legal authority or responsibility than its 1903 predecessor; and, although its 1935 publication, *Joint Action Board of the Army and Navy*, gave some guidance for the unified operations of World War II, the board itself was not influential in the war. The board was officially disbanded in 1947.

    e. <u>Goldwater–Nichols Department of Defense Reorganization Act of October 4, 1986</u>. The Goldwater–Nichols Department of Defense Reorganization Act of October 4, 1986 Pub. L. 99–433, made the most sweeping changes to the U.S. Department of Defense (DOD) since the department was established in the National Security Act of 1947 by reworking the command structure of the U.S. military.[1] The Goldwater–Nichols Act was an attempt to fix problems caused by inter-service rivalry, which had emerged during the Vietnam War, contributed to the catastrophic failure of the Iranian hostage rescue mission in 1980, and which were still evident in the invasion of Grenada in 1983.[2] It increased the powers of the Chairman of the Joint Chiefs of Staff (CJCS) and streamlined the military chain of command, which now runs from the President through the Secretary of Defense (SecDef) directly to combatant commanders (CCDRs), bypassing the Service Chiefs. The Act further outlined the responsibilities of those CCDRs, giving them total authority to accomplish assigned missions within their geographic areas of responsibility.[3] The Service Chiefs were assigned to an advisory role to the President and the SecDef as well as given the responsibility for training and equipping personnel for the unified combatant commands.

    Five years after the Goldwater-Nichols Legislation the U.S. military successfully conducted Operation Desert Storm and other associated operations (such as Operation Provide Comfort). The clarification of the operational chain of command, as well as the advances in jointness that were made as a result of the Goldwater-Nichols legislation, were viewed by many as instrumental to that success.[4]

    f. <u>Strategic Context</u>. Operation IRAQI FREEDOM (OIF) marked the most integrated joint force and joint campaign American armed forces had ever conducted up to that time. The OIF campaign is marked with a number of firsts. Arguably, it is the first "jointly" coherent campaign since the Korean War. American joint forces executed a large-scale, complex operation while simultaneously continuing active operations in Afghanistan, the Balkans, and in support of Homeland Defense. In OIF, a combined and joint land component commanders (CDR) directed all ground operations for the first time since the Eighth Army did so in the Korean War. Not since World War II have the armed forces of the U.S. operated in multiple theaters of war while simultaneously conducting security operations and support operations in several other theaters.

[1] *The Perfect Storm, The Goldwater-Nichols Act and Its Effect on Navy Acquisition/ Charles Nemfakos • Irv Blickstein • Aine Seitz McCarthy • Jerry M. Sollinger*

[2] *Cole, Ronald H. (1999). "Grenada, Panama, and Haiti: Joint Operational Reform" (PDF). Joint Force Quarterly (20 (Autumn/Winter 1998-99)): 57–74*

[3] *Richard W. Stewart, ed. (2005). "Chapter 12: Rebuilding the Army Vietnam to Desert Storm." American Military History, Volume II. United States Army Center of Military History*

[4] *Leighton W. Smith, "A Commander's Perspective," as found in, Dennis J. Quinn (ed), The Goldwater-Nichols DOD Reorganization Act: A Ten-Year Retrospective (Washington, DC: National Defense University Press, 1999) p. 29. See also Clark A. Murdock, Beyond Goldwater-Nichols: Defense Reform for a New Strategic Era, Phase 1 Report (Washington, DC: Center for Strategic and International Studies Press, 2004). http://csis.org/files/media/ csis/pubs/ bgn_ph1_report.pdf, p. 14*

OIF forces also employed emerging new concepts that had just been incorporated into the body of joint doctrine. Noteworthy joint coordination from OIF not imagined a decade earlier includes: the unprecedented degree of air-ground coordination and integration, coalition air forces shaping the fight allowing for rapid dominance on the ground, the establishment of the Coalition Forces Land Component Command (CFLCC), the "running start," integration of precision munitions with ground operations, supported by a largely space-based command and control network, effective integration of artillery and attack aviation, and air- and sea-launched precision-guided munitions (PGMs) and cruise missile strikes responding rapidly to targets developed by improved intelligence, surveillance, and reconnaissance systems. These all represent the maturation of *joint doctrine* developed since Goldwater-Nichols and tested through joint simulations and training. The Goldwater-Nichols Act of 1986 enabled combat operations to occur in 2003 in ways only imagined a short decade earlier.

2. The Strategic and Security Environment

The strategic environment has shifted dramatically. Since the enactment of the Goldwater-Nichols legislation, a number of important historical events have taken place, starting with the end of the Cold War. Subsequently, the U.S. performed crisis management and contingency operations globally, in theaters including Iraq, the Balkans, Somalia, and Colombia. After the September 11, 2001, terror attacks, the U.S. undertook major counter-insurgency campaigns in Iraq and Afghanistan, as well as a number of smaller operations as part of its "global war on terror."

The international security environment was already demanding when the Goldwater-Nichols legislation was enacted, yet most observers agree it has become significantly more complex and unpredictable in recent years.[5] This is challenging the U.S.to respond to an increasingly diverse set of requirements.[6] As evidence, observers point to a number of events, including (but not limited to) the rise of the Islamic State, including its military successes in northern Iraq and Syria; the strength of drug cartels in South and Central America; Russian warfare in Ukraine; heightened North Korean aggression; Chinese "island building" in the South China Sea; terror attacks in Europe; the ongoing civil war in Syria and its attendant refugee crisis to name a few.

Today's security environment is not unlike those of historic times. The CDRs during those eras also considered the enemy extremely complex and fluid with continually changing coalitions, alliances, partnerships, and new threats constantly appearing and disappearing.

With the national and transnational threats we face today our political and military leaders conduct operations in an ever-more complex, interconnected, and increasingly global operational environment (OE). This increase in the scope of the OE may not necessarily result from actions by the confronted adversary alone, but is likely to result from other adversaries exploiting opportunities as a consequence of an overextended or distracted U.S. or coalition. These adversaries encompass a variety of actors from transnational organizations to states or even ad hoc state coalitions and individuals.

---

[5] See, for example, *CRS Report R43838, A Shift in the International Security Environment: Potential Implications for Defense—Issues for Congress*, by Ronald O'Rourke ; James Clapper "Worldwide Threat Assessment of the U.S. Intelligence Community," Testimony before the Senate Armed Services Committee, February 9, 2016

[6] U.S. Department of Defense, *Department of Defense Press Briefing by Deputy Secretary Work and Gen. Selva on the FY2017 Defense Department Budget Request in the Pentagon Press Briefing Room*, February 9, 2016, http://www.defense.gov/News/News-Transcripts/Transcript-View/Article/653524/department-of-defense-press-briefing-by-deputy-secretary-work-and-gen-selva-on

A central challenge as noted in the Defense Strategy Review (DSR) is the reemergence of long-term, strategic competition by what the National Security Strategy (NSS) classifies as revisionist powers.[7] Along with these revisionist powers rogue regimes are destabilizing regions through their pursuit of nuclear weapons or sponsorship of terrorism. Both revisionist powers and rogue regimes are competing across all dimensions of power. They have increased efforts short of armed conflict by expanding coercion to new fronts, violating principles of sovereignty, exploiting ambiguity, and deliberately blurring the lines between civil and military goals.

The Joint Force faces two persistent realities. **First**, the security environment is always in flux. Change is relentless and occurs in all aspects of human endeavor. Ideas about how human beings should govern one another emerge, spread, and then fade away. Advances in science and technology progress and proliferate. Countries and political groups simultaneously cooperate and compete based on their relative power, capabilities, interests, and ideals. Change in the security environment occurs at an irregular pace, and over time small changes compound to shatter our assumptions. **Second**, the pursuit of political objectives through organized violence is and will remain a feature of the security environment. Strife, conflict and war are certain to endure into the foreseeable future.[8]

Today, every domain is contested—air, land, sea, space and cyberspace. We face an ever more lethal and disruptive battlefield, combined across domains, and conducted at increasing speed and reach—from close combat, throughout overseas theaters, and reaching to our homeland. Some competitors and adversaries seek to optimize their targeting of our battle networks and operational concepts, while also using other areas of competition short of open warfare to achieve their ends (e.g., information warfare, ambiguous or denied proxy operations, and subversion).[9]

To prepare the U.S. for today's threats and contingencies we have, over time, established a system of checks and balances to include numerous governmental organizations that are involved in the implementation of U.S. security policy. However, constitutionally, the ultimate authority and responsibility for the national defense rests with the President.

3. Strategy

The objective of strategy, in the modern sense, is to serve policy—the positions of governments and others cooperating, competing, or waging war in a complex environment. National policy articulates national objectives. National policy is broad guidance statements adopted by national governments in pursuit of national objectives. The ultimate goal of strategy is to achieve policy objectives by maintaining or modifying elements of the strategic environment to serve those interests.

Strategy formulation must consider the strategic environment (e.g., geography, character, and relationship of political entities and their interests, and resources) subject to norms and constants present. These factors present themselves differently in each strategic interaction and exert considerable influence on a particular strategic situation. Additionally, these factors may change during execution, necessitating revision of the strategy.

In its simplest expression, strategy determines what needs to be accomplished, the methods to accomplish it, and the resources required by those methods. A comprehensive and effective strategy answers four basic questions:[10]

- What are the desired *ends*?
- What are the *ways* to get there?

[7] *Joint Operating Environment*
[8] *National Defense Strategy*
[9] *Defense Strategy Review*
[10] *Joint Doctrine Note 2-19, Strategy*

- What *means* or resources are available?
- What are the *risks* associated with the strategy?

This ends, ways, means model is the basic construct of modern strategy, but it alone is inadequate to turn ideas into action. Strategy is both an iterative process and a product—the reflective synergy of art and science creating a coherent bridge from the present to the future, enabling the translation of ideas into action to get what you want while addressing potential risks to the nation.

a. National Strategy. National strategy articulates broadly developed interests and identifies threats, resources, and policies. National strategy usually does not address particular operational and tactical ends, and does not consider military power in isolation from other sources of national power. It defines the direction for the entire country and includes all the instruments of national power. In brief, a national strategy is a country's overarching "strategy of strategies." National Strategy secures and advances a nation's long-term, enduring, core interests over time. The most common expression of national strategy for U.S. military strategists are the President's NSS and policy guidance issued through the National Security Council (NSC). These provide a broad strategic context for employing military capabilities in concert with other instruments of national power. In the ends, ways, and means construct, the NSS provides the ends. This national strategy is anchored in the national interests that support a strategic vision of the role of the U.S. National strategy also reflects societal dynamics and their underlying enduring values and beliefs.

Emerging challenges in the security environment that require the joint force to operate in all domains and across multiple regions simultaneously have placed a new importance on national strategy. While the dictates of national policy translate into different strategic objectives for different theaters, that policy guidance requires the determination of strategic objectives at the national level.

b. Strategic Uses of Military Force. The U.S. leverages all instruments of national power to pursue its national interests. Reinforcing America's traditional tools of diplomacy, DOD provides military options to ensure the President and our diplomats negotiate from positions of strength. DOD is in a supporting role when the military instrument of national power is not the predominant instrument for the strategy. When directed or if the other instruments of national power prove insufficient, the military becomes the nation's primary instrument. In either case, the military facilitates and supports the application of the other instruments. Whether in a primary or supporting role, the military has the ability to use its strengths, assets, and capabilities in a range of strategic ways in order to achieve national aims.

(1) National Defense Strategy. Congress directs in law (Title 10, United States Code [USC], Section 113) that the SecDef develop a National Defense Strategy (NDS) every four years to describe how the DOD will contribute to the execution of the president's NSS. The NDS translates the national interests and objectives in the NSS into prioritized defense objectives for DOD and articulates DOD's approach for developing and employing military forces and Departmental resources to protect and promote U.S. national security interests.

(a) The NDS is a classified strategy with an unclassified summary, and includes:

- Assumed strategic environment, including critical and enduring threats.
- Strategies to counter threats and provide for national defense.
- Priority missions of DOD.
- The roles and missions of the Armed Forces assumed force planning scenarios and constructs.
- Force size and shape, posture, defense capabilities, readiness, infrastructure, organization, personnel, and technological innovation, including major investments for the budget.

(b) In addition to these statutory requirements, the defense strategy often covers other institutional aspects such as the defense industrial base, national-level logistics, basing, agreements, and organizational reform. The NDS is the Secretary's preeminent strategic document for DOD providing guidance on force employment, force planning, force design, posture, programming, and other activities. It provides the framework and prioritization for all subordinate DOD strategic guidance and activities, and serves as the launch point for structured DOD strategy assessments and deliberations to ensure its implementation and adjustment as the environment evolves.

(2) National Military Strategy (NMS). The CJCS produces the NMS based on consultation with the CCDRs and members of the JCS. It serves as a strategic framework for how the armed forces will execute the overall policy goals laid out in the most recent NSS and NDS. Military strategy is the creation, employment, and articulation of the military instrument of national power to achieve policy objectives. Coherent and effective military strategy is essential to achieve specific objectives or sets of objectives that protect the national interests as conveyed by national policy and national strategy. Military strategy by inferring a rational order on reality, makes action by the joint force purposeful; without it, military activities could be ad hoc, incoherent, and potentially counterproductive.

(a) Unlike national strategy, the scope of a military strategy is limited to the military instrument of national power. To be effective it still must be integrated with the diplomatic, informational, and economic instruments. Only a comprehensive approach to strategy will result in effective outcomes. The ends in military strategy are a subset of the defense strategy's objectives, while the ways and means represent how the joint force will execute the defense strategy. The frameworks in a military strategy provide a lens for subsequent campaign and contingency planning. Risk in a military strategy is localized to the ends, ways, and means of the military strategy and, while related, may differ from risk in a defense strategy.

(b) Leaders develop strategies through the exercise of strategic art and science. The essence of strategic art is inductive—organizing and articulating in clean terms the complex interrelationship between national interests, policy, strategic ends, and practice in clear terms. The conduct of strategic art occurs at the strategic level of military activity, spanning national strategy, defense strategy, military strategy, and theater strategy. The interests and objectives in the security environment, and the organizational considerations inherent to implementation.

c. Strategies articulate a story that operates in a competitive space to bridge the present to the future within the duration of the strategy. Audiences of the strategic story include friendly forces, enemies, adversaries, allies, partners, and of special significance, a variety of other relevant actors, in both public and classified venues. The "plot" provides the conceptual basis for supporting military campaigns. To be effective, a military strategy should be clear, concise, and easily understood. This enables the strategy to be successfully translated into campaign plans. The design of the operations that form those campaigns shapes the perceptions, and ultimately behaviors, of relevant actors toward strategic success. The ability to visualize and conceptualize how strategic success can be achieved or supported by military means is fundamental to the application of operational art and operational design.

(a) Policy is not strategy. Policies are the stated positions of government, usually communicated in the form of ends, that policymakers direct to be attained. Policy also includes stated assumptions, available resources, and permissions that are allowed to the military strategist. A military strategist must identify the conditions in the strategic environment that generate those policy goals. In much the same way that tactics must support operations, military strategy necessarily serves national policy, while providing insights to the costs incurred in attaining those policy goals. Policy may require a strategy which reduces strategic risk, but may incur greater tactical risk. Good tactical execution serves strategy, but it cannot substitute for sound strategy and policy.

(b) Military strategy is the art and science of achieving a political objective through the military instrument of national power. Military strategy is not planning. Similarly, a strategy is not a campaign plan. It guides and directs, but is distinct from the products and activities of a campaign plan. A campaign plan organizes day-to-day operations of the joint force to achieve national objectives. It is also distinct from the products and activities that military organizations use to organize, resource, and shape their future capabilities.

4. Strategic Direction

The common thread that integrates and synchronizes the activities of the Joint Staff (JS), Combatant Commands (CCMDs), Services, and Combat Support Agencies (CSAs) is strategic direction. As an overarching term, strategic direction encompasses the processes and products (documents) by which the President, SecDef, and CJCS provide strategic vision and direction to the Joint force. Strategic direction is normally published in key documents, generally referred to as strategic guidance. As seen in Figure A these strategic guidance documents are the principle source for DOD Global Campaign Plans (GCPs), theater strategies, CCMD Campaign Plans (CCPs), operation plans, contingency plans, base plans, and CDRs' estimates. The CCDRs, provided with national direction and guidance, each prepares their strategy and campaign plans in the context of national security and foreign policy goals.

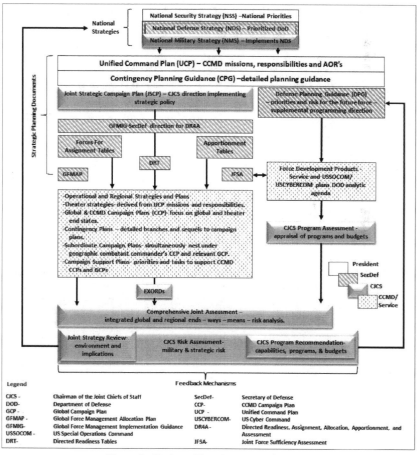

Figure A. Strategy, Planning and Resourcing

a. The President provides strategic guidance through the NSS, Presidential Policy Directive (PPD), executive orders and other strategic documents in conjunction with additional guidance and refinement from the NSC.[11]

b. The President and SecDef, through the CJCS, direct the national effort that supports combatant and subordinate CDRs. The principal forum for deliberation of national security policy issues requiring Presidential Decisions (PDs) that will directly affect the CCDR's actions is the NSC. Knowledge of the history and relationships between elements of the national security structure is essential to understanding the role of JS organizations.

c. The National Security Council System (NSC). DOD participation in the interagency process is grounded within the Constitution and established by law in the National Security Act of 1947 (NSA 47).

(1) The NSC is a product of NSA 47. NSA 47 codified and refined the interagency process used during World War II, modeled in part on Franklin D. Roosevelt's 1919 proposal for a "Joint Plan-Making Body" to deal with the overlapping authorities of the Departments of State, War, and Navy. Because of the diverse interests of individual agencies, previous attempts at interagency coordination failed due to lack of national-level perspectives, a staff for continuity, and adequate appreciation for the need of an institutionalized coordination process. Evolving from the World War II experience (during which the Secretary of State was not invited to War Council meetings), the first State-War-Navy Coordinating Committee was formed in 1945.

(a) From the earliest days of this nation the President has had the primary responsibility for national security stemming from Presidential constitutional powers both as Commander-in-Chief of the Armed Forces and Presidential authority to make treaties and appoint cabinet members and ambassadors. The intent of NSA 47 was to assist the President with respect to the integration of domestic, foreign, and military policies relating to national security. Most current U.S. Government (USG) interagency actions flow from these beginnings.

(b) Within the constitutional and statutory system, interagency actions at the national level may be based on both personality and process, consisting of persuasion, negotiation, and consensus building, as well as adherence to bureaucratic procedure.

(2) The NSC is the principal forum for deliberation of national security policy issues requiring Presidential decision. The NSC advises and assists the President in integrating all aspects of national security policy — domestic, foreign, military, intelligence, and economic (in conjunction with the National Economic Council).

(a) Together with supporting interagency working groups (some permanent and others ad hoc), high-level steering groups, executive committees, and task forces, the National Security Council System (NSCS) provides the foundation for interagency coordination in the development and implementation of national security policy. The NSC develops policy options, considers implications, coordinates operational problems that require interdepartmental consideration, develops recommendations for the President, and monitors policy implementation. The national security staff is the President's principal staff for national security issues and also serves as the President's principal arm for coordinating these policies among various government agencies. NSC documents are established to inform USG departments and agencies of Presidential actions.[12]

(b) Each administration typically adopts different names for its NSC documents. For example; the Reagan Administration used the terms NSDD (National Security Decision Directive) and NSSD (National Security Study Directive). The George H. W. Bush Administration used NSR (National Security Review) and NSD (National Security Directive), while

[11] *Joint Pub 5-0, Joint Planning*

[12] *JP 3-08, Interorganizational Coordination During Joint Operations*

the Clinton Administration used the terms Presidential Decision Directive (PDD) and Presidential Review Directive (PRD). The George W. Bush administration used National Security Presidential Directive (NSPD) and Homeland Security Presidential Directive (HSPD). The Obama Administration used the terms Presidential Policy Directive (PPD) and Presidential Study Directive (PSD). The Trump Administration used National Security Presidential Memoranda (NSPMs) and the Biden Administration utilizes National Security Memorandum (NSM) to promulgate Presidential decisions on national security matters. However, previously issued directives from prior administrations remain in effect unless and until rescinded or superseded.

(3) <u>National Security Council Membership</u>. The President chairs the NSC. The NSC has as its regular attendees (both statutory and non-statutory) the President, the Vice President, the Secretary of State, the Secretary of the Treasury, the Secretary of Defense, the Secretary of Homeland Security, the Secretary of Energy, and the Assistant to the President for National Security Affairs. The Director of Central Intelligence and the CJCS, as statutory advisors to the NSC, shall also attend NSC meetings. The Chief of Staff to the President and the Assistant to the President for Economic Policy are invited to attend any NSC meeting. The Counsel to the President is consulted regarding the agenda of NSC meetings, and attends any meeting when, in consultation with the Assistant to the President for National Security Affairs, it is deemed appropriate. The current administration has also for the first time included the ambassador to the United Nations, the director of the Office of Science and Technology Policy (OSTP) and the administrator of the U.S. Agency for International Development (USAID) as regular members.

(4) <u>NSC Organization.</u> The members of the NSC constitute the President's personal and principal staff for national security issues. The council tracks and directs the development, execution, and implementation of national security policies for the President, but does not normally implement policy. Rather, it takes a central coordinating or monitoring role in the development of policy and options, depending on the desires of the President and the National Security Advisor. There are three levels of formal interagency committees for coordinating and making decisions on national security issues; Principles, Deputies, and Interagency Policy Committees (Figure B). The advisory bodies include:

(a) <u>NSC Principals Committee (NSC/PC).</u> The NSC/PC is the senior Cabinet-level interagency forum for consideration of policy issues affecting national security. The Principals Committee meets at the call of and is chaired by the National Security Advisor.

(b) <u>NSC/Deputies Committee (NSC/DC).</u> The NSC/DC meets at the call of and is chaired by the Deputy National Security Advisor. It is the senior sub-Cabinet-level (deputy secretary-level) interagency forum for consideration of policy issues affecting national security. The NSC/DC prescribes and reviews the work of the NSC Interagency Policy Committee (NSC/IPC) and ensures that NSC/IPC issues have been properly analyzed and prepared for decision. The NSC/DC focuses on policy implementation and periodic reviews of the Administration's major foreign policy initiatives to ensure that they are being implemented in a timely and effective manner. The NSC/DC is also responsible for day-to-day crisis management, reporting to the NSC. Any NSC principal or deputy, as well as the National Security Advisor, may request a meeting of the NSC/DC in its crisis management capacity.

(c) <u>NSC/IPCs.</u> The NSC/IPCs are the main day-to-day action committees for interagency coordination of national security policy. NSC/IPCs manage the development and implementation of national security policies by multiple agencies of the USG, provide policy analysis for consideration by the more senior committees of the NSCS, and ensure timely responses to decisions made by the President. The NSC/IPCs is established at the direction of the NSC/DC, and is chaired by the NSC; at its discretion, the NSC/DC may add co-chairs to any NSC/IPC. The NSC/IPCs convenes on a regular basis to review and coordinate the implementation of Presidential decisions in their policy areas. Six NSC/IPCs are established for the following regions: Europe and Eurasia, Western Hemisphere, East Asia, South Asia, Near East and North Africa, and Africa.[13]

(d) During a rapidly developing crisis, the President may request the National Security Advisor to convene the NSC. The NSC reviews the situation, determines a preliminary COA, and tasks the Principals and Deputies Committees.

(e) Under more routine conditions, concerns focus on broader aspects of national policy and long-term strategy perspectives. National Security Presidential Directives outline specific national interests, overall national policy objectives, and tasks for the appropriate components of the executive branch.

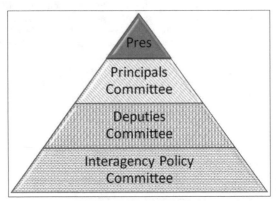

Figure B. NSC Committies

(5) <u>DOD Role in the National Security Council System</u>.

(a) Key DOD players in the NSCS come from within the Office of the Secretary of Defense (OSD) and the JS. The SecDef is a regular member of the NSC and the NSC/PC. The Deputy Secretary of Defense (DepSec) is a member of the NSC/DC. In addition to membership, an Under Secretary of Defense (USD) may chair an NSC/IPC.

(b) The NSCS is the channel for the CJCS to discharge substantial statutory responsibilities as the principal military advisor to the President, the SecDef, and the NSC. The CJCS regularly attends NSC meetings and provides advice and views in this capacity. The other members of the Joint Chiefs of Staff (JCS) may submit advice or an opinion in disagreement with that of the CJCS, or advice or an opinion in addition to the advice provided by the CJCS.

(c) The Military Departments which implement, but do not participate directly in national security policy-making activities of the interagency process, are represented by the CJCS.

(d) Of note and worth mentioning here are the geographic boundary differences be-tween the Department of State (DOS) Bureaus and the DOD geographic commands. It's important we recognize these seams and boundary differences to ensure smooth coordination between these two interagency partners, see Chapter 1-2, *Campaigning & Contingency Plans,* for details.

(6) <u>The Joint Staff Role in the National Security Council System</u>.

(a) Per Title 10, Section 155, the JS provides operational input and staff support through the Chairman (or designee) for policy decisions made by the OSD. It coordinates with the CCMDs, Services, and other agencies and prepares appropriate directives, such as warning, alert, and execute orders, for SecDef approval. This preparation includes definition of command and interagency relationships.

(b) When CCMDs require interagency coordination, the JS, in concert with the OSD, routinely facilitates that coordination.

[13] *National Security Presidential Directive 1 (NSPD-1)*

(c) Within the JS, the offices of the CJCS, Secretary of the JS, Intelligence (J2), Operations (J3), Logistics (J4), Plans and Policy (J5), and Operational Plans and Joint Force Development Directorates are focal points for NSC-related actions. The J3 provides advice on execution of military operations, the J4 assesses logistic implications of contemplated operations, and the J5 often serves to focus DOD on a particular NSC matter for policy and planning purposes and as the Joint Staff touchpoint for Promote Cooperation, a OSD/ Joint Staff program to coordinate plans with other agencies. Each of the JS directorates coordinates with the Military Departments to solicit Service input in the planning process. The Secretary may also designate one of the Services as the executive agent for direction and coordination of DOD activities in support of specific mission areas.

(7) The CCDRs' Role in the National Security Council System. Although CCDRs sometimes participate directly in the interagency process by directly communicating with committees and groups of the NSC system and by working to integrate the military with diplomatic, economic, and informational instruments of national power, the normal conduit for information between the President, SecDef, NSC, and a CCMD is the CJCS. CCDRs may communicate with the Deputies Committee during development of the POLMIL plan with the Joint Staff in a coordinating role.

5. Authorities and Responsibilities

Joint planning and execution incorporates a joint enterprise for the development, maintenance, assessment, and implementation of GCPs, CCPs and related contingency plans and orders prepared in response to Presidential, SecDef, or Chairman direction or requirements. This spans many organizational levels, including the interaction between the Secretary, CCDRs, coalition, and interagency which ultimately assists the President and Secretary to decide when, where, and how to commit U.S. military forces. Since this book examines policy and procedures that are used to develop and implement joint military plans and orders prepared in response to Presidential, SecDef, Under Secretary of Defense for Policy (USD(P)), CJCS, CCDR or other guidance it is only fitting we discuss the stakeholder responsibilities and activities. All these responsibilities and activities trace to a documented source of authority.

a. Purpose. This section describes the authorities and responsibilities of the Joint Planning and Execution Community (JPEC) stakeholders that spans the joint planning decision making.

b. United States Code (U.S.C.). The U.S.C. governs the functioning of the branches of the USG and military. The statutory authorities most relevant to joint planning and execution are as follows:[14]

(1) Title 6, U.S.C. governs Domestic Security including homeland security organization, infrastructure security, and national emergency management. It establishes the Department of Homeland Security (DHS) and prescribes its mission and authorities.

(2) Title 10, U.S.C. governs the U.S. Armed Forces and provides for the organization of the DOD, including the secretary, OSD, CJCS, JCS, JS, CCDRs, Secretaries of the Military Departments, Military Departments, Service Chiefs, Chief National Guard Bureau (NGB), and Directors of DOD Agencies and establishes statutory authorities and responsibilities. The organization of the U.S. Armed Forces and related executive departments is depicted in Figure C.

(3) Title 14, U.S.C. governs the duties of the U.S. Coast Guard (USCG). The USCG operates as a Service in the DHS, except when directed to operate as a Service in the Department of the Navy.

(4) Title 22, U.S.C. (Chapter 32, "Foreign Assistance) governs foreign relations and diplomatic conduct. It is relevant to joint planning and execution as the source of authority

---

[14] Titles 6, 10, 14, 22, 32, 42, and 50 Unites States Code

for DOS Security Assistance and Cooperation programs, which are often implemented by U.S. Armed Forces and personnel.

(5) Title 32, U.S.C. governs the National Guard organization and function. It is relevant to joint planning and execution as the source of authority for National Guard forces, not mobilized under Title 10, U.S.C., to provide support to civil authorities and homeland defense.

(6) Title 42, U.S.C. (Chapter 68, "Disaster Relief) governs disaster relief and is the statutory authority under which the DOD provides assistance (referred to as Defense Support of Civil Authorities (DSCA)) at the request of other designated Federal agencies in support of state, local, or tribal response aid recovery efforts when a Presidential major disaster or emergency declaration has been issued.

(7) Title 50, U.S.C. governs USG activities related to war and national defense. Subsections of this title govern the mobilization of U.S. industry and resources in support of national defense.

c. Stakeholders[15]

(1) **President**. Article II of the U.S. Constitution establishes the President as the Commander-in-Chief of the U.S. Armed Forces. The President provides strategic direction to the U.S. Armed Forces and is the directive authority for planning and executing military operations. See Figure C for Chain of Command graphic.

(2) **Secretary of Defense**. Title 10, U.S.C. (Section 113) authorizes the Secretary, who is the head of the DOD, to act as the principal assistant to the President in all matters relating to the DOD. Subject to the direction of the President, Title 10, U.S.C. and Section 202 of the National Security Act of 1947, the Secretary has authority, direction and control over the DOD. The Secretary is also a statutory member of the NSC. Under the authority of Title 10, U.S.C. (Section 162), the Secretary directs the assignment and allocation of specified forces to CCDRs or the CDR, U.S. Element of the North American Aerospace Defense Command (USELEMNORAD). Under authority of Title 10, U.S.C. (Section 164), and with approval of the President, the Secretary provides directive authority to the CCDRs for the planning and execution of military operations. Additionally, the Secretary is responsible for directing the OSD in the development of defense guidance and policy. The DOD Organization Structure is depicted in Figure D.

(3) **Office of the Secretary of Defense**. The OSD operates under the authority, direction, and control of the Secretary. The function of the OSD staff is to assist the Secretary in carrying out responsibilities and other duties as prescribed by law. The principle OSD positions, with respect to planning and execution are:

(a) Under Secretary of Defense for Policy (USD(P)). Per Title 10, U.S.C. (Section 134) the USD(P) is the principal staff assistant and advisor to the Secretary for all matters on the formulation of national security and defense policy and oversight of DOD policy and plans to achieve national security objectives. The USD(P) assists the Secretary in preparing written policy guidance for the preparation and review of plans. Per DoDD 5111.01, *Under Secretary of Defense for Policy*, USD(P) also serves as a member of the NSC/DC for Crisis Management, advising the Secretary on crisis prevention and management.

(b) Under Secretary of Defense for Personnel and Readiness (USD(P&R)). Per Title 10, U.S.C. (Section 136) the USD(P&R) exercises the authorities and responsibilities of the Secretary in the areas of military readiness and force management. Per DoDD 5124.02, *Under Secretary of Defense for Personnel and Readiness, the USD(P&R)* is also responsible, subject to the authority, direction, and control of the Secretary, for the monitoring of the personnel tempo and operations tempo of the U.S. Armed Forces. The USD(P&R) establishes uniform standards within the DOD for terminology and policies relating to deployment of units and personnel away from their assigned duty stations (including the length of time units or personnel may be away for such a deployment) and establishes uniform reporting systems for tracking deployments.

---

[15] *Joint Publication 1*

Chain of Command

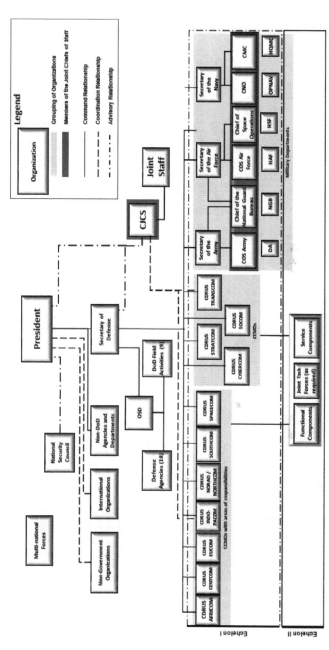

Figure C. Chain of Command

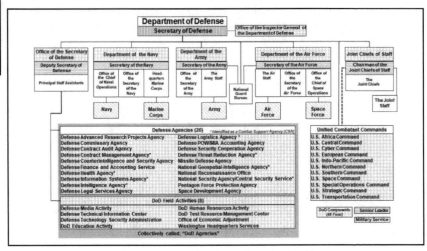

Figure D. Department of Defense Organization Structure

(c) <u>Under Secretary of Defense for Intelligence (USD(I))</u>. Per Title 10, U.S.C. (Section 137) the USD(I) performs such duties and exercises such powers as the Secretary may prescribe in the area of intelligence. Per policy direction of DoDD 5143.01, *Under Secretary of Defense for Intelligence,* USD(I) establishes priorities for the Defense Intelligence Enterprise and develops policies, plans, and programs to enable the operational application of Defense Intelligence. In the performance of these responsibilities, USD(I) exercises Secretary authority, direction, and control over the heads of the following Defense Agencies: Defense Intelligence Agency (DIA), National Geospatial-Intelligence Agency (NGA), National Reconnaissance Office (NRO), National Security Agency (NSA), Central Security Service (CSS) and Defense Security Service (DSS).

(4) **Chairman of the Joint Chiefs of Staff**. The Chairman functions under the authority, direction, and control of the President and Secretary. The Chairman heads the JCS, but does not exercise military command over the JCS or any of the U.S. Armed Forces. The primary statutory responsibilities assigned to the Chairman in Title 10, U.S.C. (Sections 151, 153 and 163) are to conduct independent assessments; act as the principal military adviser to the President, NSC, and the Secretary; and assist the President and Secretary in providing unified strategic direction to the U.S. Armed Forces. Per Title 10 Section 163 and the Unified Command Plan (UCP), the Chairman is also responsible for transmitting communications between the President, Secretary, and CCDRs, as well as for developing strategic plans and assisting in the development of policy for the preparation and review of plans (Figure E).

(a) <u>Assess</u>. The Chairman through the Secretary is required to provide Congress a report of the strategic and military risks associated with executing the NMS and a mitigation plan. The Chairman confers with and obtains information for the assessments from CCDRs, per Title 10, U.S.C. (Section 163) and evaluates and integrates that information to provide the best military advice to the Secretary. Based on assessments, the Chairman provides advice on critical deficiencies and strengths in force capabilities (including manpower, logistics, intelligence, and mobility support) with respect to their impact on meeting national security objectives and policy and on strategic plans.

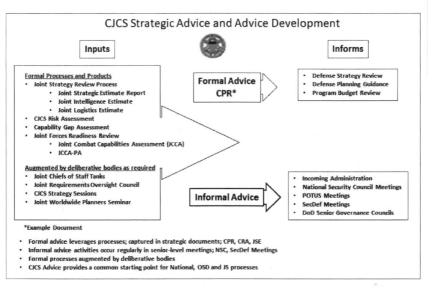

Figure E. CJCS Strategic Advice and Advice Development

(b) <u>Advise</u>. Consistent with Title 10, USC section 151, the Chairman consults with other members of the JCS and CCDRs prior to providing military advice to national leadership. In order not to unduly delay the Chairman's military advice to the national strategic leadership, the chairman has established staffing procedures with timelines to consolidate opinions and positions from the CCDRs and members of the JCS. The staffing procedures include but are not limited to Joint Staff Action Process (JSAP) Logbook, Collaborative Issue Resolution Tool (CIRT) and Secretary of Defense Orders Book (SDOB). Consolidated opinions and positions of CCDRs and other members of the JCS along with the Chairman's best military advice are provided to national strategic leaders. The Chairman also advises on the extent to which program recommendations and budget proposals of the Military Departments and other DOD components conform to the priorities established in strategic plans and with the priorities established for the requirements of the CCMDs.

(c) <u>Direct</u>. Subject to the authority, direction, and control of the President and the Secretary, the Chairman is responsible for assisting them in providing strategic direction to the U.S. Armed Forces. The NMS and the Joint Strategic Capabilities Plan (JSCP) are primary strategic guidance documents the Chairman publishes in support of this responsibility. Conforming with policy guidance from the President and SecDef, including resourcing levels projected by the SecDef, the Chairman provides strategic planning guidance and policy to the CCMDs and Services via CJCS issuances (which include the CJCS Joint planning family of documents, Joint doctrine via Joint publications, Force Apportionment Tables, and planning orders).

> *It is an old saying, 'Show me the company a man keeps, and I will tell you his character.' Why not a similar principle, "Show me the correspondence which a man receives, and I will show you what manner of man he is.'*
> *Sir John Barrow, Life of Lord Anson (1839),*
> *"The Pursuit of Victory, The Life and Achievements of Horatio Nelson,"*
> *Roger Knight*

(d) <u>Execute</u>. The Chairman issues orders, on behalf of the President or Secretary, implementing the command decisions of the President or Secretary via the release of orders or other authoritative direction.

(5) **Joint Chiefs of Staff**. Title 10, U.S.C. (Section 151) establishes the JCS. Each member of the JCS is responsible for providing military advice to the President, NSC, and Secretary. The Chairman acts as the principal advisor but the other members may submit dissenting opinions or additional advice to be included with the Chairman's presentation. The JCS consist of the following:

- The Chairman
- The Vice Chairman
- The Chief of Staff of the Army
- The Chief of Naval Operations
- The Chief of Staff of the Air Force
- The Commandant of the Marine Corps
- The Chief of Space Operations
- The Chief of the National Guard Bureau

(6) **Joint Staff (JS)**. Title 10, U.S.C. (Section 155) establishes the JS to assist the Chairman and JCS in carrying out their statutory responsibilities for the unified strategic direction of the U.S. Armed Forces. The JS is subject to the authority, direction, and control of the Chairman, and is prohibited, by statute, from being organized or operating as an overall Armed Force General Staff, and as such, exercises no executive authority.

(a) <u>Director, JS (DJS)</u>. The DJS supervises and provides guidance to the directors of the JS directorates and ensures their effective coordination with the OSD. The DJS attends JCS meetings and chairs the Operations Deputies (Ops Deps) meetings and the Global Force Management Board (GFMB). The DJS executes and manages the Joint Strategic Planning System (JSPS) in support of the Chairman's statutory responsibilities per CJCSI 3100.01, *Joint Strategic Planning System*.

(b) <u>Director for Manpower and Personnel (JS DJ-1)</u>. The JS DJ-1, is responsible for manpower management, personnel management, and personnel augmentation support. JS DJ-1 develops joint plans, policy, and guidance on manpower and personnel issues. The JS DJ-1 validates CCMD Joint Manning Documents (JMDs) augmentation requirements against criteria prescribed in the *Global Force Management Implementation Guidance* (GFMIG) and CJCSI 1301.01, *Joint Individual Augmentation Procedures*. Validated requirements are forwarded to the Joint Staff Joint Force Coordinator (JS JFC) for development of a sourcing recommendation.[16]

(c) <u>Director for Intelligence (JS DJ-2)</u>. The JS DJ-2 is a unique organization that is both an element of the JS and a major component of the DIA. The JS DJ-2 is responsible for providing intelligence support to satisfy Secretary, Chairman, National Military Command Center (NMCC), and CCMD requirements. The JS DJ-2 is the lead for crisis intelligence support to national and theater decision makers and tasks DIA and other DOD intelligence components to provide intelligence support. The JS DJ-2 coordinates, represents,

---

[16] *CJCS, through the Director, J-3 (DJ-3), serves as the Joint Force Coordinator (JFC) responsible for providing recommended sourcing solutions for all validated force and JIA requirements. In support of the DJ-3 the Joint Staff Deputy Director for Regional Operations and Force Management (J-35) assumes the responsibilities of the JFC. As such the JFC will coordinate with the Joint Staff J-3, Secretaries of the Military Departments, CCDRs, JFPs and DOD Agencies. The Joint Force Coordinator (JFC) is referred to in current DOD GFM guidance and policy as the JFC. For clarity in this text the Joint Force Commander will be annotated by the acronym (JFC) and Joint Force Coordinator will be referred to with the acronym (JS JFC) to denote the Joint Staff Joint Force Coordinator*

and advocates CCMD intelligence requirements, interests and priorities to the Chairman, OSD, and the Office of the Director, National Intelligence.

1 The JS DJ-2 leads the DOD intelligence planning process and its integration into joint planning and execution. On behalf of the CCMDs, JS DJ-2 coordinates with DIA to produce finished intelligence products in support of the CCMD Joint Intelligence Preparation of the Operational Environment (JIPOE) process. The JS J-2, in coordination with (ICW) the supported CCMD J-2, will coordinate, integrate, and synchronize the Intelligence Planning activities of the Defense Intelligence Enterprise in the development of the problem set National Intelligence Support Plan (NISP). These intelligence products support CCMD campaign plan development and assessment and provide common threat analysis to support CCMD problem set contingency plan development and maintenance.

2 The JS DJ-2 produces the Joint Strategic Intelligence Estimate (JSIE) that describes the mid- and long-range security environment from both a functional and regional approach. The JSIE informs the Chairman and CCDR's strategic planning and decision-making.

3 The JS DJ-2 prepares the Intelligence supplement to the JSCP, which provides planning guidance, objectives, and tasks for operational intelligence support to the JSCP-directed campaign and contingency plans.

(d) <u>Director for Operations (JS DJ-3)</u>. The JS DJ-3 assists the Chairman in carrying out responsibilities as the principal military advisor to the President and Secretary by developing and providing guidance to the CCMDs and relaying communications between the President, the Secretary and the CCDRs regarding current operations, plans, and readiness.

1 <u>Orders</u>. By the Chairman's direction, the JS DJ-3 develops, coordinates and prepares joint planning and execution orders, to include warning orders (WARNORDs), mobilization orders (MOBORDs), planning orders (PLANORDs), North Atlantic Treaty Organization (NATO) force preparation (FORCEPREP) messages, alert orders (ALERTORDs), deployment orders (DEPORDs) including the Global Force Management Allocation Plan (GFMAP), fragmentary orders (FRAGORDs), and execute orders (EXORDs). Subsequently, the JS DJ-3 is responsible for preparing and coordinating the SDOB to present recommendations to the Secretary for decision.[17]

2 <u>Global Force Management (GFM) Force Requirement Validation</u>. The JS DJ-3 evaluates CCMD force requirements against criteria published in the GFMIG. Validated force requirements are forwarded to the JS JFC to develop sourcing recommendations. The JS J-35 is the staff lead for force requirement validation and the preparation and coordination of the SDOB (see Chapter 3, *GFM*).

3 <u>Joint Force Coordinator (JS JFC)</u>. Per the GFMIG, as the JS JFC, the JS DJ-3 is responsible for developing recommended sourcing solutions for all validated force and individual requirements. The JS DJ-3 also performs the duties as the Joint Force Provider (JFP) for conventional forces and individual requirements. Coordinating with the Force Providers (FPs) and CCMDs, the JS DJ-3 identifies and recommends, from all conventional force across the globe, the most appropriate and responsive force or capability to meet validated force and JIA requirements. In addition, the DJ-3 provides initial force and sourcing feasibility estimates in order to provide resource-informed operation plan (OPLAN) advice to the Chairman.

(e) <u>Director for Logistics (JS DJ-4)</u>. The JS DJ-4, assists the Chairman by leveraging the Joint Logistics Enterprise (JLEnt); developing and integrating joint logistics capabilities; monitoring, assessing and shaping joint logistics readiness; supporting CCMD strategy development and operational logistics readiness; evaluating joint force logistics capability, capacity, and responsiveness; and communicating JLEnt capability gaps to heighten awareness and influence risk mitigation.

---

[17] CJCSI 3100.01 *Joint Strategic Planning System*, and CJCSI 3141.01, *Management and Review of Campaign and Contingency Plans*

<u>1</u> The JS DJ-4 prepares the logistics supplement (LOGSUP) to the JSCP which provides logistics planning guidance to CCMDs, Service HQs, and combat support agencies (CSAs) concerning the specified planning tasks assigned in the Chairmans Planning Guidance (CPG) and JSCP.

<u>2</u> The JS DJ-4 prepares the Joint Logistics Estimate (JLE) which provides an assessment of the JLEnt's ability to project, support, and sustain itself in the near-, mid-, and far-terms, in support of the full range and number of missions called for in strategic guidance.

(f) <u>Director for Strategic Plans and Policy (JS DJ-5)</u>. The JS DJ-5 is the Chairman's principal staff official for the preparation and review of strategic guidance documents and the review of campaign and contingency plans.

<u>1</u> The JS DJ-5 prepares, for leadership approval, strategic guidance, such as the UCP, NMS, and JSCP.

<u>2</u> Subject to the Chairman's authority, the JS DJ-5 coordinates with the CCMDs and the OSD staff for directed planning requirements.

<u>3</u> The JS DJ-5 coordinates the JPEC review of designated campaign and contingency plans. The JPEC review of a plan focuses on its suitability, acceptability, feasibility, completeness, and compliance with joint doctrine.

<u>4</u> The JS DJ-5 maintains the overall responsibility within the JS for interagency affairs. The J-5 directorate serves to focus the JS on particular NSC and HSC matters for policy and planning purposes. In conjunction with OUSD(P), the JS DJ-5 executes the Promote Cooperation program as DOD's primary means to gain interagency partners' input into DOD plans. It also provides a means to engage interagency partners on strategic planning and other related initiatives.

<u>5</u> The JS DJ-5 leads deliberate and continuous assessment processes, to include the CJA, the Joint Strategy Review (JSR) process, and the Joint Combat Capability Assessment (JCCA) plan assessment. These provide the Chairman's assessment of risks to national security interests relative to current and potential future military activities.

<u>6</u> The JS DJ-5 also develops recommendations on matters of security assistance, stability operations, stabilization and reconstruction, and security arrangements with allies.

(g) <u>Director for Command, Control, Communication, and Computers (C4)/Cyber (JS DJ-6)</u>. The JS DJ-6 leads the Chairman's efforts for C4/Cyber capability development, integration, and assessment. The directorate identifies, evaluates, and prioritizes joint military C4 /Cyber requirements.

(h) <u>Director for Force Development (JS DJ-7)</u>. The JS DJ-7 assists the Chairman in carrying out statutory responsibilities to develop joint doctrine, concepts and exercises, formulate policies for joint training, and coordinate joint military education and training.

(i) <u>Director for Force Structure, Resources, and Assessment (JS DJ-8)</u>. The JS DJ-8 provides resource and force structure analysis and program advice to the Chairman. Responsibilities include key support to the Chairman's JSPS responsibilities:

<u>1</u> Preparation of the Assignment Tables in the GFMIG and *Forces For Unified Command Memorandum ("Forces For")*, which documents the Secretary's direction to the Service Secretaries for the assignment of specified forces to CCDRs IAW Title 10, U.S.C. (Section 162).

<u>2</u> Preparation of the Force Apportionment Tables, for Chairman approval, documenting the quantity of units, by force element, the Services estimate can be reasonably made available to respond globally, along general timelines, in order to inform CCMD planning efforts.

<u>3</u> Analysis and assessments of joint force capabilities and readiness that inform Chairman recommendations to the Secretary regarding force planning, funding and acquisition.

(7) **Secretaries of the Military Departments**. Title 10, U.S.C. (Sections 111, 3013, 5013, and 8013) specifies the Secretaries of the Military Departments as part of the DOD. They are subject to the authority, direction and control of the Secretary and responsible for the organization, development, and programming for their respective Military Departments, consistent with policy and national security objectives. The responsibilities of the Military Department Secretaries include:

(a) Developing and maintaining their authorized force structure to support their employment by CCDRs to achieve national security objectives. They are responsible for their Department's acquisition, auditing, comptroller, information management, inspector general, legislative affairs, and public affairs.

(b) Assigning specified forces to CCDRs or to USELEMNORAD as directed by the Secretary in the assignment tables in the GFMIG and "Forces For" to perform missions assigned to those commands per Title10, U.S.C. (Section 162).

(c) The administration and support of Department forces assigned to a CCMD per Title 10, U.S.C. (Section 165). These administration and support responsibilities are typically effected through a CCMD Service component command.

(d) Department Secretaries may direct their Military Departments to campaign support plans (CSPs) in conjunction with their aligned CCMD Service components. These CSPs inform the supported CCDRs of the Service's ways and means for supporting their campaign plan activities.

(e) Synchronizing their deployed Service training and security cooperation/assistance activities with CCDR campaign planning and execution in order to ensure unity of effort.

(8) **Service Headquarters**. Title 10, U.S.C. (Sections 3031, 5031,504I, and 8031) specify the establishment of Service HQs under the Military Departments (Department of the Navy comprises the OPNAV Staff and HQs Marine Corps). Service HQs, led by their Service Chiefs, are subject to the authority and direction of their Military Department secretary. The U.S. Coast Guard also functions as a Service HQs under separate authorities and is discussed later in this chapter. The specified responsibilities of the Service HQs include:

(a) *Prepare* their Service for employment to include activities of recruiting, organizing, supplying, equipping, training, mobilizing, administering, and maintaining.

(b) *Assess* and report upon the efficiency of the Service and its preparation to support CCDR operations. They provide force readiness data that informs GFM allocation risk analysis and supports updates to the force apportionment tables.

(c) *Develop* Service plans and recommendations to the Secretary and supervise their implementation when approved. Service HQs may develop CSPs that align with campaign plans.

(9) **Force Provider (FP)**. The term "Force Provider" is used throughout this book to indicate the Secretaries of the Military Departments, the USCG, CCMDs with assigned forces, DOD Agencies, and OSD organizations that provide force-sourcing solutions to CCMD force requirements. Per the GFMIG, the FPs coordinate directly with CCMDs, JFPs, and the JS J-3 (as the Joint Force Coordinator) to develop recommended global sourcing solutions. **Service Force Providers (SFPs)** are commands designated by the Service Chief to coordinate with the JS JFC to provide recommended GFM allocation sourcing nominations to validated CCMD force requirements to include: U.S. Army (USA) Forces Command, U.S. Navy (USN) Fleet Forces Command, U.S. Air Force (USAF) Air Combat Command (ACC), U.S. Marine Corps (USMC) Forces Command and U.S. Space Forces (USSF) Command.

1 FPs may assist and advise CCMDs with identifying preferred forces.

2 FPs provide contingency sourcing solutions to the JFPs and JS JFC.

3 FPs submit execution sourcing nominations to the JS JFC and JFPs. FPs also provide deployment and movement data for ordered forces.

(10) **Combatant Commanders**. Per Title 10, U.S.C. (Sections 161,162, 164) and the UCP, a CCDR is responsible to the President and Secretary for the performance of mis-

sions assigned to their geographic or functional CCMD. CCDRs perform their duties under the authority, direction, and control of the Secretary and provide authoritative direction to the subordinate commands and forces assigned to their respective CCMD. CCDRs have the authority to employ forces within their commands to carry out missions assigned to the command. CCDRs act as the supported CDR for the planning and execution of their assigned missions, which involves the responsibility of synchronizing the range of activities required to achieve the desired objectives. They may simultaneously be a supporting CDR to other CCDR's for planning and executing the other CCDR's missions. The responsibilities of CCDRs include the following:

(a) Conduct missions and tasks within their assigned geographic area of responsibility or functional area and synchronize the efforts of forces and resources to achieve unified action.

(b) Plan, conduct, and assess security cooperation activities as authorized by the Secretary. Coordinate activities with the specific country team and Chief of Mission.

(c) Conduct planning and execution to achieve national policy objectives and protect national security interests. Responsibility includes synchronization of CCPs.

(d) Conduct planning and execution as required to respond to potential or emergent threats to national security interests. Responsibility includes synchronization of supporting plans and resources.

(e) Maintain the preparedness of the command to carry out assigned missions.

(f) Develop budget proposals for funding of CCMD activities to include joint exercises, force training, contingencies, or other selected operations.

(g) Force Provider. Per the GFMIG, CCMDs with assigned forces are designated FPs and may be directed via the GFM allocation process to provide trained and ready forces to satisfy other CCDR force requirements.

(h) Joint Force Providers. Per the UCP and GFMIG, Commander U.S. Special Operations Command (CDR, USSOCOM), Commander U.S. Transportation Command (CDR, USTRANSCOM), Commander, U.S. Cyber Command (CDRUSCYBERCOM) and Commander U.S. Space Command (CDRUSSPACECOM) are the Special Operations, Mobility, Cyber and Space operation force JFPs, respectively. Each is responsible for identifying and recommending, via the JS JFC, force execution and contingency sourcing solutions ICW the Military Departments, CCMDs, DOD Agencies and other JFPs for all Special Operations, Mobility, Cyber and Space operation forces and individual requirements. JFPs are granted specific limited authorities by the Secretary to deploy SOF, mobility, cyber and space operation forces detailed in EXORDs.

(11) **CCMD Components and Theater Special Operations Commands (TSOCs)**

(a) Service Components. CCMD Service components consist of the Service component HQs assigned to each CCMD. CCMD Service components provide Service-specific administrative support for Service forces and personnel assigned or allocated to the CCMD and will coordinate with their aligned Service HQs. CCMD Service components may also be designated as a Joint Task Force (JTF) HQs or a functional component.

(b) Functional Components. Joint Force Commanders (JFCs) have the authority to organize commands and forces within their respective command to carry out missions assigned and may establish functional component commands to control military operations. The alignment of these functional components typically corresponds to operating domains (air, land, maritime, space, and cyberspace).

(c) Theater Special Operations Commands. TSOCs are regional sub-unified commands assigned to USSOCOM, over which the respective geographic combatant commander (GCC) exercises OPCON. They enable the integration of SOF missions and forces with the GCCs planning and execution.

(12) **Secretary of State**. Title 22, U.S.C. authorizes the Secretary of State to act as the principal advisor to the President on U.S. foreign policy. Responsibilities include the implementation of security assistance activities as directed by the President and approved

by the Congress. These security assistance programs may be a source of planning and execution authority for military forces or personnel to work with partner nation militaries.

(a) The DOS and U.S. Agency for International Development (USAID) prepare a JSP as well as regional, functional, and country focused strategies. These provide diplomatic context for military strategy and plan development necessary for unity of effort in pursuing national objectives.

(b) Chiefs of Mission serve as the President's direct representative to the country appointed. Per Title 22, U.S.C. (Section 3927) the U.S. Chief of Mission to a specific country shall have full responsibility for the direction, coordination, and supervision of all USG executive branch employees in that country (except Voice of America correspondents on official orders and employees under the command of a U.S. area military CDR).

(13) Secretary of Homeland Security. Per Title 6, U.S.C. (Sections 111-115), the Secretary of Homeland Security is the principal Federal official for domestic incident management and is responsible for coordinating Federal operations within the U.S. and U.S. Territories to prepare for, respond to, and recover from terrorist attacks, major disasters, and other emergencies. Under Title 42, U.S.C. (Chapter 68, "Disaster Relief) the DOD may provide DSCA at the request of other designated Federal agencies in support of state, local, or tribal response and recovery efforts.

(a) Coast Guard. Per Title 14, U.S.C. the USCG is a Service in the DHS principally serving as a maritime law enforcement, regulatory, and humanitarian agency with roles, missions, and duties specified in Title 6, U.S.C. (Section 468) and Title 14, U.S.C. (Section 2). Additionally, per Title 14, U.S.C. the USCG is also a military service and a branch of the U.S. Armed Forces and may be directed by Congress or the President to operate in the Service of the Navy under the provisions of Title 14, U.S.C. (Section 3).

(b) MOA DOD/DHS FP. IAW Title 14, U.S.C. (Section 145) the Secretaries of the Navy and Homeland Security may make available to each other personnel, vessels, facilities, and equipment, and agree to undertake such assignments and functions for each other as they agree necessary. This interagency cooperation is detailed by a Memorandum of Agreement (MOA) between the DOD and Department of Homeland Security (DHS) on the use of USCG capabilities and resources in support of the NMS. Per the GFMIG, the USCG serves as a FP in the GFM allocation process to meet CCDR force requirements in support of current operations and CCPs. As directed, USCG units deploy and operate under the appropriate command authority of the CCDRs in support of DOD operations worldwide.

(14) Governors of the States and Territories. Responsible for executive direction of state/territorial government agencies and resources IAW state and Federal statutes. Per Title 32, U.S.C. (Sections 901-908) in the event of domestic emergencies and/or disasters, governors (or equivalent officials in U.S. territories) will retain command and control of all National Guard forces when operating in either state active duty or a Title 32 duty status within their respective states conducting civil support activities. Governors may submit a request for assistance to the primary Federal agency (typically DHS and Federal Emergency Management Agency (FEMA)) for support.

Further discussion of the DOD role in providing DSCA can be found in the current DSCA EXORD. State/territorial Adjutants General implement the governor's executive direction to their National Guard forces and serve as the state's representative to and coordinator with the NGB.

(15) Chief, National Guard Bureau. Established by Title 10, U.S.C. (Section 151) the Chief, NGB is a member of the JCS and is under the authority, direction, and control of the Secretary as described in DoDD 5105.77, *National Guard Bureau.* The Chief, NGB serves as the principal advisor to the Secretary, through the Chairman, on matters involving non-federalized National Guard forces. He also advises the Secretaries of the Army and Air Force on National Guard matters. National Guard forces or personnel may operate under Titles 10, U.S.C. authorities in support of CCDRs or Title 32, U.S.C. authorities in support of state or territory governors. The NGB coordinates with state/territory governors to synchronize state/territorial activities with potential Title 10, U.S.C. *Federal Missions.*

(a) Per Title 10, U.S.C. (Section 10501), the NGB is a joint activity within the DOD that is the channel of communications on all matters pertaining to the National Guard, Army National Guard, and Air National Guard between the Departments of the Army and Air Force and the States/territories.

(b) The NGB is responsible for coordinating and planning National Guard Civil Support acting in support of State/territory responses to manmade and natural disasters.

(c) The NGB may prepare a CSP outlining National Guard support to CCDR campaign plans. Support includes National Guard activities such as the State Partnership Program developing foreign military-to-military and selected military-to-civilian relationships that may be leveraged by CCMDs and Services during planning and execution.

(16) **Department of Defense Agencies**. DOD Agencies (comprised of Defense Agencies and DOD Field Activities) are established as DOD components IAW Title 10, U.S.C. (Sections 101, 191(a), and 192). The President, or Secretary, may authorize a supply or service activity that is common to more than one Military Department be supported by a single agency of the DOD. Each DOD Agency operates under the authority, direction, control, and supervision of the Secretary, through a designated civilian officer within OSD or the Chairman. Further information detailing DOD Agency functions can be found in DoDD 5100.01, *Functions of the DOD and its Major Components.*

(a) Combat Support Agencies (CSAs) are a subset of Defense Agencies assigned combat support or combat service support functions as designated in Title 10, U.S.C. (Section193). Per Title 10, U.S.C., CSAs consist of the following Defense Agencies: Defense Information Systems Agency (DISA), DIA, Defense Logistics Agency (DLA), NGA, and any other Defense Agency designated as a CSA by the Secretary.[18]

(b) The National Security Agency/Central Security Service (NSA/CSS), while not a DOD Component, is also designated a CSA, consistent with Title 10, U.S.C. (Section 193), with respect to those combat support activities it performs for the DOD.

(c) CSAs fulfill combat support or combat service support functions across the range of military operations in support of the CCDRs executing military operations. CCDRs ensure that CCMD and subordinate command plans adequately specify CSA missions and tasks, and establish appropriate relationships for the CSAs supporting operations within their area of responsibility or over which they exercise tactical control for force protection. Each CCDR will provide planning and operational guidance to the CSAs and ensure that requirements for CSA support are adequately identified in requests for capabilities.

6. Pursuing National Interests and Strategic Guidance

The President, Secretary, and Chairman provide their orders, intent, strategy, direction and guidance via strategic direction to the military to pursue national interests within legal and constitutional limitations. Strategic direction is typically published in key documents, generally referred to as strategic guidance, but it may be communicated through any means available. Strategic direction may change rapidly in response to changes in the global environment, whereas strategic guidance documents are typically updated cyclically and may not reflect the most current strategic direction. Documents containing strategic guidance include:

a. **President**: The President provides strategic guidance through the NSS, executive orders/Presidential Directives, and other strategic documents in conjunction with additional guidance and refinement from the NSC. The President also signs the UCP and the CPG, which are both developed by DOD.

(1) National Security Strategy (NSS). The NSS provides the President's strategic vision and outlines the major national security concerns. The NSS is a comprehensive report *required annually* by Title 50, USC, Section 3043. It is prepared by the executive branch

[18] DoDD 3000.06, *Combat Support Agencies* CSAs

of the USG for Congress and outlines the major national security concerns of the U.S. and how the administration plans to address them using all instruments of national power. The document is purposely general in content, and its implementation by DOD relies on elaborating guidance provided in supporting documents (e.g., such as the Defense Strategy Review (DSR), CPG, NMS, etc.).

(2) Unified Command Plan (UCP). The Director for Strategy, Plans, and Policy, J-5, is responsible for developing, staffing, reviewing, and preparing the UCP for the CJCS and the SecDef's review and the President's approval. The UCP establishes CCMD missions and CCDR responsibilities, addresses assignment of forces, delineates geographic AORs for geographic combatant commanders (GCCs); and specifies responsibilities for functional combatant commanders (FCCs).

(3) Contingency Planning Guidance (CPG). The CPG contains detailed planning guidance from POTUS on specific contingency plans that CCDRs must fully develop. The CPG conveys the President's guidance for force management, security cooperation, and posture planning. The CPG translates NSS objectives into prioritized and comprehensive military planning guidance for the employment of DOD forces. The CPG issues the President's guidance for contingency planning and by explicitly listing the DOD's top-priority contingency planning efforts. The CPG is produced in conjunction with the NDS, NMS and JSCP, which provides additional planning guidance and direction for the development of contingency plans, and with DOS input.

(4) Presidential Directives (PDs). PDs are authoritative documents used to promulgate Presidential decisions or policy on national security matters. They may be classified or unclassified depending upon the subject matter. Presidential directives, such as proclamations and executive orders, are a tool used by Presidents to announce official policy and make declarations in their roles as leader of the executive branch, Commander-in-Chief of the Armed Forces, and head of state. Presidents have used directives throughout American history for a wide variety of purposes, but they have taken on a more central policy role in recent Administrations. In 1789, during the first months of government under the new Constitution, President Washington issued both the first presidential proclamation, declaring a National Day of Thanksgiving in November 1789, and the first precursor of the modern executive order. In that directive, Washington ordered executive branch officers who had also served under the Articles of Confederation to report to the new President on the ongoing affairs of the government and to explain their duties. The Constitution does not explicitly recognize a presidential power to issue directives. However, there has been general acceptance since the Washington Administration that this authority is necessary for the President to manage the executive branch and that some authority to issue directives is inherent in the executive's Article II powers. Congress may also grant the President discretion to make certain decisions and determinations that have legal effect but may be affected by subsequent legislative action.[19]

b. **Secretary**: DPG, NDS and GFMIG.

(1) Defense Planning Guidance (DPG). The DPG establishes DOD's force development planning and resource priorities in order to prevail in the Nation's current operations and to develop a balanced joint force to meet future contingencies. It consolidates and integrates force development planning priorities into a single overarching document for DOD planners to use. Product of the Planning, Programming, Budgeting and Execution (PPBE) process' planning phase. The DPG reflects the President's NSS, the Secretary's NDS and the Chairmans NMS. It also reflects results of the NDS, and the annual Chairman's Program Recommendations (CPR). The DPG drives the development of the Program Objective Memoranda (POM) and Budget Estimate Submissions (BES).

(2) National Defense Strategy (NDS). The NDS addresses how the Armed Forces of the U.S. will fight and win America's wars and describes how DOD will support the objectives outlined in the NSS. It also provides a framework for other DOD strategic guidance, specifically on contingency planning, force development, and intelligence.

[19] *Congressional Research Service: Presidential Directives, https://crsreports.congress.gov*

(3) <u>Global Force Management Implementation Guidance (GFMIG)</u>. The GFMIG integrates complementary assignment, apportionment, and allocation information into a single GFM document. GFM aligns force assignment, apportionment, and allocation methodologies in support of the DSR, CPG, joint force availability requirements, and joint force assessments. It provides comprehensive insights into the global availability of U.S. military resources and provides senior decision makers a process to quickly and accurately assess the impact and risk of proposed changes in forces assignment, apportionment, and allocation.

(a) <u>Assignment Tables</u>. The Assignment Tables provide the Secretary's direction for the assignment of forces from the individual service secretaries to the CCDRs. These tables are contained in the GFMIG and, in years the GFMIG is not updated, the *Forces For Unified Commands Memorandum*, also known as "Forces For" are available on the J8 Forces Division Secret Internet Protocol Router Network (SIPRNET).

(b) The Secretary's approved GFMAP is the global deployment order for all allocated forces for the fiscal year developed under the procedures prescribed in the GFMIG. The GFMAP documents the Secretary's decisions for allocating forces to CCDRs to conduct military activities globally, and as such, communicates near-term priorities. GFMAP documents are available on the J35 GFM Division SIPRNET website.

(c) <u>Standing EXORDs</u>. There are Secretary-approved standing EXORDs that delegate to one or more CCDRs specific execution authorities. Standing EXORDs are relevant to joint planning and execution and can be found on the JS J-35 GFM Division website (SIPRNET).

c. **Chairman**: JSPS, NMS, JSCP, and Force Apportionment Tables (published in GFMIG). The CJCS serves as principal military advisor to the President, SecDef, and other members of the NSC and assists the President and SecDef with providing unified strategic direction to the Armed Forces of the United States. The CJCS uses the JSPS as the formal mechanism to fulfill responsibilities under Title 10, United States Code (USC), Section 153, to maintain a global perspective, conduct assessments, develop the force, and develop military advice for SecDef and the President. The JSPS also supports the CJCS's interactions with Congress, the Services, and the CCMDs. The CJCS provides additional strategic planning guidance and policy to the CCMDs and Services from CJCS directives, joint doctrine, force apportionment tables, and PLANORDs. The CJCS also issues orders on behalf of the President or SecDef.

(1) <u>Joint Strategic Planning System (JSPS)</u>. Titles 6, 10, 22, and 50, U.S.C. assign four primary responsibilities to the Chairman: assess, advise, direct, and execute. The Chairman uses the JSPS to effectively perform these statutory responsibilities. With the help of the JS, the Chairman assesses the strategic environment. In consultation with the JCS and CCDRs, the Chairman advises the President, NSC, Secretary, and Congress. The Chairman directs the Joint Force through strategic guidance and joint policy and doctrine. Finally, the Chairman executes the command decisions of the President or Secretary via orders and other authoritative directives. The JSPS enables the CJCS to:

- Conduct independent assessments.
- Provide independent advice to the President, SecDef, NSC, and Homeland Security Council (HSC).
- Assist the President and SecDef in providing unified strategic direction to the Armed Forces (includes execution).

The JSPS also aligns its products and processes to one of the Chairman's six primary statutory functions which are:

- Providing strategic direction for the Armed Forces
- Conducting strategic and contingency planning
- Assessing comprehensive joint readiness
- Fostering joint capability development

- Managing Joint Force Development
- Advising on global military integration

Figure F on the following page presents the relationships among these six functions and the framework for strategic planning and direction of the armed forces.[20]

<u>1</u> The JSPS provides formal structure to the CJCSs statutory responsibilities and considers the strategic environment and the alignment of ends, ways, means, risk, and risk mitigation over time to provide the best possible assessments, advice, and direction of the Armed Forces in support of senior leaders and processes at the national and OSD level. JSPS constitutes a continuing process in which each document, program, or plan is an outgrowth of preceding cycles and of documents formulated earlier and in which development proceeds concurrently.

<u>2</u> Joint strategic planning begins the process which creates the forces whose capabilities form the basis for CCPs. It culminates with planning guidance for the CCDRs to develop campaign and contingency plans. Figure G, *Product, Linkages and Dependencies* also depicts the JSPS cross-functional internal linkages and external product dependencies. The CJCS:[21]

- Requires development of and reviews strategic plans.
- Prepares and reviews contingency plans. Advises the President and SecDef on requirements, programs, and budgets.
- Provides net assessments on the capabilities of the Armed Forces of the U.S. and its allies relative to potential adversaries.

<u>3</u> JSPS is a flexible and interactive system intended to provide supporting military advice to the DOD PPBES and strategic direction for use in joint planning and execution. Through the JSPS, the JCS and the CCDRs:

- Review the national security environment and U.S. national security objectives and evaluate the threat.
- Assess current strategy and existing or proposed programs and budgets.
- Propose military strategy, programs, and forces necessary to achieve those national security objectives in a resource-limited environment consistent with policies and priorities established by the President and SecDef.

<u>4</u> Although all JSPS documents are prepared in consultation with other members of the JCS and the CCDRs, the final approval authority for all JSPS documents is the CJCS. Most JSPS documents are published biennially; however, all documents are subject to annual review and may be changed as required. The products of JSPS that gives direction to strategic and operational planning are the NMS and JSCP.

(2) <u>National Military Strategy</u>. The NMS is the CJCS's central strategy document. Title 10, USC, Section 153, directs the CJCS to determine for each even-numbered year whether to prepare a new NMS or update an existing strategy. The NMS provides the CJCS's amplifying guidance for planning, force employment, posture, and future force development. It provides the strategic framework to prioritize planning, resource allocation, and risk management. As such, this classified military strategy serves as the starting point for all other JSPS actions and constitutes the CJCS's military advice to SecDef and the President. The NMS defines the national military objectives (**ends**) and how to achieve these objectives (**ways**) and addresses the military capabilities (**means**) required to execute the strategy.

(3) <u>Joint Strategic Campaign Plan (JSCP)</u>.

(a) The JSCP is a five-year global strategic plan (reviewed every two years) and is the primary vehicle through which the CJCS carries out statutory responsibility for providing

---

[20] *CJCSI 3100.1 Joint Strategic Planning System*

[21] *Ibid*

Joint Strategic Planning System (JSPS)

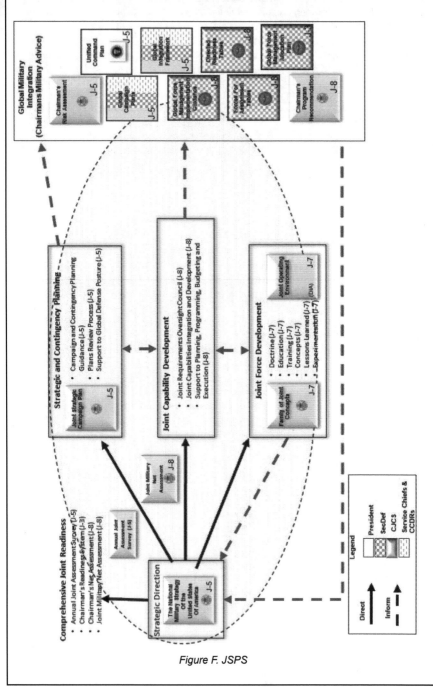

Figure F. JSPS

JSPS Product Linkages and Dependencies

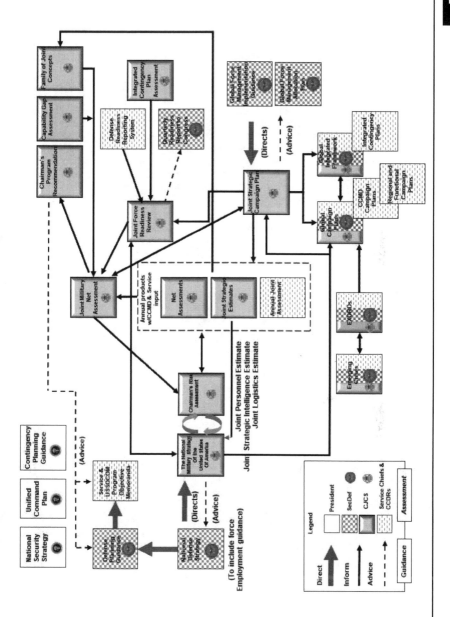

Figure G. JSPS Product, Linkages and Dependencies

unified strategic direction to the Armed Forces. It fulfills the CJCS Title 10 requirement for the preparation of a strategic framework and strategic plans; providing for the preparation of contingency plans; and the relationship between strategy and the GCPs, CCPs, other campaign and contingency plans, and operations.[22]

(b) The JSCP is the Chairman's implementation of the Secretary's planning guidance to synchronize CCMD campaign, contingency, and posture planning. The JSCP tasks CCMDs, Services, JS, and certain agencies to prepare campaign, contingency, and posture plans. Further, it provides guidance for support relationships and synchronization of directed planning efforts that span multiple CCMDs. Supplements to the JSCP provide additional planning guidance, each focused on a particular functional area.

(c) The JSCP provides CJCS strategic guidance as the "execution document" to implement strategic guidance from the President and Secretary. The JSCP nests with the strategic direction delineated by the NSS, CPG, NDS, and the DOD's planning and resourcing guidance provided in the DPG. It establishes a common set of processes, products, priorities, roles and responsibilities to integrate the Joint Force's global operations, activities, and investments from day-to-day campaigning to contingencies. It provides military strategic and operational guidance and measurable intermediate military objectives (IMOs) and guidance for CCMDs, Services, CSAs, and applicable defense agencies focused on preparation of plans and employing the Joint Force at current resource levels.[23]

(d) The JSCP directs campaign, contingency, and support plans and provides integrated planning guidance and direction for planners. The three types of campaign plans addressed in the JSCP are GCPs, FCPs and CCPs. Although no longer directed in the JSCP, another type of campaign plan is a regional campaign plan (RCP). Regional planning guidance addresses regional threats or challenges that require coordination across multiple combatant commands. Generally, issues that require RCPs are not as significant a threat to U.S. interests as global campaign plans, but they require attention to ensure they do not evolve into a more significant crisis. The JSCP has shifted from a CCMD-focused "capabilities" plan to a Strategic "Campaign" Plan. The JSCP directs:

1 Integrated Contingency Planning (ICP). The JSCP also directs contingency planning consistent with the CPG. It expands on the CPG with specific objectives, tasks and linkages between campaign and contingency plans.[24] Individual contingency plans (numbered plans developed as branches to campaign plans that are planned for potential threats, catastrophic events, and contingent missions) that deal with like problem sets are grouped together into an ICP. ICPs are the primary branch plans and war plans associated with a GCP. The ICP brings together contingency plans from multiple organizations to achieve increased unity-of-effort and closer linkages between complementary contingency plans for a specific problem set. When threats emerge, crises occur, or escalation warrants, a GCP will transition into a contingency plan for execution.[25]

2 Global Campaign Plans (GCP). GCPs address threats or challenges that significantly affect U.S. interests across the globe and require coordinated planning across all, or nearly all, CCMDs. The CJCS manages these plans on behalf of the SecDef. The CJCS approves GCPs after SecDef endorsement.

a Regional Campaign Plans (RCPs) - RCPs are plans written for regional challenges that do not rise to the interest/threat level of GCPs. RCPs are assigned to a Coordinating Authority (CA) and employ Collaborators to deal with cross-AOR elements of the challenge and/or solution.

b Functional Campaign Plans (FCPs) - FCPs are plans written for global challenges that do not rise to the interest/threat level of GCPs, and deal primarily with a function instead of a region. FCPs are also assigned to a CA and employ "Collaborators" to deal with

[22] *CJCSI 3110.01, Joint Strategic Campaign Plan*

[23] *CJCSI 3100.01 Joint Strategic Planning System*

[24] *CJCSI 3110.01, Joint Strategic Campaign Plan*

[25] *Joint Pub 5-0, Joint Planning*

cross-AOR elements of the challenge or solution. Cyber might be one area where an FCP would be produced.

c Combatant Command Campaign Plans (CCPs) - CCDRs maintain responsibility for developing campaign plans that address their respective area and functional responsibilities. A CCP incorporates intermediate objectives and tasks from GCPs, RCPs, and FCPs. The CCP is, therefore, the principal operational plan for execution of a CCMD's theater and global responsibilities for all priority challenges. The CCP balances the risks and opportunities of the command and simultaneously accounts for all assigned theater and problem-focused tasks to provide a campaign plan that fully integrates operations, activities, and investments (OAIs) spanning the CCMDs' assigned responsibilities. CCPs are the centerpiece of the CCMDs' planning construct, and executes JSCP direction and CCMD strategies. CCPs align the command's day-to-day activities (which include ongoing operations, military engagement, security cooperation, deterrence, and other shaping or preventive activities) with resources to achieve the CCMD's objectives. CCPs should address the following additional topics: NMS mission area linkages; objectives and tasks assigned to the CCMD from GCPs, RCPs, and FCPs; key support plan linkages; key contingency plan linkages; and risks and priorities considering the CCMDs balancing of global, theater, regional, and functional assigned objectives and tasks. A CCP has a five-year planning horizon.[26]

(3) Force Apportionment Tables. The Services provide an estimate of the number of forces, over a period of time, which can reasonably be expected to be available, globally along general timelines. On behalf of the Chairman, the JS J-8 consolidates and publishes these estimates in the Force Apportionment Tables as planning assumptions for the CCMDs. The Force Apportionment Tables depict the total number of specific types of forces the Services can make available to respond to contingencies, including supporting and supported plans, worldwide. Forces are not apportioned to individual plans or CCDRs. The Force Apportionment Tables can be found on the JS J-8 Forces Division SIPRNET website and are addressed in the GFMIG.

d. Department of State: Quadrennial Diplomacy and Development Review (QDDR) and Joint Strategic Plan (JSP). CCMD planning may also be informed by specific Joint Regional Strategies (JRSs) and Integrated Country Strategies (ICSs) relevant to a given area of operation. While not directive of the U.S. Armed Forces, strategic guidance developed within the DOS should inform military planners' efforts and enable unity of effort between the military and diplomatic instruments of national power and, like DOD strategic guidance, support achieving enduring national interests.

(1) Quadrennial Diplomacy and Development Review. The QDDR is not congressionally mandated, but it is similar to the DSR from a DOS perspective. Developed every four years by the DOS and USAID, the QDDR provides integrated planning guidance for diplomacy and development missions.

(2) Joint Strategic Plan. The JSP provides strategic direction at a similar level to the NMS, but from a DOS perspective. Developed jointly by DOS and USAID, this document outlines the highest-level foreign policy goals and objectives of both organizations.

(3) Joint Regional Strategies. Formerly called Bureau Strategic and Resource Plans, JRSs are developed jointly by the regional bureaus of DOS and USAID and identify U.S. foreign policy and development priorities for a given region. The regional bureaus of the DOS roughly correspond to the GCCs. The geographic and functional seams should be coordinated with GCC theater strategies to achieve unity of effort. JRSs are considered Sensitive but Unclassified and can be found on the NIPRNET (See Chapter 1-II, *Campaigning and Contingency Plans.*)

(4) Integrated Country Strategies. Formerly called Mission Strategic and Resource Plans, these documents are developed in each country by the embassy's Country Team and identify U.S. foreign policy and development priorities for that country. They are in-

[26] *CJCSI 3100.01 Joint Strategic Planning System*

formed by the appropriate JRS for the region of that country. ICSs are considered Sensitive but Unclassified and can be found on the NIPRNET.

(5) <u>Functional Strategies</u>. Strategies are developed on functional topics (e.g., anti-terror countering weapons of mass destruction). These functional strategies should be considered during planning as appropriate.

e. <u>Planning Policy and Joint Doctrine</u>. Per Title 10, U.S.C. and IAW guidance and direction by the President, Secretary, and USD(P) provide guidance for the preparation of military plans. In accordance with this guidance and direction, the Chairman directs the development of plans and publishes additional planning policy documents in order to meet this statutory responsibility. This policy is necessary to maintain consistency of planning products and supports unity of effort across the JPEC. The Chairman also publishes policy issuances on joint planning and execution processes along with publishing policy issuances and approving the joint doctrine in Joint Publications.

7. Strategic and Operational Forums

a. <u>National Security Council</u>. As discussed in paragraph 4.c. the NSC is the principal interagency forum established by Title 50, U.S.C. (Section 3021) for developing policy options, considering implications, coordinating operational problems that require interdepartmental consideration, developing recommendations for the President, and monitoring policy implementation. The Secretary, as a statutory member, and Chairman, as the principal military advisor, regularly attend NSC meetings. Per statutory responsibility, the Chairman presents the views of the JCS and CCDRs in providing the best military advice. The procedures of the NSC are further codified by Presidential Directive usually issued at the start of an administration.

(1) The NSC addresses policy issues with consideration for the proper application of all instruments of national power. These considerations and capabilities are integrated throughout planning and are designed to lead to more effective coordination of whole-of-government efforts during execution.

(2) Well-crafted plans, developed with frequent dialogue between stakeholders, provide an effective shared civilian-military understanding of existing or emerging threats. Should a crisis occur, this shared understanding developed during the NSC forums provides the foundation to address emergent events that impact national security interests.

b. <u>Promote Cooperation</u>. JS J-5 in conjunction with OSD(P) executes the Promote Cooperation program to facilitate further discussions and coordination between CCMD planners and interagency partners in the National Capital Region. It is intended to be an ongoing forum that leverages the expertise of interagency partners to inform development of CCMD plans. It enables interaction and coordination with stakeholders external to DOD and provides an opportunity to collaborate on planning. Promote Cooperation is initiated by CCMD planners and can take the form of consultations, seminars, workshops, or tabletop exercises depending upon requirements. These are typically held at the action officer level, with the option of holding a Deputy Assistant Secretary level out brief at the conclusion of the event. Promote Cooperation events provide an important opportunity for collaboration on planning.

c. <u>Diplomacy, Development, and Defense (3D) Planning Group</u>. Select representatives from DOS, USAID, and DOD constitute the Diplomacy, Development, and Defense (3D) Planning Group. It is an ongoing community of interest designed to improve collaboration and coordination on planning efforts. The 3D Planning Group typically meets monthly and provides a critical venue to share information, ensure alignment across planning priorities, and support and coordinate on shared initiatives when required.

d. <u>Deputy's Management Action Group (DMAG)</u>. The DMAG is a management forum chaired by the Deputy Secretary of Defense (DepSecDef) with the intent of supporting the Secretary by prioritizing issues with resource, management, and broad strategic and/or policy implications. DMAG is cochaired by the Vice Chairman and membership includes the Military Department Secretaries, Under Secretaries of Defense, Deputy Chief Management Officer, Chiefs of Military Services, Chief of the NGB, CDR USSOCOM, and Director of Cost Assessment and Program Evaluation. The DMAG is also supported by a senior steering group that convenes regularly to review the DMAG schedule, assess its effectiveness, and review progress on DMAG action items.

e. Secretary. Chairman. and Operations Deputies Meetings. Senior-level meetings are conducted regularly, and as required, in the JCS Conference Room (also known as the 'Tank"). The intent of these meetings is to ensure full coordination across the CCMDs and Services. The meetings are typically conducted at one of three levels: OpsDeps, JCS, and Secretary. CJCSI 5002.01, *Meetings in the JCS Conference Room*, discusses the scheduling and use of the room in further detail.

(1) Operations Deputies Meetings. The OpsDeps meetings are used to review plans and discuss DOD issues that can be resolved at the 3-star level. The DJS convenes and presides over the meetings. In the absence of the DJS, the JS DJ-3 presides. The meetings are attended by the DJS, Joint Staff Directors (as required), Service OpsDeps, Joint Staff Vice Director, Secretary of the Joint Staff JS, and other personnel (e.g., CCMDs, DOD Agencies) needed to support an agenda item. CCDRs or Service action officers may attend at the discretion of their principal or as directed by the Director, Joint Staff. If there are any irreconcilable or unresolved differences or non-concurs from the OpsDeps meeting, they are framed for review and resolution at the 4-star level JCS Tank.

(2) JCS Meetings. The purpose of the JCS Tank is to provide a forum for the Chairman to discuss with the other members of the JCS issues and identify dissenting views or differences of opinion. The Chairman convenes and presides over the meetings and attendance is restricted to the Chairman, Vice Chairman, JCS, Director, Joint Staff Assistant to the Chairman, and other individuals specifically designated by the Chairman or Director, Joint Staff. If the Chairman cannot achieve consensus among the JCS on an action that he provides advice on to the President, NSC, or the Secretary, the opposing opinions will be presented along with the Chairman's advice.

(3) Secretary Meetings. Secretary meetings are private executive conferences. Attendance is restricted to the Secretary, DepSecDef, Chairman, Vice Chairman, JCS, DJS, and other individuals specifically designated by the Secretary, Chairman, or DJS.

f. Global Force Management Board (GFMB). The GFMB is a committee of general, flag officers and SES chaired by the DJS to provide senior DOD leadership the means to assess operational outcomes of GFM decisions and provide strategic planning guidance. Joint Staff Directorates; CCMDs; Military Departments/ Services; NGB; and OSD (to include OUSD(P), and OUSD(P&R)), participate in the GFMB at a minimum, with others attending as required. DOD Agencies representatives may attend as required. The GFMB provides GFM direction and reviews and endorses sourcing recommendations for senior leaders. Further details of the GFMB are contained in the GFMIG.

g. Secretary of Defense Orders Book (SDOB). Per the GFMIG, the SDOB is a 3-ring binder used to present orders to the Secretary for approval. It is also a staffing process for all orders to be approved by the Secretary. The JS J-3 is responsible for ensuring the SDOB is properly prepared and staffed through Military Departments/ Services, CCDRs, JS, Chairman, and OSD reflecting stakeholder risks and recommendations. Occurring every 3 weeks, or as required, the SDOB process presents "recommendations for decision" to the Secretary.

h. In-Progress Reviews. In-progress reviews (IPRs) are an iterative dialogue among senior leaders in order to support strategic decision making. The purpose of planning IPRs is to stimulate strategic dialogue between the CCDRs, Secretary, USD(P), OSD, Chairman, JCS, and other appropriate senior leaders. IPRs help shape the plan, discuss interagency or multinational participation, develop a shared understanding of risks, and identify anticipated Secretary decision points. The purpose is not the approval of specific aspects of the plan, but to provide a venue for DOD senior leadership to discuss current planning. The result of the IPR should include OSD and JS endorsement of the planning to date or acknowledgement of friction points and guidance to shape continued planning (Chapter 6-II, *Plan Review Process*, provides further details of the plan review and IPR process.)

 i. Joint Planning Board (JPB). JPBs are hosted by the JS J-5 and OUSD(P) to support CCMD plan development. The JPBs aid in the socializations of CCMD plans with DOD and across the JPEC and can be used to discuss planning issues and conflicts or provide lower-level guidance early in the planning process.

 j. Joint Planning Group (JPG). A JPG is an organization established by a JFC to integrate planning efforts under crisis or non-crisis conditions. A JPG includes representative from all principal stall sections, and component and subordinate commands (see Chapter 5-0, *Joint Planning Process* for more discussion on JPGs).

8. Global Context

Strategic guidance can at times be overwhelming. There are multiple National Strategies and constantly revised Regional Strategies/Plans that require our attention. In 1999 the U.S. had one National Strategy with ten supporting strategies from the OSD and JS. The year 2008 found us with twelve National Strategies and sixteen OSD and JS supporting strategies for a total of twenty-eight strategies. Today there are many, many more. See www.jcs.mil/Library/CJCS-instructions for a current list of PDF downloadable Strategies.

9. Summary

Joint plans and orders are developed with the strategic and military objectives and endstates in mind. The CDR and planners derive their understanding of those endstates from strategic guidance. Strategic guidance comes in many forms and provides the purpose and focus of joint operation planning. Joint operation planners must know where to look for the guidance to ensure that plans are consistent with national priorities and are directed toward achieving national security goals and objectives. The iterative and recursive Joint planning and execution processes work both up and down the chain of command. When necessary senior leaders adjust their strategic direction, that direction can come in many forms: memos, verbal, etc. Joint officers need to know where to look for the additional strategic direction from senior leaders and adjust their plans accordingly. Likewise, Joint officers need to offer military advice up the chain of command with fully considered detailed and risk-informed options and recommendations. It is an incumbent responsibility of military leaders to communicate this understanding of military realities to civilian leadership. This CIV-MIL dialogue may result in additional changes to strategic direction.

> *Simplicity and common sense should characterize planning and strategic direction.*
> *Ingvar Kamprad*

II. Campaigning & Contingency Plans

I. Integration

1. Introduction

CCDRs can be tasked to address missions that cross geographic CCMD boundaries. They develop campaigns to support the global campaign and shape the OE in a manner that supports strategic objectives by integrating posture, resources, and activities to achieve objectives and tasks identified by the CJCS in the global, functional, and regional campaign plans and complement other government efforts related to a geographic region or functional area. CCDRs conduct their campaigns primarily through military engagement, operations, posture, and other activities that seek to achieve U.S. national objectives, protect U.S. national interests, and prevent the need to resort to armed conflict while setting conditions to transition to contingency operations when required.

Campaign plans are developed within the context of existing U.S. national security and foreign policies, and are the primary vehicle for designing, organizing, integrating, and executing security cooperation activities and routine military operations, integrating their posture and contingency plans, and synchronizing these DOD plans and activities with U.S. diplomatic and development efforts. Theater campaign plans also reflect each CCDR's overarching strategy and implement the military portion of national policy and defense strategy by identifying those actions the CCMDs will conduct on a daily basis. Campaign plans are intended to focus and direct steady-state activities that can prevent or mitigate conflict and set the conditions necessary for successful execution of contingency plans.[1]

2. Integrated Planning

The intent of integrated planning is to produce globally integrated plans to advance U.S. interests and achieve U.S. strategic objectives. Addressed in this chapter are the full range of campaign plans (e.g., Global Campaign Plans (GCPs), CCMD campaign plans (CCPs), Functional Campaign Plans (FCPs) and Regional Campaign Plans (RCPs). These plans provide the SecDef and President the best possible information and options to address the complex and uncertain global environment.

3. Global Integration

a. Global Integration is the arrangement of cohesive Joint Force actions in time, space, and purpose, executed as a whole to address trans-regional, all domain, and multifunctional challenges. Global integration ensures the Joint Force maintains a shared understanding of the global OE; collaborates to address threats and challenges; provides the information needed to assess and refine strategies and operations, activities, and investments (OAIs); and ensures the CJCS is able to make informed decisions and provide military advice.[2]

b. The key roles within Global Integration are as follows:

(1) <u>Global Integrator</u>. The SecDef designated the CJCS as the Global Integrator with responsibilities defined in 10 U.S.C. 153. The Chairman's responsibilities as Global Integrator with respect to planning include developing strategic frameworks, preparing strategic plans, providing for the preparation and review of contingency plans, and advising the

---

[1] JP 5-0, Joint Planning

[2] CJCSM 3141.01 Management and Review of Campaign and Contingency Plans

SecDef on allocation and transfer of forces among geographic and functional CCMDs to address trans-regional, all domain, and multifunctional threats. Additional roles include assessing risk, priorities, readiness, preparedness, and budgets.

(2) Coordinating Authority (CA). The Chairman, as Global Integrator, may designate CAs to integrate CCMD planning and campaigning. A CA is generally a CCDR with the preponderance of responsibility for a problem set, but does not receive additional command authority beyond that already assigned in the UCP. CAs will perform three key major functions: planning, assessing, and recommending changes to plans. The CA convenes collaborative forums to develop integrated plans among CCMDs, CSAs, Services, other government agencies, allies, and partner nations and is the designated lead for representing a problem set including topics such as planning, risk, prioritization, resourcing, synchronization of activities in plans, and transition to contingencies. CAs determine relative risk and prioritization of objectives and tasks, and identify additional authorities or execute orders, as required.

(3) Priority Challenge Cross-Functional Team (CFT). Priority challenge cross-functional teams (CFTs) consist of Joint Staff members from each functional and regional area and members from CCMDs and other government agencies, as required. CFTs are charged with maintaining a shared understanding of the strategic and operational environment through activities such as the development of strategies (NMS) and plans (GCPs) with respect to one of the Chairman's priority challenges as designated in the NMS. CFTs develop guidance for the Global Integrator and support globally integrated planning.[3]

(4) Collaborator. A collaborator is a Joint Force organization assigned by the Global Integrator to support integrated planning for a problem set. The collaborator is responsible for working with the CA to develop and assess globally integrated plans. A collaborator is also responsible for providing support plans to the CA when required by the JSCP or other strategic guidance.

4. Global Integrated Plans

a. Global Integration seeks to increase collaboration through intentional JPEC engagement across the whole-of-government to address priority challenges. To accomplish this, the traditional planning framework requires a greater degree of integration. The integrated planning framework, therefore, requires the two traditional plan types—campaigns and contingencies—to closely align in purpose and activities to execute a strategy spanning the spectrum of conflict.

> *Priority Challenge: An actual or potential adversary with the will and capability to undermine U.S. national interests. CJCS priority challenges are designated in the NMS.*

b. Plan Framework

(1) The JSCP is the CJCS's primary document to guide and direct the preparation and integration of joint force campaign plans and associated contingency plans. The three types of campaign plans addressed in the JSCP are: GCPs, FCPs, and CCPs. Although no longer directed in the JSCP, another type of campaign plan is a Regional Campaign Plan (RCP). Regional planning guidance addresses regional threats or challenges that require coordination across multiple CCMDs. Generally, issues that require RCPs are not as significant a threat to U.S. interests as GCPs, but they require attention to ensure they do not evolve into a more significant crisis. If necessary, the SecDef, through the CJCS, could direct a RCP with a designated CA.

GCPs, RCPs, and FCPs are generally problem-focused plans that focus the efforts of multiple organizations on specific problem sets that span organizational and geographic boundaries. GCPs focus on competing with a single priority challenge, while RCPs and FCPs focus on addressing crosscutting challenges, not necessarily one priority challenge.

[3] CJCSM 5115.01, *Priority Challenge Cross Functional Teams*

CCPs are generally organization-focused and serve to guide day-to-day campaigning (incorporating requirements from GCPs and FCPs) and operational execution to achieve U.S. strategic objectives short of war.

(2) <u>Global Integration Frameworks (GIF)</u>. The JSCP directs contingency planning consistent with the CPG. It expands on the CPG with specific objectives, tasks, and linkages between campaign and contingency plans. The JSCP directs the development of Integrated Contingency Plans (ICPs) and Global Integration Frameworks (GIFs). While GCPs guide day-to-day Joint operations, activities, and investments, GIFs provide strategic frameworks to enable a coordinated Joint Force response to crisis or conflict associated with a **priority challenge**. The Chairman recommends which challenges require GIFs based on the SecDef's priorities in the NDS. GIFs are strategic frameworks that enable the Chairman's advice and the SecDef's decisions on strategic risks and trade-offs across and within campaigns and contingencies during crisis or conflict with a priority challenge. GIFs provide a global look at crisis and conflict with one of the priority challenges beyond the scope of a single CCMD. GIFs are informed by GCPs and existing contingency plans.[4]

(3) <u>Contingency Plans</u>. Contingency plans serve as branches or sequels to campaign plans. The Joint Force executes them in a synchronized manner as an ICP or independently for limited purposes. Overlaps between plans represent a convergence of objectives, organizational responsibilities, resources, and readiness. Planners must integrate objectives between campaign plans and contingency plans to employ the campaign plan in a way that seeks to prevent contingencies and posture the Joint Force for successful contingency execution when necessary.

(a) **ICPs are the primary branch plans and war plans associated with a GCP**. The ICP brings together contingency plans from multiple organizations to achieve increased unity-of-effort and closer linkages between complementary contingency plans for a specific problem set.

(b) **Stand-alone contingency plans** will remain necessary for situations not tied to conflict with priority challenges. Organizations may use support plans written for campaign plans or stand-alone contingencies to support ICPs, provided the plans meet the CA's requirements.

5. Problem-Focused Plans (GCP/FCP/RCP)

a. A problem set is an array of threats or adversary capabilities unified in its actions against U.S. interests. Rather than a theater or AOR-centric view, the JPEC starts with a problem-centric view across AORs and functional boundaries, assigning planning responsibilities to CCMDs, CSAs, and other Defense agencies capable of addressing them. To ensure successful integration of planning, the Global Integrator assigns a single CA to each global, regional, or functional problem set. There may be instances, however, when a problem set is divided between more than one CA (e.g., responsibilities for homeland defense within and outside of the Continental United States (CONUS)).

(1) Problem-focused campaign plans include guidance and direction from the Global Integrator, integrated planning by the CA, and support plans developed by collaborators. For GCPs, the Joint Staff develops and maintains the plan while the CA implements, assesses, and recommends updates to it. For FCPs and RCPs, the CA develops, maintains, and updates the plans. The CA regularly coordinates with collaborators to provide feedback to the Global Integrator. As part of this process, the CA may also draft establishing directives to recommend support relationships for approval by the SecDef.

(2) Problem-focused campaign plans enable aligning operations, forces, footprints, agreements, authorities, permissions, and capabilities necessary to promote and protect national interests using the Joint Force. Problem-focused campaign plans provide a description of the strategic environment and situation, campaign approach and intent, related contingency plans, intermediate military objectives, and high-level tasks.

---

[4] *CJCSI 3141.01 Management and Review of Campaign and Contingency Plans*

b. <u>Global Campaign Plans (GCP)</u>. GCPs address threats or challenges that significantly affect U.S. interests across the globe and require coordinated planning across all, or nearly all, CCMDs. GCPs globally integrate the activities of the Joint Force to campaign against the priority challenges. The CJCS manages these plans on behalf of the SecDef. The CJCS approves GCPs after SecDef endorsement.

(1) The Director for Strategy, Plans, and Policy, J-5, is responsible for developing, staffing, reviewing, and preparing GCPs for the CJCS and the SecDef's approval. The GCPs are integrated plans that address the most pressing trans-regional, multi-functional strategic challenges across all domains. The CJCS, as the global integrator, determines which challenges require GCPs.

(2) As problem-focused plans, GCPs look across GCC and FCC seams and simultaneously provide direction to the CCDRs and military advice to the SecDef. GCPs are the focal point for integrated assessment and resource decisions regarding prioritization, posture, capabilities, risk, and risk mitigation measures. The Chairman's military advice, derived from GCP assessments, can take the form of a GCP memorandum focused on a single challenge or be contained within a broader JSPS product. GCPs contain linkages to key contingency plans, identify responsibilities, define objectives, and assign tasks. The CCDR with the preponderance of responsibility for a GCP generally serves as the CA.[5]

c. <u>Regional Campaign Plans (RCP)</u>. RCPs address regional threats or challenges that require coordinated planning across multiple CCMDs. CAs develop, approve, and manage these plans.

d. <u>Functional Campaign Plans (FCP)</u>. FCPs address functional threats or challenges that are not geographically constrained and require coordinated planning across multiple CCMDs. CAs develop, approve, and manage these plans. Functional planning guidance addresses security challenges that are often global in nature or affect more than one GCC. Though functional planning guidance often leads to planning by FCCs, GCCs must ensure their CCPs support achievement of strategic end states and objectives.

The JSCP tasks CCDRs to develop FCPs when achieving strategic objectives requires joint operations and activities conducted in multiple area of responsibilities (AORs). FCPs establish the strategic and operational framework within which subordinate campaign plans are developed. The FCP's framework also facilitates coordinating and synchronizing the many interdependent, cross-AOR missions such as security cooperation, intelligence collection, and coalition support.

6. Combatant Command Campaign Plans (CCPs)

a. CCDRs maintain responsibility for developing campaign plans that address their respective area and functional responsibilities. A CCP is a CCDR-approved plan that incorporates intermediate objectives and tasks from GCPs, RCPs, and FCPs. The CCP is, therefore, the principal operational plan for execution of a CCMD's theater and global responsibilities for all priority challenges. The CCP balances the risks and opportunities of the command and simultaneously accounts for all assigned theater and problem-focused tasks to provide a campaign plan that fully integrates OAIs spanning the CCMDs' assigned responsibilities.

(1) CCP and FCPs implement the military portion of national policy and defense strategy by identifying those actions the CCMDs will conduct on a daily basis. Campaign plans are intended to focus and direct steady-state activities that can prevent or mitigate conflict and set the conditions necessary for successful execution of contingency plans. In linking steady-state objectives with resources and activities, campaign plans enable resource-informed planning and permit prioritization across DOD. The UCP, CPG and JSCP are the core strategic guidance directives for campaign planning. The CPG and the JSCP

---

[5] *CJCSM 3130.01 Campaign Planning Procedures and Responsiblities*

direct that campaign plans focus on AOR strategy implementation through the command's steady-state, foundational activities, which include ongoing operations, security cooperation, and other shaping, preventative, or deterrence actions.

(2) Campaign planning operationalizes a CCMD's theater and functional strategy by comprehensively and coherently integrating all of its directed steady-state (actual) and contingency (potential) operations and activities. The CCMD's strategy and resultant campaign should be designed to achieve the prioritized theater or functional objectives provided in national guidance and serve as the integrating framework that informs and synchronizes all subordinate and supporting planning and operations. CCPs also address objectives directed by GCPs, RCPs, and FCPs.

(a) The CCP's long-term, persistent and preventative activities are intended to identify and deter, counter, or otherwise mitigate an adversary's actions before escalation to combat. Many of these activities are conducted with DOD in support of the diplomatic, economic, and informational efforts of USG partners and partner nations.

(b) The CCP flows from the CDR's strategy for the AOR and provides the action plan to implement that strategy. While each CCMD's campaign plan may approach the task of executing the strategy differently, the plan will address the CDR's AOR in an interconnected and holistic manner and seek to avoid what can be a myopic focus on one or two stove-piped contingency plans. The current construct for nesting plans is first to build the GCPs, RCPs and FCPs, then to build a CCP that implements the activities required to achieve the desired conditions for the theater while dealing with deviations from the strategy through branch plans. Branch plans are brought back into a global planning framework by the creation of ICPs. Supporting activities (to ICPs and to the GCPs/RCPs/FCPs) are contained in Campaign Support Plans (CSPs).

(c) The format of a campaign plan is the basic operation plan format tailored by the CCMD with supporting, posture, and country plans organized and formatted as annexes and appendices of the overarching campaign plan. The CCP should:

- Describe the relevant environment(s).

- Describe the desired military and associated conditions for the environment in the timeframe covered by the strategy.

- -This will include conditions associated with the GCP, RCP and FCPs that apply to the command.

- Address the use of all instruments of power, but be specific about the role of the military instrument in the strategy.

- Describe the military objectives that will support achieving the desired conditions for the relevant environment(s).

- Describe the current and required force posture for the theater, and identify elements of risk in the gap between current and required forces.

- Prioritize activity among subordinate components.

- Link security cooperation activities to specific objectives.

- Describe branches to the campaign plan that require contingency plans and describe the connectivity between the day-to-day activities of the plan and each contingency plan's shaping activities, such as setting the theater for successful contingency plan execution should it be required.

b. CCPs should address the following additional topics: NMS mission area linkages; objectives and tasks assigned to the CCMD from GCPs, RCPs, and FCPs; key support plan linkages; key contingency plan linkages; and risks and priorities considering the CCMDs balancing of global, theater, regional, and functional assigned objectives and tasks.

7. Combatant Command Strategy

a. **Combatant Command Strategy.** A strategy is a broad statement of the CCDR's long-term vision for the AOR and the FCCs long-term vision for the global employment of functional capabilities. It is the bridge between national strategic guidance and the joint planning required to achieve national and command objectives and attain endstates. Specifically, it links CCMD activities, operations, and resources to USG policy and strategic guidance. A strategy should describe the ends as directed in strategic guidance and the ways and means to attain them and it should begin with the *strategic estimate*. Although there is no prescribed format for a strategy, it may include the CDR's vision, mission, challenges, trends, assumptions, objectives, and resources. CCDRs employ strategies to align and focus efforts and resources to mitigate and prepare for conflict and contingencies, and support and advance U.S. interests. To support this, strategies normally emphasize security cooperation activities, force posture, and preparation for contingencies. Strategies typically employ military engagement, close cooperation with DOS, embassies, and other USG departments and agencies. A strategy should be informed by the means or resources available to support the attainment of designated end states and may include military resources, programs, policies, and available funding. CCDRs publish strategies to provide guidance to subordinates and supporting commands/agencies and improve coordination with other USG departments and agencies and regional partners. A CCDR operationalizes a strategy through a campaign plan.[6]

(1) <u>Strategic Estimate</u>. The CCDR and staff, with input from subordinate and supporting commands and agencies, prepare a strategic estimate by analyzing and describing the political, military and economic factors, and the threats and opportunities that facilitate or hinder achievement of the objective over the timeframe of the strategy. The CCMD's input to the Chairman's Comprehensive Joint Assessment (CJA) is produced annually and informs the strategic estimate and its periodic updates.

(2) <u>Policy-Strategy Dynamic</u>. Strategy is always subordinate to policy. However, there is a two-way dependent relationship between policy and strategy. CCDRs bridge the inevitable friction that policy and politics create when developing their AOR strategy. Military strategy must be clear, achievable, and flexible to react to changing policy. Policy may evolve as the AOR strategy is implemented in a dynamic operational environment. Also, policy may change in reaction to unanticipated opportunities or in reaction to unanticipated challenges. The CCDR's role is to keep national policymakers informed about changes in the theater's operational environment that effect such policy decisions and provide advice on the potential outcomes of proposed policy changes.

(3) <u>AOR Strategy Components</u>. The AOR strategy consists of a description of key factors about the environment that provide context for the strategy and affect the achievement of the desired objectives in the theater, a description of the desired strategic objectives (ends), a strategic approach to apply military power in concert with the other elements of national power over time to achieve the desired objectives (ways), a description of resources needed to source the operational approach (means), and a description of the risks in implementing the strategy.

(a) <u>Environment.</u> Planners describe the current environment of the theater, as well as the desired future environment that meets national and regional policy objectives. This provides context for the strategy. While strategy is subordinate to policy, so is it subordinate to the environment – as the environment changes, so must the strategy. The CCDR and staff conduct a strategic estimate which describes the broad strategic factors that influence the theater strategic environment. This continually updated estimate helps to determine missions, objectives, and potential activities required in the CCP.

(b) <u>Ends</u>. The ends for both the strategy and campaign are the strategic-directed objectives. CCDRs use the Secretary's prioritization to guide the order in which they employ

---

[6] *JP 5-0, Joint Planning*

limited resources, accepting risk on lower priority objectives before accepting risk on higher priority objectives.

(c) <u>Ways.</u> The strategic approach describes the ways that the CCDR will employ the command's total joint force along with other elements of national power to advance toward its objective. Although military operations, activities, and investments may achieve some objectives without the involvement of non-DOD agencies, the CDR's strategic approach should be complementary with partner agencies' national security and foreign policy efforts.

(d) <u>Means.</u> The strategy's means are the resources and authorities required to conduct the strategic approach. If there is a reasonable expectation that required means will not become available, then the CCMD must develop an alternative approach within the means that are available or can reasonably be expected to become available. The CCDR takes unresolved issues of means to the DoDs' senior leaders to identify shortfalls.

(e) <u>Risk</u>. CCDRs assess how U.S. national interests are effected within their AOR, how those interests can be threatened, and their ability to execute assigned missions to protect them. This is documented in the CCDR's strategic estimate and in the annual Chairman's CJA. CCDRs' and DoDs' senior leaders work together to reach a common understanding of campaign risk, decide what risk is acceptable, and minimize the effects of accepted risk by establishing appropriate risk controls.

b. <u>Sources of Guidance and Direction for CCMD Strategies</u>. The CCMD staff translates national policy and strategy into military operations, actions, and activities. The guidance to the CCDR formulating the AOR strategy comes from a variety of formal and informal sources. Very often, the national policy and corresponding guidance is not explicit. This places a premium on the CCDR's ability to interpret, analyze, and synthesize the many sources of national intent, and then communicate this synthesis back to the national policymakers to ensure that he/she is in sync with their vision (in fact, the CCDR may actually shape their vision).

(1) Chapter 1-I, *Strategic Organization*, describes the CPG, NDS, NMS, and JSCP, as sources of formal guidance. However, in a dynamic strategic environment, policy may evolve and the CCDR must stay attuned to evolving descriptions and applications of national interests as described by the President, SecDef, and other senior government officials through less formal means such as speeches, social media, and verbal guidance. Though not directive in nature, guidance contained in various U.S. interagency and even international directives, such as UN Security Council Resolutions, will also impact campaign end states and objectives. Perhaps most importantly, the CCDR must continually analyze the dynamic relationships within the theater to describe the desired end state and present limitations on ways to achieve that end state.

(2) <u>Identifying and collaborating with stakeholders</u>. CCDRs must coordinate and synchronize their strategies and implementation activities with other stakeholders, to include non-DOD government agencies and other nations. One critical partner is the DOS, which provides some guidance and many of the resources for the CCDR's security cooperation program, which is vital to the implementation of the AOR strategy. Similarly, other agencies, such as the U.S. Agency for International Development (USAID), routinely conduct developmental activities in countries of the region, requiring the CCDR to ensure compatibility between military activities and USAID activities. The CCDR and staff may have to find ways to work through some policy interpretations that might inhibit formal coordination with non-DOD executive branch agencies. The CCDR should coordinate closely with international partners, to include nations, international organizations, and non-governmental and private organizations. Though it is not always realistic to align goals and activities among all stakeholders, it is important to understand the purpose of the other activities, and to work towards mutual benefit when possible. On the other hand, the CCDR should be aware of competing agendas and activities by other non-U.S. organizations (and, in rare cases, U.S. organizations) that may present obstacles to achievement of the AOR strategy objectives. Formally, the CCDR works through OSD to reconcile and synchronize activities with other organizations, but an informal coordination network is also crucial to success. It is important

to consider that non-military and international actors have legitimate agendas and will be active (sometimes the lead) players to a greater or lesser extent across the full spectrum of conflict.

    c. Operationalizing Strategy. The CPG also gives regional planning guidance for CCMD campaign and related contingency planning efforts. CCMD campaign planning operationalizes the CCMD's strategy by integrating all of its directed steady-state (actual) and contingency (potential) operations and activities. The resultant CCMD campaign serves as the integrating framework that informs and synchronizes all regional planning and operations. The theater campaign's operational approach and design provides the structure and linkage to integrate all subordinates and supporting U.S. military regional and functional operations, activities, events, and investments in support of the AOR strategy. Strategy implementation, and the broader theater engagement, ebbs and flows within a dynamic operational environment as U.S. military operations, activities, and investments are executed in concert with other USG activities, create effects, and those effects are assessed by the CCMD against tangible goals linked to desired objectives. Contingency plans are developed for anticipated (potential) threats and catastrophic events as branch plans under the theater campaign. As events or circumstances dictate, the CCDRs may direct through an appropriate order (e.g., PLANORD, WARNORD, ALERTORD) a given contingency plan be placed under enhanced plan management because political circumstances or indications and warning have changed its priority, visibility, likelihood, or risk to national security. Issuance of an EXORD signals the transition of that contingency plan to an OPORD for implementation as a branch plan under the theater campaign.

    (1) Geographic Combatant Commands (GCCs) prepare CCPs to achieve CPG-directed objectives for their UCP-assigned AORs. The GCCs are responsible for integrating the planning of designated missions assigned to FCCs into their theater campaigns. FCCs prepare FCPs to achieve CPG-directed objectives for their UCP-assigned missions and responsibilities. FCCs are responsible as directed for synchronizing planning across CCMDs, Services, and Defense Agencies for designated missions. CCDRs document the full scope of their theater or functional campaigns in the set of plans that includes the overarching theater or FCP, and all of its subordinate, supporting, posture, country (for GCCs), operation (for operations currently in execution), and contingency plans.

    (2) CCDRs use their campaign plans to articulate resource requirements in a comprehensive manner vice an incremental basis and also to provide a vehicle for conducting a comprehensive assessment of how the CCMD activities are contributing to the achievement of IMOs and strategic objectives (see Chapter 6-III, *Plan Assessment*).

    d. Nested Plans. The nesting relationship of GCPs, CCPs, FCPs, subordinate campaign plans, and contingency plans is portrayed in Figure A. GCCs must ensure their CCP and contingency plans support the achievement of strategic objectives. Similarly, FCCs must ensure their functional plans support the achievement of strategic objectives. Subordinate campaign plans are nested under the CCP and synchronized with the FCPs. Contingency plans are branches to CCPs and/or subordinate campaign plans. The relationship of CCPs, FCPs, subordinate, supporting and campaign support plans, and contingency plans is unique to each CCDR. GCP's impact multiple theaters; therefore, GCCs develop subordinate plans in support of the GCPs. These subordinate campaign plans are then embedded in the CCP.

    (1) Strategic guidance requires all regional and contingency planning to nest under a CCMD campaign as depicted in Figure B. CCMD campaign planning is mostly art; requiring the judicious application of operational design to arrive at a concept for theater engagement that provides the conceptual linkage of strategic-directed objectives, through resource informed ways, and means, to regional operational actions. CCMD campaign planning directs and justifies the resources required by the CCDR to accomplish assigned theater objectives.

Plans Relationship and Nesting

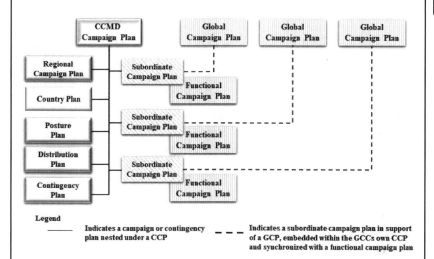

Legend

─────── Indicates a campaign or contingency plan nested under a CCP

─ ─ ─ Indicates a subordinate campaign plan in support of a GCP, embedded within the GCCs own CCP and synchronized with a functional campaign plan

Figure A. Plans Relationship (CJCSM 3141.01)

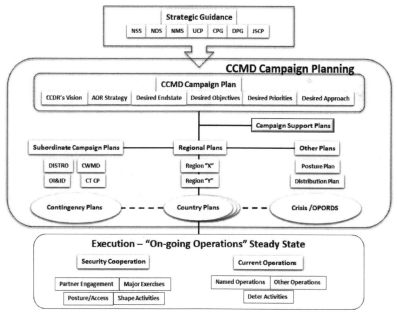

Figure B. Nested Plans in Generic CCP Framework (CJCSG 3130)

(2) CCMD campaign planning is inherently intergovernmental. It is informed by the strategic planning of other USG agencies, in particular the DOS and USAID. The intent is for the CCMD campaign design to complement and support the DOS's broader foreign policy objectives and, to the extent possible, not undermine or work at cross purposes to the goal and activities of other USG agencies in the region. CCMD theater campaign planning is also heavily informed by detailed country planning. Country-level plans help quantify and justify aggregate theater military resource requirements. In peacetime, regional military actions occur in a world where the U.S. Ambassadors' objectives have primacy. Therefore, regional U.S. military operations, activities, events, and investments are prioritized, aligned, and/or integrated with U.S. developmental and diplomatic actions at the country level to achieve unity of effort and husband-scarce resources. In the end, CCMD campaign planners seek to synchronize and nest the planned operations, activities, events, and investments across posture planning, country planning, security cooperation planning, contingency planning, shaping phase integration, strategic communication planning, interagency planning, and multi-national planning in CCPs to promote overall regional unity of effort.

8. Elements of a Combatant Command Campaign Plan

a. Campaign Plan. The CCP consists of all plans contained within the established theater or functional responsibilities to include contingency plans, subordinate and supporting plans, posture plans, country-specific security cooperation sections/country plans (for geographic commands), and operations in execution.

(1) The campaign plan operationalizes the CCDR's strategy by organizing operations, activities, and investments within the assigned and allocated resources to achieve the CPG- and JSCP-directed objectives, as well as additional CCDR-determined objectives within the timeframe established by the CPG or JSCP.

(2) The campaign plan should show the linkages between operations, activities, investments, and expenditures and the campaign objective and associated end states that available resources will support. The campaign plan should identify the assessment process by which the command ascertains progress toward or regression from the national security objectives.[7]

b. Posture Plan. The posture plan is the CCMD's proposal for forces, footprint, and agreements required and authorized to achieve the command's objectives and set conditions for accomplishing assigned missions. GCCs prepare Posture Plans which outline their posture strategy, link national and theater objectives with the means to achieve them, and identify posture requirements and initiatives to meet CCP objectives. The Posture Plans is the single source document used to advocate for changes to posture and to support resource decisions.

(1) Posture Planning. Posture Plans propose a set of posture initiatives and other posture changes, along with the corresponding cost data necessary to support the DoDs activities as described in the CCP. Posture Plans also must account for the desires of the FCCs, other GCCs, and Services, then balance these possibly-competing desires.

(2) CCMD planners must ensure theater objectives that run counter to global and regional objectives are properly aligned and prioritized to ensure that those objectives with the highest priority are elevated and the risk associated with the theater objectives that are counter are well understood. Also, planners must understand that Posture Plans are integrally linked to the Services ability to resource them both from a fiscal and a force requirements perspective.

c. Theater Logistics and Distribution Plans (TDP) The TDP provides detailed theater mobility and distribution analysis to ensure sufficient capacity or planned enhanced capability throughout the theater and synchronization of distribution planning throughout the global distribution network. The TDP includes a comprehensive list of references, country data,

---

[7] *CJCSM 3130.01 Campaign Planning Procedures and Responsiblities*

and information requirements necessary to plan, assess, and conduct theater distribution and joint reception, staging, onward movement, and integration (JRSOI) operations. The GCCs develop their TDPs using the format in USTRANSCOM's Campaign Plan for Global Distribution, JSCP, and JSCP Logistics Supplement. TDPs and Theater Posture Plans (TPPs) complement each other by posturing forces, footprints, and agreements that will interface with the GCC's theater distribution network in order to provide a continuous flow of material and equipment into the AOR. This synchronization enables a GCC's theater distribution pipeline to have sufficient capacity and capability to support development of CCPs, OPLANs, and CONPLANs.[8]

(1) <u>Theater Logistics Overview (TLO)</u>. The TLO codifies the GCC's Theater Logistics Analysis (TLA) within the TPP. The TLO provides a narrative overview, with supporting matrices of key findings and capabilities from the TLA, which is included in the TPP as an appendix.

(2) <u>Theater Logistics Analysis (TLA)</u>. The TLA provides detailed, country-by-country analysis of key infrastructure by location or installation (main operating base [MOB]/forward operating site [FOS]/cooperative security location [CSL]), footprint projections, HN agreements, existing contracts, and task orders required to logistically support theater campaigns and their embedded contingency operations.

d. <u>Regional and Country-Specific Security Cooperation Sections (CSCS)/Country Plans</u>. As needed or directed, CCDRs prepare country-specific security cooperation plans (codified in CSCS) within their campaign plans for each country where the CCMD intends to apply significant time, money, and/or effort. CCDRs may also prepare separate regional plans. These are useful to identify and call out activities directed toward specific regional or country objectives and provide focus for the command.

(1) <u>Country Plans</u>. GCCs prepare selected country plans in collaboration with respective security cooperation organizations. Country plans are nested within CCPs by designing country objectives to support CCP IMOs, link to strategic-directed objectives, and complement the CCP's activities. A country plan describes how the CCMD, working with the U.S. country team, will engage with the partner country to achieve both U.S. and partner country security objectives, and the role the partner has agreed to play or is expected to play in the CCMD campaign; and, it must be consistent with the bilateral security agreements that govern the U.S.-partner country relationship.

(2) <u>Country Planning</u>. This is usually developed in parallel with CCPs which allows appropriate nesting of country plans within CCPs and allows country planners to inform and be informed by the higher-level discussions between CCMD and relevant steady-state actors. Country planners ultimately make a determination on "what the U.S. wants," and "does not want," each country to do in support of U.S. strategy. These country-level assessments inform the development of a country engagement concept (for country-level plans) and the larger theater campaign concept of engagement. Country planning begins with an assessment of the partner country's security environment. In addition to understanding geopolitical trends or conditions that influence a partner nation, planners also assess significant internal and external threats to both the partner nation and neighboring countries in the region. CCMD theater planners evaluate the capabilities and resources of partner nations and their ability and willingness to participate with U.S. forces in coordinated operations, activities, and events. CCMD theater planners also identify key security-related opportunities in context with the goals and activities of other USG agencies in the region. Planners must also review the DOS and USAID strategies to gain appreciation of diplomatic and developmental objectives and goals for a particular country.[9]

(3) Regional and country-specific security cooperation sections/country plans can also serve to better harmonize activities and investments with other agencies. By isolating the

---

[8] *JP 4-0, Joint Logistics*

[9] JP 3-20, Security Cooperation and CJCSM 3130.03, Planning and Execution Formats and Guidance

desired objectives, planners can more easily identify supporting efforts and specific assessment measures toward achieving U.S. objectives.

(4) Where the U.S. has identified specific objectives with a country or region (through strategic guidance or policy), separate regional or country-specific security cooperation sections/country plans help to identify resource requirements and risk associated with resource limitations that may be imposed.

e. Subordinate, Supporting, and Campaign Support Plans. Once the Global Integrator assigns a CA and issues guidance and direction for a problem set, the CA will refine the campaign plan or contingency plan with the collaborators. Collaborators prepare support plans to document assigned tasks and how to address them. CAs use support plans to develop the overall concept of operations for the campaign or contingency, synchronizing the actions of the Joint Force in time and space. Collaborators work with the CA to ensure their support plans effectively address the problem set and integrate with the OAIs of other organizations. Support plans do not change supported/supporting command relationships. An organization must submit a support plan if a specific plan in the JSCP designates it as a collaborator and the CA requests a support plan. If the CA does not request a support plan, a collaborator is not obligated to submit one. If the JSCP does not specify that an organization is a collaborator, the CA may still request one and negotiate the details with the organization. CAs are not required to create support plans for the plans they lead, but planning must capture (at a minimum) all collaborators' force and logistics resource/capabilities requirements.

(1) Subordinate Campaign Plan. JFCs subordinate to a CCDR or other JFC may develop subordinate campaign plans in support of the higher plan to better synchronize operations in time and space. It may, depending upon the circumstances, transition to a supported or supporting plan in execution.

(2) Supporting Plans. Prepared by a supporting CDR, a subordinate CDR, or the head of a department or agency to satisfy the requests or requirements of the supported CDR's plan.

(3) Campaign Support Plans (CSPs). Developed by the Services, NGB, and DOD agencies that integrate the appropriate USG activities and programs, describes how they will support the CCMD campaigns, and articulate institutional or component-specific guidance.

f. Security Cooperation Plans. Security cooperation encompasses activities undertaken by the DOD to assist and enable international partners to apply capability and capacity; provide U.S. access to territory, information, and resources; and/or take a political action in support of U.S. goals. It includes, but not limited to, DOD interactions with foreign defense and security establishments and DOD-administered security assistance programs. Security cooperation activities are in CCMDs campaign, posture, and country plans and the Services, Defense agencies, and NGBs CSPs. Examples of security assistance programs are the Foreign Military Sales Program, Economic Support Fund, and the International Military Education and Training Program, to name a few.

g. Contingency Plans. Contingency plans are branch plans to the campaign plan that are based upon hypothetical situations for designated threats, catastrophic events, and contingent missions outside of crisis conditions (see Section II, The Contingency Plan).

(1) The campaign plan should address those known issues in the contingencies that can be addressed prior to execution to establish conditions, conduct deterrence, or address assumptions. As planners develop contingency plans, issues and concerns in the contingency should be included as an element of the campaign.

(2) Deliberately Planning for a Contingency. Contingency planning is conducted to respond to defined planning tasks identified in the CPG and the JSCP or as directed by the CCDR. It is based on derived assumptions and forces apportioned from Section V of the GFMIG. Contingency plans are built to account for the possibility the CCP steady-state

activities could fail to prevent aggression, preclude large-scale instability in a key state or region, or mitigate the effects of a major disaster. Under the CPG/JSCP campaign planning direction, contingency plans are conceptually considered branches of the overarching CCPs. Contingency planning informs the theater campaign and prepares the CCMD for transition to crisis planning when necessary. Planners developing contingency plans identify shaping, condition setting, prevention, or preparation for entry task requirements to theater campaign planners that can be accomplished within the scope of the CCP's steady-state activities. Contingency planners may also identify shaping, condition setting, prevention, or preparation for entry task requirements specific to their plan that would only be implemented in the event of an emerging crisis.

> *"Perception is strong and sight weak. In strategy it is important to see distant things as if they were close and to take a distanced view of close things."*
> *Miyamoto Musashi, legendary Japanese swordsman*

h. Interagency Coordination. The quality of DOD CCMD campaign planning improves with early and regular participation of other U.S. departments and agencies. Interagency coordination is the interaction that occurs among U.S. agencies, including DOD, for the purpose of accomplishing CCMD theater objectives. In general, security cooperation will align with broader USG policy. Interagency coordination forges the vital link between the U.S. military and the other instruments of national power.

(1) Of note and worth mentioning are the geographic boundary differences between the DOS Bureaus (Figure C) and the DOD geographic commands (Figure D) as depicted in the figures on the following pages. It's important we recognize these seams and boundary differences to ensure smooth coordination between these two interagency partners.

(2) When overlaid with each other we see the potential coordination challenges that face both the DOS and DOD when working across boundaries. Close coordination is required between DOS Bureaus and DOD geographic CCMD's to ensure national security issues and priorities are addressed.

i. Multinational Plans. CCMD campaign planning considers the capabilities and activities of allies and partners including regional security organizations and other multinational organizations to complement U.S. efforts to achieve regional objectives. Multinational integration in theater campaign planning is accomplished in national and international channels. Collective security goals, strategies, and combined plans are developed in accordance with individual treaty or alliance procedures. Host nation support and contingency mutual support agreements are usually developed through national planning channels.

II. THE CONTINGENCY PLAN

1. Contingency Plans

a. Campaign plans provide the vehicle for linking steady-state shaping activities to current operations and contingencies. They establish operations, activities, and investments the command undertakes to achieve specific objectives (set conditions) in support of national policy and objectives. Contingency plans identify how the command might respond in the event of a crisis or the inability to achieve objectives. Contingency plans specifically seek to favorably resolve a crisis that either was not or could not be deterred or avoided by directing operations toward achieving specified objectives (campaign plan).

(1) Contingency plans have specified end states that seek to re-establish conditions favorable to the U.S.. They react to conditions beyond the scope of the CCP.

(2) Contingency plans have an identified military objective and set of termination criteria. Upon terminating a contingency plan, military operations return to campaign plan execution. However, the post-contingency OE may require different or additional military activities to sustain new security conditions.

Regional Priorities

Department of State

The DOS has six Bureaus covering regional priorities.[10]

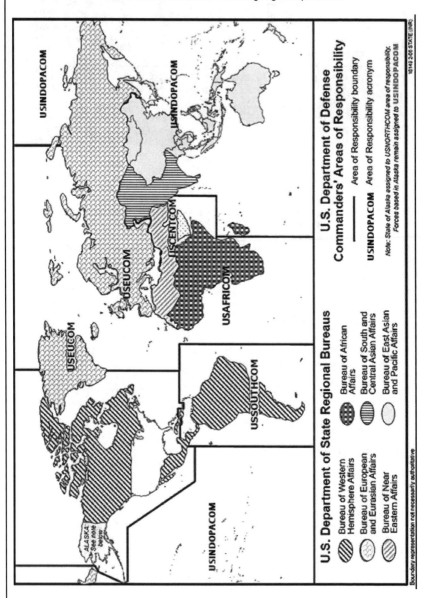

Figure C. U.S. Department of State Regional Bureaus Overlaid on CCDR's AOR's

[10]*CJCSM 3130.01 series*

Geographic Commands

The DOD has seven geographic commands:[11]

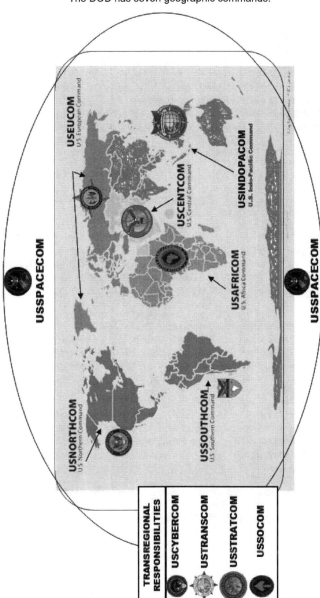

Figure D. DOD Geographic and Functional Commands

[11]CJCSM 3130.01 series

(3) Although campaign plan operations, activities, and investments can have deterrent effects, contingency plan deterrence activities specifically denote actions for which separate and unique resourcing and planning are required. These actions are executed on order of the President or SecDef and generally entail specific orders for their execution and require additional resources allocated through GFM processes.

b. Contingency plans become "branch" plans to the overarching CCMD campaign plan. A contingency is a situation that likely would involve military forces in response to natural and man-made disasters, terrorists, subversives, military operations by foreign powers, or other situations as directed by the President or SecDef. Contingency plans are built to account for the possibility that steady-state shaping measures, security cooperation activities, and operations could fail to prevent aggression, preclude large-scale instability in a key state or region, or mitigate the effects of a major disaster. Contingency plans address scenarios that put one or more U.S. strategic objectives or end states in jeopardy and leave the U.S. no other recourse than to address the problem at hand through military operations. Military operations can be in response to many scenarios, including armed aggression, regional instability, a humanitarian crisis, or a natural disaster. Contingency plans should provide a range of military options coordinated with total U.S. Government response.

2. Contingency Planning

Contingency planning is the deliberate process in which the JPP is utilized to develop responses to potential scenarios which occurs in non-crisis situations based on hypothetical contingencies. Following the guidance provided by the JSCP, CCDRs prepare, submit, and continuously refine their plans and continue until the requirement no longer exists. Planning guidance is provided in the CPG, JSCP, Strategic Guidance Statements (SGS) and through SecDef and CCDR IPRs, which exist to stimulate a synchronized dialogue between the supported CCDR, the SecDef and other appropriate senior leaders. Figure E on the following page provides a notional depiction of joint planning for a contingency plan.

a. Contingency planning is an iterative process and is adaptive to situational changes within the operational and planning environments. The process allows for changes in plan priorities, changes to the review and approval process of either a single plan or a category of plans, and contains the flexibility to adjust the specified development timeline for producing and refining plans. Contingency planning also facilitates the transition to crisis planning by having a library of information available (intelligence, logistics, forces, etc.) for disparate regions of the globe.

b. The JSCP links the JSPS to joint planning, identifies broad scenarios for plan development, specifies the type of joint plan required, and provides additional planning guidance as necessary. A CCDR may also initiate a contingency plan by preparing plans not specifically assigned but considered necessary to discharge command responsibilities. If a situation has developed that warrants contingency planning but was not anticipated in strategic guidance, the SecDef, through the CJCS, tasks the appropriate supported CCDR and applicable supporting CCDRs, Services, and CSAs to begin planning in response to the new contingency. The primary mechanism for tasking a contingency plan outside of the strategic cycle will be through SGS from the SecDef and endorsed by message from the CJCS to the CCDRs.

c. Plans are produced and updated periodically to ensure relevancy. Contingency planning most often addresses military options requiring combat operations; however, plans must account for other types of joint operations across the range of military operations. For example, stability operations, are very complex and require extensive planning and coordination with non-DOD organizations with the military in support of other agencies. Contingency planning occurs in prescribed cycles in accordance with formally established procedures that complement and support other DOD planning cycles. In coordination with

Joint Planning: Contingency (Notional Depiction)

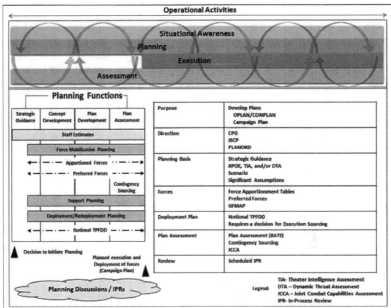

Figure E. Contingency Notional Depection

Purpose. Contingency plans or campaign plans are the principal outputs of this type of planning. These plans can provide the basis for subsequent order development should the plan transition to execution.

Direction. Contingency planning is usually initiated via planning tasks in the CPG or JSCP. Planning requirements that emerge outside of the strategic guidance update cycle may also be directed via a CJCS PLANORD. Additionally, CDRs at all levels may initiate planning on their own authority when they identify a planning requirement not directed by higher authority.

Planning Basis. The basis for all planning including joint planning is the planning framework provided by strategic guidance intended to advance U.S. national interests. This framework is further informed by an analysis of the OE based upon the outputs of the JIPOE process, TIA, and DTA where applicable. Contingency planning that is based upon a hypothetical scenario usually requires significant planning assumptions to complete detailed planning.

Forces. The forces available for planning are a key planning assumption. The quantities of forces in the Force Apportionment Tables provide an estimate of the Military Departments'/Services' capacity to generate force elements along general timelines. These apportioned quantities may be an initial starting point for planning. From this starting point planning may be refined by preferred force identification to provide higher fidelity force planning assumptions necessary for plan feasibility analysis. For campaign plans, the appropriate FY GFMAP may be considered as a projection of forces available to conduct planned campaign activities.

Deployment Plan. Time-phased force lists during plan development may be documented in a notional TPFDD. The JSCP prescribes which specific contingency plans are required to be developed with a notional TPFDD. The units identified in a notional TPFDD are planning assumptions and are not execution sourced forces. A notional TPFDD developed during planning still requires execution sourcing in order to be executed.

Plan Assessment. Developed plans may be assessed periodically to determine if the plan needs to be refined or adapted, terminated or executed based on changes in the OE. The assessment of contingency plans may include contingency sourcing as part of a JCCA to provide a more detailed assessment of the ability to execute a selected plan under prescribed conditions.

Review. Developed plans allow for sustained informal dialog between planners and senior leadership during planning. Some CPG- or JSCP-directed CCMD planning requirements are prescribed JPEC review and scheduled IPRs with the Secretary or designated representative.

the JPEC, the JS develops and issues a planning schedule that coordinates plan development activities and establishes submission dates for joint plans.

d. Plan Management. The CJCS reviews planning accomplished for contingency plans specified in the JSCP, combined military plans, military plans of international treaty organizations, and as otherwise specifically directed by the SecDef.

(1) The JS Director for Operational Plans and Joint Force Development (JS J7) is responsible for the plan management processes for all contingency plans, to include plans maintained by the JS Director of Operations (DJ-3) that are not in execution. DJ-3 is responsible for managing the process of developing operations plans in a crisis environment, overseeing the execution of operations, and maintaining subject matter experts (SME) on all J3 developed plans.

(2) The J-5/Joint Operational War Plans Division (JOWPD) serves as the office of primary responsibility (OPR) within the JS for all contingency plan matters, to include bilateral military plans and military plans of international treaty organizations not specifically designated otherwise. This consists of both the management of contingency plans and the plan review process, including but not limited to review of the notional Time Phased Force Deployment Data (TPFDD), final plan, and facilitation of contingency plan IPRs with the SecDef.

(3) JOWPD is the primary liaison for the CCDR with both the Office of the Chairman of the Joint Chiefs of Staff (OCJCS) and the Office of the Under Secretary of Defense for Policy (OUSD(P)) for development of contingency plans and the plan review process.

(4) To achieve rapid planning with greater efficiency, this process features early planning guidance with frequent dialogue in the form of IPRs between the JS and planners and the CDRs and DOD senior leaders. This dialogue promotes an understanding of, and agreement on the mission, planning assumptions, threat definitions, interagency, allied planning cooperation, risks, courses of action, and other key factors (see Chapter 6-II, *Plan Review Process*).

3. Contingency Planning and the GFM Process

a. GFM Process during Contingency Planning. The Apportionment Tables maintained in Section V of the GFMIG (*Apportionment*) provides the number of forces reasonably expected to be available for planning. These tables should be used as a beginning assumption in planning (for greater details see Chapter 3, *GFM*).

b. Force Identification Planning. The purpose of force identification planning is to identify all forces needed to accomplish the supported CDR's CONOPS and phase the forces into the theater. It consists of determining the force requirements by operation phase, mission, mission priority, mission sequence, and operations area (OA). It includes major force phasing; integration planning; force list structure development; followed by force list development. Force identification planning is the responsibility of the CCDR supported by the Service component CDRs, Services, JS, and JS JFC/JFPs. The primary objectives of force identification planning are to apply the right force to the mission while ensuring force visibility, force mobility, and adaptability. Force identification planning begins during CONOPS development. During the planning stage, these forces are planning assumptions called preferred forces. The supported CDR determines force requirements, identifies and resolves or reports shortfalls, develops a TPFDD and letter of instruction (LOI), and designs force modules to align and time-phase the forces in accordance with (IAW) the CONOPS. Major combat forces are selected from those apportioned for planning and included in the supported CDR's CONOPS by operation phase, mission, and mission priority. The supported CCDR Service components then collaboratively make tentative assessments of the combat support (CS) and combat service support (CSS) required IAW the CONOPS.

c. For some high-priority plans, as the plan is either approved or nearing approval, the CJCS may direct the JS JFC/JFPs to contingency source a plan to support CJCS and/

or SecDef's strategic risk assessments. Contingency sourcing allows greater fidelity in force planning. Contingency sourcing is based on planning assumptions and contingency sourcing guidance. To the degree that the contingency sourcing planning assumptions and sourcing guidance mirrors the conditions at execution, contingency sourcing may validate the supported CCDR-preferred forces.

d. If the plan is directed to be executed, the force requirements are forwarded to the JS for validation and tasked to the JS JFC/JFPs to provide sourcing recommendations to the SecDef. When the SecDef accepts the sourcing recommendation, the FP is ordered to deploy the force to meet the requested force requirement. After the actual forces are sourced, the CCDR refines the force plan to ensure it supports the CONOPS, provides force visibility, and enables flexibility.

III. PLANNING IN A CRISIS
1. Crisis

A crisis is an incident or situation involving a threat to the U.S. rapidly and creates a condition of such diplomatic, economic, political, or military importance that the President or SecDef considers a commitment of U.S. military forces and resources to achieve national objectives. It may occur with little or no warning (e.g., Indonesian Tsunami or Haitian Earthquake) and it is fast-breaking and requires accelerated decision-making and timelines. Sometimes a single crisis may spawn another crisis elsewhere. Figure F provides a notional depiction of Joint planning under crisis conditions (see Chapter 3-II, *Force Identification and Sourcing*.)

2. Crisis Planning Relationship to Contingency Planning

> *"It is a common mistake in going to war to begin at the wrong end, to act first, and wait for disasters to discuss the matter."*
> – Thucydides, Greek historian. "History of the Peloponnesian War"

Crisis planning is accomplished utilizing the JPP which can respond to crises spanning the full range of military operations. Strategic planning supports the JPP by anticipating potential contingencies and developing plans that facilitate execution planning during crises. As discussed earlier, contingency planning prepares for a hypothetical military contingency based on the best available intelligence, while using forces and resources projected to be available for the period during which the plan will be effective. It relies heavily on assumptions regarding the political and military circumstances that will exist when the plan is implemented. Even though every crisis situation cannot be anticipated, the distributed collaborative environment, detailed analysis, and coordination which occur during contingency planning may facilitate effective decision-making, execution, and redeployment planning as a crisis unfolds. In a crisis situation assumptions and projections made during contingency planning are replaced with facts and actual conditions. Figure G compares contingency and crisis planning with time, environment, forces, etc.

a. Planning in a crisis situation encompasses the activities associated with the time-sensitive development of OPORDs for the deployment, employment, and sustainment of assigned, attached, and allocated forces and resources in response to an actual situation that may result in actual military operations. While contingency planning normally is conducted in anticipation of future events, crisis planning is based on circumstances that exist at the time planning occurs. There are always situations arising in the present that might require U.S. military response. Such situations may approximate those previously planned for in contingency planning, though it is unlikely they would be identical, and sometimes they will be completely unanticipated. The time available to plan responses to such real-time events

Joint Planning: Crisis Conditions (Notional Depiction)

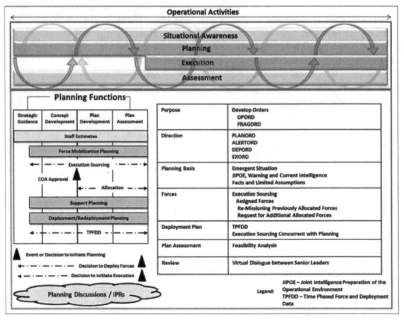

Purpose	Develop Orders OPORD FRAGORD
Direction	PLANORD ALERTORD DEPORD EXORD
Planning Basis	Emergent Situation JIPOE, Warning and Current Intelligence Facts and Limited Assumptions
Forces	Execution Sourcing Assigned Forces Re-Missioning Previously Allocated Forces Request for Additional Allocated Forces
Deployment Plan	TPFDD Execution Sourcing Concurrent with Planning
Plan Assessment	Feasibility Analysis
Review	Virtual Dialogue between Senior Leaders

Legend: JIPOE – Joint Intelligence Preparation of the Operational Environment
TPFDD – Time Phased Force and Deployment Data

Purpose. Planning under crisis conditions is intended to produce an order as the principal output. Typically, OPORDs or FRAGORDs are developed during crisis to direct the execution of military activities.

Direction. Planning is initiated when an emergent situation or crisis is identified that may warrant a near-term military response. Planning under crisis conditions may be directed by a PLANORD or ALERTORD. Planning and deployment or execution may be directed by a DEPORD or EXORD. During a crisis the direction to deploy forces or execute operations may occur at any time during planning with continued planning and initial execution functions conducted in parallel.

Planning Basis. During crisis the basis for planning is the ongoing situational awareness and assessment of the emergent situation. The current JIPOE, warning and current intelligence provide the context for planning. Planning during crisis will usually require less assumptions than non-crisis planning due to more facts being available for an actual vice hypothetical event.

Forces. Force requirements identified during crisis planning may be execution-sourced concurrent to plan development. CCDRs may execution source assigned forces to the operations they are directed to conduct. With the approval of a COA, the Secretary may direct the allocation of additional forces to a CCDR. Plan development may then be conducted based upon actual sourced units vice planning assumptions to facilitate the transition to execution.

Deployment Plan. During a crisis deployment, plans are usually documented as a TPFDD. To meet requirements, the CCDR may task assigned forces. In the case of force shortfalls, the CCDR may request allocated forces. Force requirements are execution sourced concurrent to planning and unit and movement data are entered. The sourced TPFDD is a deployment plan that can be executed by the supported CCMD with support from USTRANSCOM and FPs.

Plan Assessment. Given the limited planning time under crisis conditions, the feasibility analysis done during plan development may effectively be the plan assessment. Under crisis conditions, plan feasibility is assessed and the plan refined in the context of the current conditions of the OE as time allows.

Review. Planning under crisis conditions is often shaped by frequent, potentially virtual dialog between senior civilian and military leaders. Decisions and approval of planning may be initially conveyed verbally and subsequently codified in orders or directives in order to expedite the transition to execution.

Figure F. Crisis Notional Depiction

Contingency and Crisis Comparison

Planning initiated in response to an emergent event or crisis uses the same construct as all other planning. However, steps may be compressed to enable the time-sensitive development of OPLANs or OPORDs for the deployment, employment, and sustainment of forces and capabilities in response to a situation that may result in actual military operations. While planning for contingencies is based on hypothetical situations and normally is conducted in anticipation of future events, planning in a crisis is based on circumstances that exist at the time planning occurs.

	Planning for a Contingency	Planning in a Crisis
Time Available to Plan	As defined in authoritative directives (normally 6(+) months)	Situation dependent (hours, days, or up to 12 months)
Environment	Distributed, collaborative planning	Distributed, collaborative planning and execution
JPEC Involvement	Full JPEC participation. Note: JPEC participation may be limited for security reasons.	Full JPEC participation. Note: JPEC participation may be limited for security reasons
APEX Operational Activities	Situational Awareness Planning Assessment	Situational Awareness Planning Execution Assessment
APEX Functions	Strategic Guidance Concept Development Plan Development Plan Assessment	Strategic Guidance Concept Development Plan Development Plan Assessment
Document Assigning Planning Task	CJCS issues (1) JSCP, (2) PLANDIR, or (3) WARNORD for short-suspense planning	CJCS issues WARNORD, PLANORD or SecDef approved ALERTORD
Forces for Planning	Apportioned in JSCP	Allocated in WARNORD, PLANORD, or ALERTORD
Planning Guidance	CJCS issues JSCP or WARNORD. CCDR issues planning directive and TPFDD LOI	CJCS issues WARNORD, PLANORD, or ALERTORD. CCDR issues WARNORD, PLANORD or ALERTORD and TPFDD LOI to subordinates, supporting commands and supporting agencies
COA Selection	CCDR prepares COAs and submits to CJCS and SecDef for review. Specific COA may/or may not be selected	CCDR develops Commanders Estimate with recommended COA
CONOPS Approval	SecDef approves planning or directs additional planning or changes	President/SecDef approve COA, disapproves or approves further planning
Final Planning Product	Campaign Plan Level 1-4 Contingency Plan	OPORD
Final Planning Product Approval	CCDR submits final plan to CJCS for review and SecDef for approval	CCDR submits final plat1 to President/SecDef for approval
Execution Document	NA	CJCS issues SecDef-approved EXORD. CCDR issues EXORD
Output	Plan	Execution

Legend

ALERTORD	Alert Order		JSCP	Joint Strategic Campaign Plan	
APEX	Adaptive Planning and Execution		LOI	Letter of instruction	
CCDR	Combatant Commander		OPORD	Operations Order	
CJCS	Chairman of the Joint Chiefs of Staff		PLANDIR	Planning Directive	
COA	Course of Action		PLANORD	Planning Order	
CONOPS	Concept of Operations		SecDef	Secretary of Defense	
EXORD	Execute Order		TPFDD	Time-phased Force and Deployment Data	
JPEC	Joint Planning and Execution Community		WARNORD	Warning Order	

Figure G. Contingency and Crisis Comparison

is short. In as little as a few days, CCDRs and staffs must develop and approve a feasible COA, publish the plan or order, prepare forces, ensure sufficient communications systems support, and arrange sustainment for the employment of U.S. military forces.

b. In a crisis, situational awareness is continuously fed by the latest intelligence and operations reports. An adequate and feasible military response in a crisis demands flexible procedures that consider time available, rapid and effective communications, and relevant previous planning products whenever possible.

c. In a crisis or time-sensitive situation, the CCDR uses JPP to adjust previous contingency planning and converts these plans to executable OPORDs or develops OPORDs from scratch when no useful contingency plans exist. To maintain plan viability, it is imperative that all Steps of the JPP are conducted and thought through, although some may be done sequentially. Time-sensitivities are associated with a crisis situation and the JPP may be abbreviated for time.

d. JPP activities in a crisis are similar to contingency planning activities, but a crisis is based on dynamic, real-world conditions vice assumptions. The JPP in a crisis provides for the rapid and effective exchange of information and analysis, the timely preparation of military COAs for consideration by the President or SecDef, and the prompt transmission of their decisions to the JPEC. The JPP may be performed sequentially or in parallel, with supporting and subordinate plans or OPORDs being developed concurrently. The exact flow of the procedures is largely determined by the time available to complete the planning and by the significance of the crisis. Capabilities such as collaboration and decision-support tools will increase the ability of the planning process to adapt quickly to changing situations and improve the transition from contingency to crisis planning. The following paragraphs summarize the activities and interaction that occur during a crisis situation.

(1) When the President, SecDef, or CJCS decides to develop military options, the CJCS issues a Planning Directive to the JPEC initiating the development of COAs and requesting that the supported CCDR submit a CDR's Estimate of the situation with a recommended COA to resolve the situation. Normally, the directive will be a WARNORD, but a PLANORD or ALERTORD may be used if the nature and timing of the crisis warrant accelerated planning. In a quickly evolving crisis, the initial WARNORD may be communicated vocally with a follow-on record copy to ensure that the JPEC is kept informed. If the directive contains force deployment preparation or deployment orders, SecDef approval is required.

(2) The WARNORD describes the situation, establishes command relationships, and identifies the mission and any planning constraints. It may identify forces and strategic mobility resources, or it may request that the supported CCDR develop these factors. It may establish tentative dates and times to commence mobilization, deployment, employment, or it may solicit the recommendations of the supported CCDR regarding these dates and times. If the President, SecDef, or CJCS directs development of a specific COA, the WARNORD will describe the COA and request the supported CCDR's assessment. A WARNORD sample can be found in CJCSM 3122.02 (*JOPES Vol III*).

(3) In response to the WARNORD, the supported CCDR, in collaboration with subordinate and supporting CCDRs and the rest of the JPEC, reviews existing joint plans for applicability and develops, analyzes, and compares COAs. Based on the supported CCDR's guidance, supporting CCDRs begin their planning activities.

(4) Although an existing plan almost never completely aligns with an emerging crisis, it can be used to facilitate rapid COA development. An existing plan can be modified to fit the specific situation. An existing CONPLAN can be fully developed beyond the stage of an approved CONOPS. The TPFDD related to specific plans are stored in the JOPES database and available to the JPEC for review.

(5) The CJCS, in consultation with other members of the JCS and CCDRs, reviews and evaluates the supported CCDR's estimate and provides recommendations and advice

to the President and SecDef for COA selection. The supported CCDR's COAs may be refined or revised, or new COAs may have to be developed to accommodate a changing situation. The President or SecDef selects a COA and directs that detailed planning be initiated.

(6) On receiving the decision of the President or SecDef, the CJCS issues an ALERTORD to the JPEC to announce the decision. The SecDef approves the ALERTORD. The order is a record communication that the President or SecDef has approved the detailed development of a military plan to help resolve the crisis. The contents of an ALERTORD may vary, and sections may be deleted if the information has already been published, but it should always describe the selected COA in sufficient detail to allow the supported CCDR, in collaboration with other members of the JPEC, to conduct the detailed planning required to deploy, employ, and sustain forces. However, the ALERTORD does not authorize execution of the approved COA.

(7) The supported CCDR develops the OPORD and supporting TPFDD using an approved COA. Understandably, the speed of completion is greatly affected by the amount of prior planning and the planning time available. The supported CCDR and subordinate describe the CONOPS in OPORD format. They update and adjust planning accomplished during COA development for any new force and sustainment requirements and source forces and lift resources. All members of the JPEC identify and resolve shortfalls and limitations.

(8) The supported CCDR submits the completed OPORD for approval to the SecDef or President via the CJCS. After an OPORD is approved, the President or SecDef may decide to begin deployment in anticipation of executing the operation or as a show of resolve, execute the operation, place planning on hold, or cancel planning pending resolution by some other means. Detailed planning may transition to execution as directed or become realigned with continuous situational awareness, which may prompt planning product adjustments and/or updates.

(9) In crisis planning, plan development continues after the President decides to execute the OPORD or to return to the pre-crisis situation. When the crisis does not lead to execution, the CJCS provides guidance regarding continued planning under either a crisis or contingency environment.

e. The JPP during a crisis provides the CJCS and CCDRs with a process for getting vital decision-making information up the chain of command to the President and SecDef. The JPP facilitates information sharing among the members of the JPEC and the integration of military advice from the CJCS in the analysis of military options. Additionally, the JPP enables the President and SecDef to communicate their decisions rapidly and accurately through the CJCS to the CCDRs, subordinate and supporting CCDRs, the Services, and CSAs to initiate detailed military planning, change deployment posture of the identified force, and execute military options. It also outlines the mechanisms for monitoring the execution of the operation.

3. GFM Process During a Crisis

During crisis planning, preferred force identification is used the same as it was during contingency planning. Contingency sourcing is rarely used during crisis planning due to the time constraints involved, but if time allows, the option exists for the CJCS to direct JS JFC/JFPs to contingency source for crisis planning.

a. In contingency and crisis planning the difference in force planning is the level of detail done with the force requirements for the plan. For contingency planning, the number of planning assumptions prevents generating the detailed force requirements needed by the JS JFC/JFPs to begin execution sourcing. During a crisis, a known event has occurred and there are fewer assumptions. The focus of crisis planning is usually on transitioning to execution quickly. The detailed information requirements specified to support the execution sourcing process, either emergent or annual, preclude completion until most assumptions are validated.

b. During crisis planning, the CCDR will utilize assigned and already allocated forces to respond to the situation. If the CCDR identifies additional forces are required to respond, then a force request is submitted. Once this request is endorsed by the CCDR, that force request is considered a CCDR requirement to execute the operation as planned. The force request is sent from the CCDR to the SecDef via the JS. The vehicle for the force request is a message called a Request for Forces (RFF) to the SecDef and JS info the JS JFC/JFPs, FPs, OSD, and all other CCDRs as specified in CJCSM 3130.06, *GFM Allocation Policies and Procedures*. Each individual force requested is serialized with a Force Tracking Number (FTN). An RFF message may contain one or more FTNs. To request Joint Individual Augmentations (JIAs) for a Joint Task Force Headquarters, the message is called a Joint Manning Document (JMD) Emergent JIA Request. The initial force or JIA request to perform a mission is an emergent request (see the GFMIG, Section IV).

If the plan is directed to be executed, the force requirements are forwarded to the JS for validation and tasked to the JS JFC/JFPs to provide sourcing recommendations to the SecDef. When the SecDef accepts the sourcing recommendation, the FP is ordered to deploy the force to meet the requested force requirement via a deployment order (see Chapter 3-III, *Force Planning*.)

4. Summary

Day-to-day campaigns span the range from competition through armed conflict. Global campaigns are directed by the JSCP which provides CJCS direction on how to execute the military strategy and identifies global, functional, and regional campaigns required to support national strategy objectives. GCPs address adversaries and competitors identified in national strategy by integrating joint force actions across geographic boundaries. CCPs implement a CCDR's strategy and seek to shape the operational environment by integrating posture, resources, and activities to achieve objectives and complement other government efforts related to a geographic region or functional area. CCPs also consolidate operational objectives and tasks identified by the CJCS in the GCP, FCPs and RCPs as they pertain to the CDR's specific authorities and responsibilities.

Contingency plans are executed in response to changes in the strategic environment that require a branch from the GCPs, CCPs, FCPs or RCPs. National and defense strategies, through the CPG, direct contingency plans to address designated threats, potential catastrophic events, and contingent missions without a crisis that put one or more national interests at risk in ways that warrant military response options. Since many contingencies are branches from day-to-day campaign plans, an integrated contingency plan should capture modifications to day-to-day campaign objectives, resources, and forces so that SecDef, CJCS, and the CCDRs can coordinate joint force contingency response activities across all CCMDs. Contingency plans conclude upon achievement of identified military objectives and military operations return to day-to-day campaign plan execution, often under new or re-characterized conditions.

"Do what you can, with what you have, where you are."
Theodore Roosevelt

III. Sequencing Actions

> *"If I were given one hour to save the planet, I would spend fifty-nine minutes defining the problem and one minute resolving it."*
>
> -Albert Einstein

This section gives a broad overview of sequencing actions and phasing.

1. Sequencing Actions and Phasing

Part of the art of planning is determining the sequence of actions that best accomplishes the mission. The concept of operations describes in sequence the start of the operation to the projected status of the force at the operation's end, or endstate. If the situation dictates a significant change in mission, tasks, task organization, or priorities of support during the operation, the CDR may phase the operation.[1] A phase is a planning and execution tool used to divide an operation in duration or activity.

a. <u>Phasing</u>. A phase is a definitive stage of an operation or campaign during which a large portion of the forces and capabilities are involved in similar or mutually supporting activities for a common purpose. Phasing, which can be used in any operation regardless of size, helps the CDRs organize operations by integrating and synchronizing subordinate operations. Phasing helps CDRs and staffs visualize, design, and plan the entire operation or campaign and define requirements in terms of forces, resources, time, space, and purpose. It helps them systematically achieve military objectives that cannot be attained all at once by arranging smaller, related operations in a logical sequence. Phasing also helps CDRs mitigate risk in the more dangerous or difficult portions of an operation.

(1) Each phase is designed to nest with the intent for the overall campaign and sequenced to achieve an endstate that will set conditions for commencement of the next phase. The CDR will declare his/her intent for each phase that supports his overall intent for the operation or campaign. Each phase must have a specified set of conditions for both the beginning and intended endstate. Leaders should recognize that lines of operation (LOO) or effort (LOE) (see Chapter 5-3a, *Concept Development*) are likely to run throughout the phases to provide the logical framework for the entire operation or campaign. Each operation or campaign is unique and the phasing must make sense for the campaign. While phases should ideally be flexibly event-oriented, the staff must also consider the time-oriented resourcing requirements for the activities of each phase.

(2) For each phase, the campaign's CONOPS should describe the following elements:

(a) <u>Intent and schemes of movement and maneuver</u>. The CDR's intent for the phase must be clear. Describe the purpose, endstate, and the operational risk to the campaign during this phase. The schemes of movement and maneuver may be narratives of the various LOO and LOE as they are executed during this particular phase. The flow of forces and capability into theater are broadly described as are subsequent joint force maneuver schemes to achieve the various operational objectives. In campaigns where LOEs are used (as opposed to LOOs) and/or where positional advantage may not be consistently critical to success, the scheme of maneuver uses the logic of purpose and may describe how and

---

[1] *ADRP 3-0*

when certain objectives within each LOE must be achieved, especially in relation to the objectives on the other LOEs of the campaign.

(b) <u>Objectives and effects (desired and undesired)</u>. Describe the objectives for each phase, and the major effects that must be achieved to realize those objectives. Describe how the force's objectives are related to those of the next higher organization and to other organizations (especially if the military is a supporting effort).

(c) <u>Tasks to subordinate and supporting commands and agencies</u>. The CDR assigns tasks to subordinate CDRs, along with the capabilities and support necessary to achieve them. Area tasks and responsibilities focus on that specific area to control or conduct operations. Functional tasks and responsibilities focus on the performance of continuing efforts that involve the forces of two or more Military Departments operating in the same domain (air, land, sea, or space) or where there is a need to accomplish a distinct aspect of the assigned mission. Include identification of requests for support to organizations outside of DOD.

(d) <u>Command and control (C2) organization and geometry of the area of operations (AO)</u>. Note any changes to C2 structure or to the geometry of the AOR (for CCMDs) or joint operations area (JOA) (for subordinate joint forces) or AOs (for subordinate non-joint forces).

(e) <u>Assessment methodology</u>. Identify the basic methodology for assessing accomplishment of objectives. Include assessments to help gauge if the objectives actually support achievement of the endstate.

(f) <u>Risk mitigation</u>. Identify the areas of risk concern to the CDR and outline how the risk may be mitigated.

(g) <u>Commander Critical Information Requirement (CCIR) and associated decision points</u>.

(h) <u>Transition to the next phase</u>. Describe how the joint force will move to the next phase. Describe the endstate conditions for the phase, which should tie directly to the initiation conditions for the next phase. Include a description of transition of control from the joint force to other parties for aspects of the overall campaign.

(3) While phasing has traditionally been described in a 6-phase model, this model has been problematic in describing operations that are not predominately military. While it works well for operations such as Desert Storm, it breaks down in describing some of the operations, activities and actions associated with long-term campaigns and competition activities that occur below the level of armed conflict (e.g., U.S. actions toward Russia in Ukraine). JP 3-0, *Joint Campaigns and Operations*, models several phasing constructs that may apply. The bottom line is that the phases should be adapted to the environment, the problem, and the operational approach – not vice versa.

Inducement: Increases the benefits of and/or reduces the cost of compliance (increasing overall utility of complying with our demands).

Persuasion: Alters the preferences against which the costs and benefits are evaluated (changing the decision context).

---

(4) <u>Phasing Model.</u>

(a) Phasing is critical to arranging all tasks of an operation that cannot be conducted simultaneously. It describes how the CDR envisions the overall operation unfolding. It is the logical expression of the CDR's visualization in time. Within a phase, a large portion of the force executes similar or mutually supporting activities. Achieving a specified condition or set of conditions typically marks the end of a phase.

(b) Figure A is a notional phasing model and displays six phases: shape, deter, seize the initiative, dominate, stabilize the environment, and enable civil authority. Each phase may be considered during planning and assessment. This construct is meant to provide planners a template while not imparting constraints on the flexibility of CCDRs.

This notional six-phase model is not intended to be a universally prescriptive template for all conceivable joint operations and is expected to be tailored to the character and duration of the operation to which it applies.

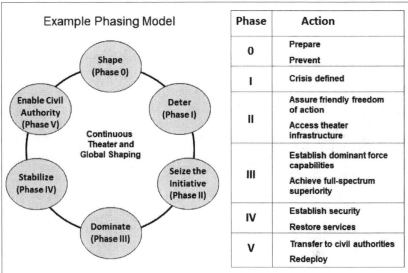

Phase	Action
0	Prepare Prevent
I	Crisis defined
II	Assure friendly freedom of action Access theater infrastructure
III	Establish dominant force capabilities Achieve full-spectrum superiority
IV	Establish security Restore services
V	Transfer to civil authorities Redeploy

Figure A. Notional Phasing Model

   (5) A phase can be characterized by the "focus" that is placed on it. Phases are distinct in time, space, and/or purpose from one another, but must be planned in support of each other and should represent a natural progression and subdivision of the campaign or operation. Each phase should have a set of starting conditions (that define the start of the phase) and ending conditions (that define the end of the phase). The ending conditions of one phase are the starting conditions for the next phase. Phases are necessarily linked and gain significance in the larger context of the campaign.

   The nature of operations and activities during a typical joint combat operation will change from its beginning (when the CJCS issues the execute order) to the operation's end (when the joint force disbands and components return to a pre-operation status). Shaping activities usually precede the operation and may continue during and after the operation. The purpose of shaping activities is to help set the conditions for successful execution of the operation. Figure B on the following page shows that from deter through enable civil authority, the operations and activities in these groups vary in magnitude—time, intensity, forces, etc., — as the operation progresses. At various points in time, each specific group might characterize the main effort of the joint force.

   For example, dominate activities would characterize the main effort after the joint force seizes the initiative until the enemy no longer is able to effectively resist. Even so, activities in the other groups would usually occur concurrently at some level of effort. The following illustration and paragraphs provide more information on the nature of these activities.

> *"It ain't what you don't know that gets you into trouble. It's what you know for sure that just ain't so."*
>
> Mark Twain

A Notional Joint Combat Operation Model

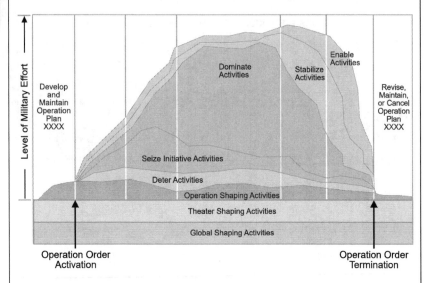

Figure B. Notional Joint Combat Operation Model

- The model depicts six general groups of military activities that typically comprise a single joint combat operation. The model applies to a large-scale combat operation as well as to a combat operation relatively limited in scope and duration. It shows that emphasis on activity types shifts as an operation progresses.

- Operation shaping activities may begin during plan development to help set conditions for successful execution. They may continue after the operation ends if the command continues to maintain an operation plan.

- Theater and global shaping activities occur continuously to support theater and global requirements. Specific theater and global shaping activities may support a specific joint operation plan during its execution.

2. The Six-Phase Construct

a. The six-phase construct is described as follows:

(1) <u>Shape</u>. Shaping Operations are focused on partners, potential partners and those that might impede our efforts or provide indirect support to adversaries. Shaping supports deterrence by showing resolve, strengthening partnership and fostering regional security. Insofar as the influencing of potential adversaries is concerned, shaping utilizes inducement and persuasion. Shaping activities set the foundations for operational access as well as develop the relationships and organizational precursors that enable effective partnerships in time of crisis.

(a) Participation in effective regional security frameworks with other instruments of national and multi-national power is critical. Pre-crisis shaping activities by their nature rely heavily on the non-military contributors to unified action; for example, the State Department as the lead agency for U.S. foreign policy leads the individual country teams, funds security assistance and is responsible for the integration of information as an instrument of national power. Also, the State Department's Office of the Coordinator for Reconstruction and Stabilization (S/CRS) has the mission to lead, coordinate and institutionalize USG civilian capacity to prevent or prepare for post-conflict situations, and to help stabilize and reconstruct societies in transition from conflict or civil strife.

(b) Ultimately, shaping operations will support the achievement of an endstate that provides a global security environment favorable to U.S. interests. Figure C is an example of a phasing construct for Humanitarian Assistance. Note that this construct sup-ports the phasing model; however, the phases have been modified to support an endstate in which the DOD is in support of other governmental organizations (OGOs, DOS/USAID).

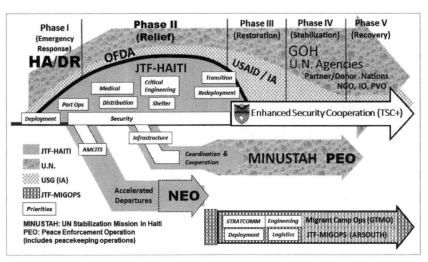

Figure C. Example Humanitarian Phasing Construct

> *"You will never reach your destination if you stop and throw stones at every dog that barks."*
> *Winston Churchill*

(c) The Joint Force, as part of a larger multinational and interagency effort, conducts continuous, anticipatory-shaping operations that build partnerships with governmental, non-governmental, regional and international organizations, and reduces the causes of conflict and instability in order to prevent or mitigate conflict or other crises and set the conditions for success in other operations - all aimed at a secure global environment favorable to U.S. interests.[2]

(d) Joint, interagency and multinational operations are executed continuously with the intent to enhance international legitimacy and gain multinational cooperation in support of defined national strategic and strategic military objectives. They are designed to assure success by shaping perceptions and influencing the behavior of both adversaries and al-lies, developing allied and friendly military capabilities for self-defense and coalition opera-tions, improving information exchange and intelligence sharing, and providing U.S. forces with peacetime and contingency access. Shape phase activities must adapt to a particular theater environment and may be executed in one theater in order to create effects and/or achieve objectives in another. Planning that supports most "shaping" requirements typically occurs in the context of day-to-day security cooperation, and CCMDs may incorporate shaping activities and tasks into the Security Cooperation Plan (SCP)/CCP. Contingency and crises requirements also occur while global and theater-shaping activities are ongoing, and these requirements are satisfied in accordance with the CJCSM 3122 series. The JPP steps described in Chapter 5, *Joint Planning Process*, are useful in planning security coop-eration activities as well as developing OPLANs and OPORDs.

[2]*Military Support to Shaping Joint Operating Concept*

(2) <u>Deter</u>. The intent of this phase is to deter undesirable adversary action by demonstrating the capabilities and resolve of the joint force. It differs from deterrence that occurs in the shape phase in that it is largely characterized by preparatory actions that specifically support or facilitate the execution of subsequent phases of the operation/campaign. Deterrence supports shaping by helping to reassure states that cooperative partnership with the U.S. will not result in an unacceptable threat. Insofar as the influencing of potential adversaries is concerned, deterrence deals with coercive forms of influence.[3]

(a) Deterrence operations are designed to convince adversaries not to take actions that threaten the vital interests of the U.S. by means of decisive influence over their decision-making. Decisive influence is achieved by credibly threatening to deny benefits and/or impose costs, while encouraging restraint by convincing the actor that restraint will result in an acceptable outcome. Because of the uncertain future security environment, specific vital interests may arise that are identified by senior national leadership. Deterrence strategy and planning must be sufficiently robust and flexible to accommodate these changes when they occur.

(b) CCDRs continue to engage multinational partners, thereby providing the basis for further crisis response. Liaison teams and coordination with other agencies assist in setting conditions for execution of subsequent phases of the campaign or operation. Many actions in the deter phase build on security cooperation activities from the previous phase and are conducted as part of security cooperation plans and activities. They can also be part of stand-alone operations.

(3) <u>Seize the Initiative</u>. CCDRs and their subordinate JFCs seek to seize the initiative in combat and noncombat situations through the application of appropriate joint force capabilities. In combat operations this involves executing offensive operations at the earliest possible time, forcing the adversary to offensive culmination and setting the conditions for decisive operations. Rapid application of joint combat power may be required to delay, impede, or halt the adversary's initial aggression and to deny the initial objectives. If an adversary has achieved its initial objectives, the early and rapid application of offensive combat power can dislodge adversary forces from their position, creating conditions for the exploitation, pursuit, and ultimate destruction of both those forces and their will to fight during the dominate phase. During this phase, operations to gain access to theater infrastructure and to expand friendly freedom of action continue while the CCDR seeks to degrade adversary capabilities with the intent of resolving the crisis at the earliest opportunity. In all operations, the CCDR establishes conditions for stability by providing immediate assistance to relieve conditions that precipitated the crisis.

(4) <u>Dominate</u>. The dominate phase focuses on breaking the enemy's will for organized resistance or, in noncombat situations, control of the operational environment. Success in this phase depends upon overmatching joint force capability at the critical time and place. This phase includes full employment of joint force capabilities and continues the appropriate sequencing of forces into the OA as quickly as possible. When a campaign or operation is focused on conventional enemy forces, the dominate phase normally concludes with decisive operations that drive an adversary to culmination and achieve the CCDR's operational objectives. Against unconventional adversaries, decisive operations are characterized by dominating and controlling the operational environment through a combination of conventional, unconventional, information, and stability operations. Stability operations are conducted as needed to ensure a smooth transition to the next phase and relieve suffering. In noncombat situations, the joint force's activities seek to control the situation or operational environment. Dominate phase activities may establish the conditions for an early favorable conclusion of operations or set the conditions for transition to the next phase.

---

[3] *Deterrence Operations, Joint Operating Concept*

(5) <u>Stabilize the Environment</u>. The stabilize phase is required when there is no fully functional, legitimate civil governing authority present. The joint force may be required to perform limited local governance, integrating the efforts of other supporting/contributing multinational, Inter-Governmental Organizations (IGOs), Non-Governmental Organizations (NGOs), or USG agency participants until legitimate local entities are functioning. This includes providing or assisting in the provision of basic services to the population. The stabilize phase is typically characterized by a change from sustained combat operations to stability operations. Stability operations are necessary to ensure that the threat (military and/or political) is reduced to a manageable level that can be controlled by the potential civil authority or, in noncombat situations, to ensure that the situation leading to the original crisis does not reoccur and/or its effects are mitigated. Redeployment operations may begin during this phase and should be identified as early as possible. Throughout this segment, the CCDR continuously assesses the impact of current operations on the ability to transfer overall regional authority to a legitimate civil entity, which marks the end of the phase.

(6) <u>Enable Civil Authority</u>. This phase is predominantly characterized by joint force support to legitimate civil governance in theater. Depending upon the level of indigenous state capacity, joint force activities may be at the behest of that authority or they may be under its direction. The goal is for the joint force to enable the viability of the civil authority and its provision of essential services to the largest number of people in the region. This includes coordination of joint force actions with supporting or supported multinational, agency, and other organization participants; establishment of measures of effectiveness (MOEs); and influencing the attitude of the population favorably regarding the U.S. and local civil authority's objectives. DOD policy is to support indigenous persons or groups promoting freedom, rule of law, and an entrepreneurial economy and opposing extremism and the murder of civilians. The joint force will be in a supporting role to the legitimate civil authority in the region throughout the enable civil authority phase. Redeployment operations, particularly for combat units, will often begin during this phase and should be identified as early as possible. The military endstate is achieved during this phase, signaling the end of the campaign or operation. Operations are concluded when redeployment is complete. CCMD involvement with other nations and agencies, beyond the termination of the joint operation, may be required to achieve the national strategic endstate.

b. <u>Transitions</u>. Transitions between phases are designed to be distinct shifts in focus by the joint force, often accompanied by changes in command or support relationships. The activities that predominate during a given phase, however, rarely align with neatly definable breakpoints. The need to move into another phase normally is identified by assessing that a set of objectives are achieved or that the enemy has acted in a manner that requires a major change in focus for the joint force and is therefore usually event driven, not time driven. Changing the focus of the operation takes time and may require changing CDR's objectives, desired effects, MOEs, priorities, command relationships, force allocation, or even the design of the OA. An example is the shift of focus from sustained combat operations in the dominate phase to a preponderance of stability operations in the stabilize and enable civil authority phases. Hostilities gradually lessen as the joint force begins to reestablish order, commerce, and local government and deters adversaries from resuming hostile actions while the U.S. and international community take steps to establish or restore the conditions necessary for long-term stability. This challenge demands an agile shift in joint force skill sets, actions, organizational behaviors, and mental outlooks, and inter-organizational coordination with a wider range of interagency and multinational partners and other participants to provide the capabilities necessary to address the mission-specific factors.[4]

---

[4] *JP 5-0, Joint Planning*

3. Summary

As a general rule, the phasing of the campaign or operation should be conceived in condition-driven rather than time-driven terms. However, resource availability depends in large part on time-constrained activities and factors—such as sustainment or deployment rates—rather than the events associated with the operation. The challenge for planners, then, is to reconcile the reality of time-oriented deployment of forces and sustainment with the event-driven phasing of operations.

Effective phasing must address how the joint force will avoid reaching a culminating point. If resources are insufficient to sustain the force until achieving the endstate, planners should consider phasing the campaign or operation to account for necessary operational pauses between phases. Such phasing enables the reconstitution of the joint force during joint operations, but the CDR must understand that this may provide the adversary an opportunity to reconstitute as well. In some cases, sustainment requirements, diplomatic factors, and political factors within the host nation may even dictate the purpose of certain phases as well as the sequence of those phases. For example, phases may shift the main effort among Service and functional components to maintain momentum while one component is being reconstituted.

Coercion: Increases the cost and/or reduces the benefits of defiance (decreasing the overall utility of defying our demands).

Deterrence: Demand that the adversary refrain from undertaking a particular action linked to a threat to use force if it does not comply.

Compellence: Demand that the adversary undertake a particular action linked to a threat to use force if it does not comply.

"These phases of a plan do not comprise rigid instructions, they are merely guideposts. Rigidity inevitably defeats itself, and the analysts who point to a changed detail as evidence of a plan's weakness are completely unaware of the characteristics of the battlefield."

General Dwight D. Eisenhower

I. Planning

> *The primary goal of planning is not the development of elaborate plans that inevitably must be changed; a more enduring goal is the development of planners who can cope with the inevitable change.*

1. Planning

Planning is the **process** of thinking about and organizing the activities required to achieve a desired goal (forethought). It is an anticipatory decision-making process that helps in coping with complexities and combines forecasting of developments with the preparation of scenarios and how to react to them. It is conducted for different planning horizons, from long-range to short-range. Depending on the echelon and circumstances, units may plan in years, months, or weeks, or in days, hours, and minutes. The defining challenges to effective planning are uncertainty and time. Uncertainty increases with the length of the planning horizon and the rate of change in an OE. A tension exists between the desire to plan far into the future to facilitate preparation and the fact that the farther into the future the CDR plans, the less certain the plan will remain relevant. Given the uncertain nature of the OE, the object of planning is not to eliminate uncertainty, but to develop a framework for action in the midst of such uncertainty.[1]

Planning is the **art and science** of understanding a situation, envisioning a desired future, and determining effective ways to bring that future about. Planning helps leaders understand situations; develop solutions to problems; direct, coordinate, and synchronize actions; prioritize efforts; and anticipate events. In its simplest form, planning helps leaders determine how to move from the current state of affairs to a more desirable future state while identifying potential opportunities and threats along the way. It is a **continuous learning activity**. While planning may start an iteration of the operations process, planning does not stop with the production of an order. During preparation and execution, the CDR and staff continuously refine the order to account for changes in the situation. Subordinates and others provide assessments about what works, what does not work, and how the force can do things better. In some circumstances, CDRs may determine that the current order (to include associated branches and sequels) no longer applies. In these instances, instead of modifying the current order, CDRs reframe the problem and develop a new plan.

a. <u>The Functions of Planning</u>. Imperfect knowledge and assumptions about the future are inherent in all planning. Planning cannot predict with precision how the enemies will react or how civilians will respond during operations. Nonetheless, the understanding and learning that occurs during planning have great value. Even if units do not execute the plan exactly as envisioned—and few ever do—planning results in an improved understanding of the situation that facilitates future decision making.[2] Planning and plans help leaders—

- Understand situations and develop solutions to problems.
- Task-organize the force and prioritize efforts.
- Direct, coordinate, and synchronize action.
- Anticipate events and adapt to changing circumstances.

[1] *ADP 5-0, The Operations Process*
[2] *Ibid*

b. Planning keeps us oriented on future objectives despite the requirements of current operations. By anticipating events beforehand, planning helps the CDR seize, retain, or exploit the initiative. As a result, the force anticipates events and acts purposefully and effectively before the adversary can act or before situations deteriorate. In addition, planning helps anticipate favorable turns of events that could be exploited during shaping operations.

c. A product of planning is a plan or order—a directive for future action. CDRs issue plans and orders to subordinates to communicate their visualization of the operations and to direct action. Plans and orders synchronize the action of forces in time, space, and purpose to achieve objectives and accomplish the mission. They inform others outside the organization on how to cooperate and provide support. These plans and orders describe a situation, establish a task organization, lay out a concept of operations, assign tasks to subordinate units, and provide essential coordinating instructions. The plan serves as a foundation for which the force can rapidly adjust from based on changing circumstance. The measure of a good plan is not whether execution transpires as planned, but whether the plan facilitates effective action in the face of unforeseen events.

d. Planning provides an informed forecast of how future events may unfold. It entails identifying and evaluating potential decisions and actions in advance to include thinking through consequences of certain actions. Planning involves thinking about ways to influence the future as well as how to respond to potential events. ***Put simply, planning is thinking critically and creatively about what to do and how to do it, while anticipating changes along the way.***

2. Conceptual and Detailed Planning

Planning consists of two separate, but closely related, components: a conceptual component and a detailed component as shown in the figure below. Conceptual planning involves understanding the OE and the problem, determining the operation's end state, and visualizing an operational approach. Conceptual planning generally corresponds to **operational art** and is the focus of the CDR with staff support. Detailed planning translates the broad operational approach into a complete and practical plan. Generally, detailed planning is associated with the **science of operations** including the synchronization of the forces in time, space, and purpose. Detailed planning works out the scheduling, coordination, or technical problems involved with moving, sustaining, and synchronizing the actions of force as a whole toward a common goal. Effective planning requires the integration of both the conceptual and detailed components of planning.[3]

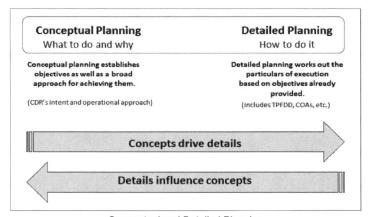

Conceptual Planning	Detailed Planning
What to do and why	How to do it
Conceptual planning establishes objectives as well as a broad approach for achieving them.	Detailed planning works out the particulars of execution based on objectives already provided.
(CDR's intent and operational approach)	(includes TPFDD, COAs, etc.)

Concepts drive details

Details influence concepts

Conceptual and Detailed Planning

[3] *ADP 5-0, The Operations Process*

3. Operational Art and Planning

Operational art and design provide context for decision-making and how the many facets of the problem are likely to interact, enabling CDRs and planners to identify hazards, threats, consequences, opportunities and risks. Planning is both a science and an art.[4]

a. <u>Cognitive Approach - Art</u>. Conceptual planning is directly associated with operational art which is the cognitive approach used by CDRs and staffs, supported by their skill, knowledge, experience, creativity and judgement to develop strategies, campaigns and operations to organize and employ military forces by *integrating ends, ways, means and risks.* Operational art is a thought process that guides conceptual and detailed planning to produce executable plans and orders. Operational art relies on the ability of the CDR and planners to identify what tools are required to address the planning problems. Different CDRs and planners will need different tools in their tool box to help them, as each person has inherent strengths and weaknesses. CDRs apply judgment based on their knowledge and experience to select the right time and place to act, assign tasks, prioritize actions, and allocate resources. Similarly, every problem is different and may require different tools to analyze and address them. The choice of COA, combination of forces, threats, choice of tactics, and arrangement of activities etc., will be different for every OE and problem. One size does not fit all. These belong to the art of planning. The art of planning requires understanding the dynamic relationships among friendly forces, the threat, and other aspects of the OE. It includes making decisions based on skilled judgment acquired from experience, training, study, imagination, and critical and creative thinking. The art of planning involves the CDR's willingness to accept risk.

b. <u>Analytical Framework - Design</u>. Design is a methodology for *applying critical and creative thinking* to understand, visualize, and describe complex, ill-structured problems and develop approaches to solve them. Critical thinking captures the reflective and continuous learning essential to design. Creative thinking involves thinking in new, innovative ways while capitalizing on imagination, insight, and novel ideas. Design is a way of organizing the activities of understanding, visualizing, and describing within an organization. Design occurs throughout the operations process before and during detailed planning, through preparation, and during execution and assessment. Operational design is that analytical framework that underpins planning and supports the CDRs and planners in organizing and understanding the OE as a complex interactive system. Many aspects of military operations, such as movement rates, fuel consumption, and weapons effects, are quantifiable. They are part of the science of planning, they can be measured and analyzed and while not easy, the science of planning is fairly straightforward.[5]

c. As CDRs conceptualize the operation, their vision guides the staff through design and into detailed planning. Design is continuous throughout planning and "evolves" with increased understanding throughout the operations process. Design underpins the role of the CDR in the operations process, guiding the iterative and often cyclic application of understanding, visualizing, and describing. As these iterations occur, the design concept—the tangible link to detailed planning—is forged. Design provides an approach for how to generate change from an existing situation to a desired objective or condition. Effective planners are grounded in both the science and the art of planning.

4. Defining Challenges

a. Planning is also the art and science of understanding a situation, envisioning a desired future, and laying out an operational approach (a broad description of the mission, operational concepts, tasks, and actions required to accomplish the mission) to achieve that future. Planning is both a continuous and a cyclical activity of the operations process

[4] *JP 3-0, Joint Campaigns and Operations*
[5] *ADP 5-0, The Operations Process*

which translates strategic guidance and direction. Based on this understanding and operational approach, planning continues with the development of a fully synchronized campaign plan, operation plan or order that arranges potential actions in time, space, and purpose to guide the force during execution.[6]

b. While planning may start an iteration of the operations process, planning does not stop with production of a plan or an order. During preparation and execution, the challenges to the plan are continuously refined as assessments and situational understanding improves. Supporting commands, subordinates and others provide feedback as to what is working, what is not working, and how the force can do things better.

c. Planning may be based on defined tasks identified in the strategic guidance, or it may be based on the need for a military response to an unforeseen current event, emergency, or time-sensitive crisis. The value of following the well-established Joint Planning Process (JPP) has been reinforced through operational and exercise experiences. Key to the process is the detailed analysis necessary to produce the requisite plans and orders that will direct subordinates. In addition to the required analysis, planners must strive to ensure the generated solution does not further exacerbate the problem or limit future options.[7]

5. Understand and Develop Solutions to Problems

a. A problem is an issue or obstacle that makes it difficult to achieve a desired goal or objective. In a broad sense, a problem exists when an individual becomes aware of a significant difference between what actually is and what is desired. In the context of operations, an operational problem is the issue or set of issues that impede CDRs from achieving their desired endstate or objectives.[8]

b. To understand something is to grasp its nature and significance. Understanding includes establishing context—the set of circumstances that surround a particular event or situation. Throughout the operations process, CDRs develop and improve their understanding of their OE and the problem. An OE is a composite of the conditions, circumstances, and influences that affect the employment of capabilities and bear on the decisions of the CDR. Both conceptual and detailed planning assist CDRs in developing their initial understanding of the OE and the problem. Based on personal observations and inputs from others (to include running estimates), CDRs improve their understanding and modify their visualization throughout the conduct of operations.

c. Throughout operations, CDRs face various problems, often requiring unique and creative solutions. Planning helps CDRs and staffs understand problems and develop solutions. Not all problems require the same level of planning. For simple problems, CDRs often identify them and quickly decide on a solution, sometimes on the spot. Planning is critical, however, when a problem is actually a set of interrelated issues, and the solution to each affects the others. For unfamiliar situations, planning offers ways to deal with the complete set of problems as a whole. In general, the more complex a situation is, the more important and involved the planning effort becomes.

d. As discussed in this books Introduction but worth mentioning again here is a plans variance. The variance in any plan is the constant change in the OE (system). Whether a campaign, contingency or crisis scenario, we plan in a chaotic environment. In the time it takes us to plan, the likelihood that the OE has changed is a certain, whether by action or inaction, affecting the plan (i.e., assumptions change or are not validated, leaders change, the OE fluctuates, Apportionment Tables are poor assumptions, disputed borders fluctuate, weather changes the rules, plans change at contact, enemy gets a vote, etc.).

[6] JP 3-0, Joint Campaigns and Operations

[7] ADP 5-0, The Operations Process

[8] Ibid

(1) Variables are hard to predict because each environment and situation have their own unique challenges which can certainly affect an orderly plan. Given the size and scope of an OE a plan can only anticipate, or forecast, for a short duration without being updated. This is known as the plan's horizon. In a fluid crisis situation, the plan's horizon may be very short and contain greater risk, causing the planner to constantly re-evaluate and update the plan. Inversely, for a campaign or contingency plan, the plan's horizon may be relatively static with less risk allowing time for greater analysis. The number of variables within the OE and the interactions between those variables and known components of the OE increases exponentially with the number of variables, thus potentially allowing for many new and sometimes subtle planning changes to emerge.

As an example of a plan's horizon, or stability, let's look at an environment that constantly influences us, the weather:

> A forecaster endeavors to anticipate the path of a tropical cyclone and utilizes historical models and probabilities to predict the tropical cyclones path and warn residents. When a low-pressure area first forms and the storm begins to take shape along the equator, forecasters are working within a complex environment with constant and multiple variables (i.e., winds, temperatures, currents, pressures, etc.) and few facts (i.e., exact location at this moment, jet stream location, ocean temperatures, surface winds, etc.). As variables amplify and the storm begins to move. the storm's horizon shifts yet again, and the forecaster updates the assessment. Over days of surveillance, gathering information, updating, and studying the variables, the actual track of the storm begins to emerge and the storm's horizon becomes more durable and predictable. The forecaster continuously narrows the storm's estimated track, eventually forecasting with some certainty the tropical cyclone's landfall.

(2) Planners employ the same technique by utilizing current knowledge of the OE to anticipate events, calculate what those may be by means of an in-depth analysis, assessment, update, and plan accordingly. But always remember, plans are orderly; probabilities and variables are not. Just as a tropical storm has a self-organizing phase within its environment, so must the planner.

(3) So, the challenge is how to plan within an environment with continuously changing and emerging variables. The planner must understand that every plan is unique and never as perfect as you want it because there are too many variables. But with constant awareness each iteration of the plan will improve the prospect of success as the variables become known and are planned for.

(4) Simplicity should be the aspiration for every plan. Prepare clear, uncomplicated plans and concise orders to ensure a thorough understanding. A plan need not be more complicated than the underlying principles which generate it.

e. Just as planning is only part of the operations process, planning is only part of problem solving. In addition to planning, problem solving includes implementing the planned solution (execution), learning from implementation of the solution (assessment), and modifying or developing a new solution as required. The object of problem solving is not just to solve near-term problems, but to do so in a way that forms the basis for long-term success.

6. Joint Planning Process (JPP)

a. Joint Planning Process. Planning provides an awareness and opportunity to study potential future events amongst multiple alternatives in a controlled environment. By planning we can evaluate complex systems and environments allowing us to break these down into small, manageable segments for analysis, assisting directly in the increased probability of success. In this way, deliberately planning for contingencies allows us to manage identified risks and influence the operational environment in which we have chosen to interact, in a deliberate way. The plans generated in this process represent actions to be taken if an identified risk occurs or a trigger event has presented itself.

The JPP is the overarching process that guides us in the development of plans for the employment of forces and capabilities within the context of national strategic objectives and national defense/military strategy to shape events, meet contingencies, and respond to unforeseen crises. It translates guidance into plans or orders to initiate necessary joint force actions. The JPP underpins planning at all levels and across the competition continuum. It applies to both supported and supporting CDRs and to component and subordinate commands when they participate in joint operations. JPP helps CDRs and their staffs organize their planning activities, share a common understanding of the mission, method, purpose, end state and CDR's intent and help's develop effective plans and orders.

JPP aligns military activities and resources with achieving strategic objectives. It enables CDRs to examine cost-benefit relationships, risks, and trade-offs to determine a preferred COA to achieve those specific objectives. JPP consists of planning activities that help CCDRs and their subordinate CDRs transform national objectives into actions that mobilize, deploy, employ, sustain, redeploy, and demobilize joint forces. It ties the employment of the Armed Forces of the U.S. to the achievement of national objectives. Based on understanding gained through the application of operational design, more detailed planning takes place within the steps of JPP. JPP is an orderly, analytical set of logical steps to frame a problem; examine a mission; develop, analyze, and compare alternative COAs; select the best COA; and develop a plan or order.[9]

7. Guides to Effective Planning

Planning is an inherent and fundamental part of command and control, and CDRs are the single most important factor in effective planning. Effective planning requires dedication, study, and practice. Planners must be technically and tactically competent within their areas of expertise and understand basic planning concepts.[10] The following list may be helpful in effective planning:

- <u>CDRs focus planning.</u>

The responsibility for planning is inherent in command. CDRs are planners—they are the central figure to effective planning. Often with the most experience, CDRs are ultimately responsible for the execution of the plan.

- <u>Develop simple, flexible plans.</u>

Simplicity—prepare clear uncomplicated plans and clear concise orders to ensure thorough understanding—is a principle of war. Effective plans and orders are simple and direct. Developing shorter plans helps maintain simplicity. Shorter plans are easier to disseminate, read, and remember. Simple plans require an easily understood concept of operations. Simple plans are not simplistic plans. Simplistic refers to something made overly simple by ignoring the situation's complexity. Good plans simplify complicated situations. However, some situations require more complex plans than others do. Complex plans requiring intricate coordination or having inflexible timelines have a greater potential to fail during execution. Operations are always subject to friction beyond the control of CDRs and staffs. Elaborate or complex plans that do not incorporate tolerances for friction have more chances of something irrevocable going wrong. Flexible plans assist in adapting quickly to changing circumstances. CDRs and planners build opportunities for initiative into plans by anticipating events. This allows them to operate inside of the enemy's decision cycle or to react promptly to deteriorating situations. Incorporating options to reduce risk adds flexibility to a plan. Identifying decision points and designing branches and sequels ahead of time—combined with a clear CDR's intent—helps create flexible plans.

[9] *JP 3-0, Joint Campaigns and Operations*
[10] *CJCSM 3130.01, Campaign Planning Procedures and Responsibilities*

- Optimize available planning time.

Time is a critical variable in all operations. When allocating planning time to staffs, CDRs ensure subordinates have enough time to plan and prepare their own actions prior to execution. CDRs follow the "one-third, two-thirds rule" as a guide to allocate time available. They use one-third of the time available before execution for their own planning and allocate the remaining two-thirds of the time available before execution to their subordinates for planning and preparation. Both collaborative planning and parallel planning help optimize available planning time. Collaborative planning is two or more echelons planning together in real time, sharing information, perceptions, and ideas to develop their respective plans simultaneously. Parallel planning is two or more echelons planning for the same operations nearly simultaneously facilitated by the use of warning orders by the higher headquarters. Parallel planning is used when time is of the essence and the likelihood of execution of the plan is high. CDRs are careful not to burden subordinates with planning requirements too far into the future, instead enabling subordinates to focus on execution. Generally, the higher the headquarters, the more time and resources staff have available to plan and explore options. Higher headquarters involve subordinates with developing those plans and concepts that have the highest likelihood of being adopted.

- Focus on the right planning horizon.

Planning too far into the future may overwhelm the capabilities of planning staffs, especially subordinate staffs. Not planning far enough ahead may result in losing the initiative and being unprepared. A planning horizon is a point in time CDRs use to focus the organization's planning efforts to shape future events. To guide their planning efforts, CDRs use three planning horizons—short-range, mid-range, and long-range. These plans vary in level of detail based on assumptions about the future that address "what if?" scenarios.

- Determine relevant facts and develop assumptions.

CDRs and staffs gather key facts and develop assumptions as they build their plan. A fact is something known to exist or have happened—a statement known to be true. Facts concerning the operational and mission variables serve as the basis for developing situational understanding during planning. When listing facts, planners are careful they are directly relevant to a COA and/or they assist CDRs to make a decision. Any captured, recorded, and most importantly briefed fact must add value to the planning conversation. An assumption provides a supposition about the current situation or future course of events, presumed to be true in the absence of facts. Assumptions must be valid (logical and realistic) and necessary for planning to continue. Assumptions address gaps in knowledge that are critical for the planning process to continue. Staffs continually review assumptions to ensure validity and to challenge if they appear unrealistic. CDRs and staffs use care with assumptions to ensure they are not based on preconceptions; bias; false historical analogies; or simple, wishful thinking.

> *Nothing succeeds in war except in consequence of a well-prepared plan.*
> *Napoleon Bonaparte*

8. Planning and the Levels of Warfare

It is important to understand how planning differs at the levels of warfare. The levels of warfare are a framework for defining and clarifying the relationship among national objectives, the operational approach, and tactical tasks. The three levels are strategic, operational, and tactical. There is no hard boundary between levels of warfare, nor fixed echelon responsible for a particular level. The levels of warfare focus a headquarters on one of three broad roles—creating strategy; conducting campaigns and major operations; or sequencing battles, engagements, and actions. The levels of warfare correspond to specific levels of responsibility and planning with decisions at one level affecting other levels. They help CDRs visualize a logical arrangement and synchronization of operations, allocate resources, and assign tasks to the appropriate command. Among the levels of warfare, planning horizons differ greatly.[11]

a. <u>Strategic Level.</u> The strategic level of warfare is the level of warfare at which a nation, often as a member of a group of nations, determines national or multinational (alliance or coalition) strategic security objectives and guidance, then develops and uses national resources to achieve those objectives. The focus at this level is the development of strategy—a foundational idea or set of ideas for employing the instruments of national power in a synchronized and integrated fashion to achieve national and multinational objectives. The strategic level of war is primarily the province of national leadership in coordination with CCDRs. As discussed in Chapter 1, *Planning Fundamentals*, the NSC develops and recommends national security policy options for Presidential approval. The President, the SecDef, and the CJCS provide their orders, intent, strategy, direction, and guidance via strategic direction to the military (Services and CCMDs) to pursue national interest. They communicate strategic direction to the military through written documents referred to as strategic guidance. Key strategic guidance documents include—

- National Security Strategy of the United States (NSS).
- National Defense Strategy of the United States (NDS).
- National Military Strategy of the United States (NMS).
- Joint Strategic Campaign Plan (JSCP).
- Unified Command Plan (UCP).
- Global Force Management Implementation Guidance (GFMIG)

Based on strategic guidance, GCCs and staffs—with input from subordinate commands and supporting commands and agencies—update their strategic estimates and develop theater strategies. A theater strategy is a broad statement of a GCC's long-term vision that bridges national strategic guidance and the joint planning required to achieve national and theater objectives. The theater strategy prioritizes the ends, ways, and means within the limitations established by the budget, global force management processes, and strategic guidance. FCCs also follow this process within their functional areas.

b. <u>Operational Level.</u> The operational level of warfare is the level of warfare at which campaigns and major operations are planned, conducted, and sustained to achieve strategic objectives within theaters or other operational areas. Operational-level planning focuses on developing plans for campaigns and other joint operations. A campaign plan is a joint operation plan for a series of related major operations aimed at achieving strategic or operational objectives within a given time and space. JFCs (CCDRs and their subordinate joint task force CDRs) and their component CDRs (Service and functional) conduct operational-level planning. Planning at the operational level requires operational art to integrate ends, ways, and means while balancing risk. Operational-level planners use operational design and the JPP to develop campaign plans, OPLANs, OPORDs, and supporting plans (See Chapter 5, *Joint Planning Process*).

> *"In preparing for battle I have always found that plans are useless,*
> *but planning is indispensable."*
> *General Dwight D. Eisenhower*

---

[11] *FM 5-0, Planning and Orders Production*

The CCP operationalizes the GCC's strategy by organizing and aligning operations and activities with resources to achieve objectives in an area of responsibility. The CCP provides a framework within which the GCC conducts security cooperation activities and military engagement with regional partners. The CCP contains contingency plans that are viewed as branches within the campaign. Contingency plans identify how the command might respond in the event of a crisis. Contingency plans are often phased and have specified endstates that seek to re-establish conditions favorable to the United States. Contingency plans have an identified military objective and termination criteria. They may address limited contingency operations or large-scale combat operations.

c. Tactical Level. The tactical level of warfare is the level of warfare at which battles and engagements are planned and executed to achieve military objectives assigned to tactical units or task forces. Tactical-level planning revolves around how best to achieve objectives and accomplish tasks assigned by higher headquarters. Planning horizons for tactical-level planning are relatively shorter than planning horizons for operational-level planning. Tactical-level planning works within the framework of an operational-level plan and is addressed in Service doctrine or, in the case of multinational operations, the lead nation's doctrine. Operational- and tactical-level planning complement each other but have different aims. Operational level planning involves broader dimensions of time, space, and purpose than tactical-level planning involves. Operational-level planners need to define an operational area, estimate required forces, and evaluate requirements. In contrast, tactical-level planning proceeds from an existing operational design. Normally, AOs are prescribed, objectives and available forces are identified, and a general sequence of activities is specified for tactical-level CDRs.

9. Planning Pitfalls

CDRs and staffs recognize the value of planning and avoid common planning pitfalls. These pitfalls generally stem from a common cause: the failure to appreciate the unpredictability and uncertainty of military operations. Pointing these out is not a criticism of planning, but of planning improperly.[12] Common planning pitfalls include—

• Lacking CDR involvement in the development of the plan.

The responsibility for planning is inherent in command. CDRs are the central figures in effective planning. Since the plan describes their visualization of the end state and how the force will achieve that endstate, CDRs must devote significant time and effort in the development of plans and orders.

• Failing of the CDR to make timely decisions.

Pending or missing decisions can hinder the planning process and reduce developing a flexible and timely plan. In a rapidly changing and fluid environment, a lack of timely decisions can result in a loss of agility, initiative, or opportunities. The lack of timely decisions and timely plan development also inhibits subordinate planning and may unintentionally increase risk to the mission and force.

• Attempting to forecast and dictate events too far into the future.

Even the most effective plans cannot anticipate all the unexpected events. Often, events overcome plans much sooner than anticipated. Effective plans include sufficient branches and sequels to account for the nonlinear nature of events.

> *"Plans are of little importance, but planning is essential."*
> *Winston Churchill*

[12] *ADP 5-0, The Operations Process*

- <u>Trying to plan in too much detail.</u>

Sound plans include necessary details; however, planning in unnecessary detail consumes limited time and resources that subordinates need. In general, the less certain the situation, the fewer details a plan should include. However, planners often respond to uncertainty by planning in more detail to try to account for every possibility. Often this over-planning results in an extremely detailed plan that does not survive the friction of the situation and constricts effective action.

- <u>Using the plan as a script for execution.</u>

Attempting to prescribe the course of events with precision. When planners fail to recognize the limits of foresight and control, the plan can become a coercive and overly regulatory mechanism. CDRs, staffs, and subordinates mistakenly focus on meeting the requirements of the plan rather than deciding and acting effectively.

- <u>Institutionalizing rigid planning methods.</u>

Often leads to inflexible or overly structured thinking. Tends to make planning rigidly focused on the process and produces plans that overly emphasize detailed procedures. Effective planning provides a disciplined framework for approaching and solving complex problems. Taking that discipline to the extreme often results in subordinates not getting plans on time or getting overly detailed plans.

- <u>Lacking a sufficient level of planning detail.</u>

While planning with too much detail may consume too much time, and result in plans which are overly constraining to subordinates, plans with too little detail result in unsynchronized and uncoordinated actions of subordinate units. Determining the right balance of detail and permissiveness requires a trained and experienced staff with CDR involvement.

10. Summary

Planning provides an awareness and opportunity to study potential future events amongst multiple alternatives in a controlled environment. By planning we can evaluate complex systems and environments allowing us to break these down into small, manageable segments for analysis, assisting directly in the increased probability of success. In this way, deliberately planning allows us to manage identified risks and influence the OE in which we have chosen to interact, in a deliberate way. The plans generated in this process represent actions to be taken if an identified risk occurs or a trigger event has presented itself. Planning is the deliberate process of balancing ways, means, and risk to achieve directed objectives and attain desired endstates (ends) by synchronization and integration. It is the art and science of interpreting direction and guidance and translating it into executable activities within imposed limitations to achieve a desired objective or attain an endstate. Planning enables identification of cost-benefit relationships, risks, and trade-offs to determine a preferred course of action.

Joint planning is the deliberate process of determining how (**the ways**) to use military capabilities (**the means**) in time and space to achieve objectives (**the ends**) while considering the associated **risks**. Ideally, planning begins with specified national strategic objectives and military endstates to provide a unifying purpose around which actions and resources are focused.

> *"Initiative and its dynamic, which is imagination, are not for long the exclusive possession of one belligerent."*
> *Lee's Lieutenants, page 334, Vol I, Douglas Southall Freeman, 1942*

II. Operational Activities

1. Overview

The Joint capability to create and revise plans rapidly and systematically, as circumstances require, is the function of joint planning and execution. Joint planning and execution incorporates a joint enterprise for the development, maintenance, assessment, and implementation of global campaign plans, CCMD campaign and related contingency plans and orders prepared in response to Presidential, SecDef, or Chairman direction or requirements. Its activities span many organizational levels, including the interaction between the SecDef, CCDRs, coalition, and interagency which ultimately assists the President and SecDef to decide when, where, and how to commit U.S. military forces.

a. Strategic direction shapes joint planning and execution and it is integrated within the national strategic framework. Civilian control of the military is exercised via this strategic direction, including the delegation of authorities and allocation of resources. A sustained civilian-military dialogue provides a common understanding of the operating environment and options for military ends, ways, means, and associated risk. Within the joint planning and execution framework, this civilian-military dialogue informs and is informed by ongoing Joint planning and execution. Substantive changes in the operating environment or strategic ends, ways, and means may also drive more enduring changes to strategic direction. This mutual influence is foundational and is depicted in Figure A.

b. Joint Planning and Execution. Joint Planning and Execution encompasses the full spectrum of military doctrine, organization, training, material, leadership and education, personnel, facilities, and policy (DOTMLPF-P). It is the compilation of joint policies, processes, procedures, tools, training, and education used by the Joint Planning and Execution Community (JPEC) to monitor, plan, execute and asses the planning and execution functions associated with joint operations. Joint planning integrates strategic and operational planning with execution activities of the JPEC to meet national security objectives and facilitate seamless transition from planning to execution. Operational activities and functions span many organizations at all levels of command, including interaction between the Secretary, CCDRs, subordinate forces, allied, coalition, and interagency partners. Collaboration and an integrated approach among the supported and supporting commands, Services, and other essential stakeholders is a fundamental component to achieve unified action through an understanding of the authorities, roles, and responsibilities of the JPEC stakeholders. Through joint planning and execution, the entire chain of command is informed, including the President and SecDef, facilitating informed decisions on how, when, and where to employ the joint force. Joint planning and execution is a scalable process which can be adapted to support planning and execution with or without time constraints and under changing conditions. The planning and execution functions are depicted sequentially but can be compressed or conducted in parallel in order to meet time constraints.[1]

(1) An iterative process. Each activity and function influences and is influenced by activities and functions which are performed and reviewed at multiple echelons of commands in overlapping timeframes. Facilitating communication and understanding of strategic guidance between these echelons of command takes place in several formats: formal strategy and policy documents; the plans review process; and via specific, individual communications with CCMDs. CCDR planning may also influence strategic direction and guidance, either during planning or execution.

[1] CJCSM 3130.01, Campaign Planning Procedures and Responsibilities

Civilian-Military Dialogue

Figure A. Civilian-Military Dialogue

(2) Joint planning and execution leverages existing information technology (IT) tools and doctrinal processes. IT tools enable planner collaboration and access to shared authoritative data. Doctrinal processes provide planners a variety of flexible analytical techniques for framing problems and logically developing plans or orders to accomplish missions or objectives. The joint planning and execution enterprise, including joint doctrine, policies/ procedures, and IT capabilities, facilitates the transition from planning to the effective execution of military operations. Strict adherence to policies and procedures is required to achieve unified action.

(3) The joint planning and execution process is composed of four operational activities (**situational awareness, planning, execution, and assessment**) that provide an operating framework for one or more planning or execution efforts. The operational activities support leader decision-making cycles at all levels of command and civilian leadership. The planning and execution functions depict the elements, activities, and products that may be ongoing or under development. A sustained civilian-military dialog (Figure A) occurs in parallel to these activities and functions to inform decision making at all levels of the chain of command and ensure alignment with current strategic guidance as depicted in Figures A and B.

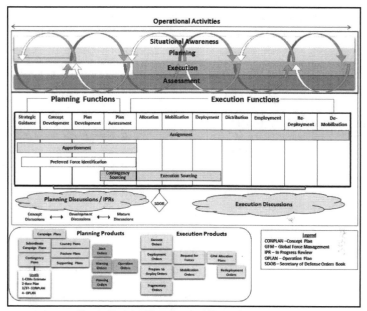

Figure B. Joint Planning and Execution Process

(4) This process leverages CCDR design, military planning and execution and the JPP framework that forms the basis for planning. The Operational Activities are discussed below. Planning Functions are discussed in Section III of this chapter and within Chapters 5 and 6. Execution Functions are detailed in Chapter 7.

2. Operational Activities

a. **Operational Activities**: Operational activities are persistent and interdependent activities performed continuously by CDRs and staffs at all levels of the chain of command. They provide a framework under which one or more planning or execution efforts are conducted. These are discussed in detail in the following appendices.[2]

[2] *CJCSM 3130.01, Campaign Planning Procedures and Responsibilities*

(1) **Situational Awareness**. Situational awareness supports the cycle of planning, execution, and assessment activities. The outputs of situational awareness inform CDRs at all levels of the chain of command, from the President to the tactical level with a current, relevant understanding of the dynamic operating environment. Situational awareness is a command-wide activity as all elements of the staff and subordinate commands report on their OE. As threats to national security interests are identified, the focus of situational awareness is adapted to the CDR's priorities. Situational awareness information is provided to the CDR through a command's operational cycle in order to inform decision making. Staff activities that inform situational awareness include:

(a) Joint Intelligence Preparation of the Operational Environment (JIPOE). JIPOE is the iterative, analytical process used by all-source joint intelligence organizations to produce and manage intelligence assessments, estimates, and other intelligence products in support of the JFC's decision-making process. It is a continuous process that involves four major steps: (1) define the OE; (2) describe the impact of the OE; (3) evaluate the adversary and other relevant actors; and (4) determine the adversary's most likely COA and the adversary's COA most dangerous to friendly forces and mission accomplishment. The CCDR's staff uses the products to produce their respective staff estimates; develop, wargame, and compare COAs; and assist in the decision regarding which COA to adopt. The CCMD Joint Intelligence Operations Centers (JIOC) have continuous JIPOE analysis and production responsibilities in support of CCMD operation planning, execution and assessment. See Chapter 4 for greater detail on JIPOE.[3,4]

(b) Strategic Estimates. The strategic estimate is a prerequisite for the development of the CCDR's theater or functional strategy to address global threats. It encompasses all the aspects that influence the CCMD's OE. Strategic estimates provide the CDR's perspective of the strategic and operational levels of the OE, desired changes required to meet specified regional or functional objectives, and the CDR's visualization of how those objectives might be achieved. CCMDs annually develop and regularly update a strategic estimate of their theater or functional area that includes a description and analysis of political, military, and economic factors and trends, and the threats and opportunities that could facilitate or hinder the achievement of strategic-directed objectives. While the strategic estimate is not specific to a planning problem, it is a starting point for conducting more detailed staff estimates and provides the CDR a baseline of understanding of the OE.[5]

(c) Staff Estimates. Staff estimates are running functional estimates, updated continuously, that support situational awareness. Staff estimates inform the CDR, staff, and subordinate commands how the functional areas support planning and execution. They should identify critical shortfalls or obstacles that impact mission accomplishment. Staff estimates may be tailored to support the unique requirements of one or more planning effort.[6]

(2) **Planning**. Planning implements strategic direction through the development of military plans and orders focused on military objectives. Planners provide military options and COAs for military actions which inform the civilian-military dialogue and decision-making and enable a shared understanding of ends, ways, means, and risk. Planning is an overarching continuous operational activity that spans the full spectrum of joint operations and may encompass multiple simultaneous planning efforts. Joint planning and execution integrates planning into one unified construct utilized during contingency or crisis situations to facilitate unity of effort and the transition from planning to execution. Planning functions can be performed in series over a period of time or they can be compressed, performed in parallel, or truncated as appropriate.

(a) Collaborative, Parallel Planning Environment. Planning at all levels involves a large collection of stakeholders and functional specialists, who require a holistic view while concentrating on specific elements of a plan. The CCMD will establish a collaborative plan-

---

[3] CJCSM 3314.01, Intelligence Planning

[4] JP 2-01.3, Joint Intelligence Preparation of the Operational Environment

[5] CJCSM 3130.01, Campaign Planning Procedures and Responsibilities

[6] JP 5-0, Operations

ning environment which facilitates plan development to include identified planning points of contact, information management procedures, and security classification guidance.

(b) <u>Resource-Informed Planning.</u> All planning should be resourced informed from the early stages of planning. Planners should consider resources required for supported and supporting plans. This includes forces, sustainment, funding, transportation, and all other resources. Plans should be developed to be executable given near-term resources reasonably expected to be available (e.g., apportioned forces, available supplies, and existing contracts or task orders).

(c) <u>Integrated Planning.</u> Integrated planning aligns the efforts of multiple CCMDs to provide the President and Secretary a clear understanding of how the total force will address a strategic problem. The integrated planning process is how the Joint Force will address complex threats that span multiple CCMD AORs and functional responsibilities. Integrated planning is directed for the development of some contingency and campaign plans, however, all planning should be integrated when practical. The integrated planning process aggregates resource requirements and integrates timelines, decision points, and authorities across multiple CCMDs to achieve strategic objectives. Integrating plan development, IPRs, and analysis/assessment provides national leadership a holistic understanding of how a particular conflict could realistically develop as well as the options for response.

<u>1 Integrated Planning for Contingency Plans.</u> Through strategic guidance, the Secretary directs integrated planning for the development of contingency plans for the high priority problem sets. Integrated planning aggregates global resource requirements and integrates timelines, decision points, and authorities across multiple CCMDs to achieve directed objectives. The results of integrated planning and analysis/assessment provides national leadership a holistic understanding of how, a particular conflict could realistically develop, its potential global consequences, and the options for response.

<u>2 Integrated Planning for Campaign Plans.</u> Selected CCDRs designated in strategic guidance are tasked to synchronize planning to better integrate specific global threats or functional perspectives into CCMD campaign plans. Globally integrated planning provides a common planning and assessment framework to integrate the operations, activities, and investments of the other CCMDs to achieve unity of effort. During execution, integrated plans present aggregated CCMD assessments and provide an evaluation of global progress against the functional strategic objectives.

(d) <u>Global Force Management.</u> GFM Integrates complementary DR4A information into force management and force planning constructs to support the Departments strategic direction. The GFM processes provide comprehensive insight into the strategic posture of forces and global availability of forces/capabilities for plans and operations. GFM also provides senior decision-makers with a construct to assess impacts and risks associated with proposed changes to the force and how the force is used. These processes are used to align U.S. forces in support of the NDS, joint force availability requirements, and joint force assessments. (See Chapter 3, *GFM* for greater details).

(e) <u>Plans.</u> Multiple types of plans with varying functions support Joint planning and execution. This list is not all inclusive:

<u>1 Campaign Plans.</u> Campaign plans document the full scope of activities a CCMD plans to execute or prepare for in order to achieve national objectives as directed in the CPG and JSCP over the next 2-5 years. Campaign plans are informed by the CCDRs theater or functional strategy and provide the overarching framework to organize operations, activities, and investments to achieve theater/functional strategic and national strategic objectives. Campaign plans focus on the command's daily activities, which include ongoing operations, security cooperation activities, intelligence collection, exercises, and other shaping or preventive activities. The intent of all campaign operations, activities, and investments is to prevent, prepare for, or mitigate contingencies.[7]

<u>2 Subordinate and Supporting Plans.</u> Supporting plans are prepared by a supporting CDR, a subordinate CDR, or a CSA to satisfy the requests or requirements of the sup-

[7] *CJCSM 3130.01, Campaign Planning Procedures and Responsibilities*

ported CDR's plan. Supporting plans should be coordinated with the Military Departments and may include organize, train, and equip responsibilities such as exercises, readiness, interoperability, augmentation, joint enablers, and capabilities development.

3 <u>Theater Distribution Plans (TDP)</u>. The TDP provides a CCMD view of the distribution network (physical, communication, information, and financial) including theater specific deployment considerations. TDPs are prepared in accordance with guidance contained within the CPG, JSCP, and JSCP LOGSUP and are usually nested within the CCMD campaign plans; however, they are developed, staffed, and submitted separately. TDPs also provide the basis to advocate for changes in distribution and support resource decisions. USTRANSCOM, as the Global Distribution Synchronizer, provides a TDP template in the Campaign Plan for Global Distribution that advises and assists the GCCs in the development and improvement of their TDPs.

4 <u>Posture Plans</u>. Posture plans establish the CCMD framework for required forces, footprints and agreements that align basing and prepositioned forces and equipment to support global operations. They are usually annexes to the CCMD campaign plan that are developed, staffed, and submitted separately. Posture plans are the vehicle for CCMDs to propose changes to approved enduring locations and resources and inform posture and resource decisions.

5 <u>Security Cooperation Planning.</u> Security cooperation planning should include establishing relationships, building capacity, increasing interoperability, developing partner nation military capabilities for self-defense and multinational operations, and gaining or maintaining access with partners that are critical to the achievement of campaign plan objectives. Security cooperation activities are aligned with the strategic objectives assigned to GCCs by strategic guidance and are nested in the CCMDs' campaign, posture, country plans, and the supporting plans of the CCMD Service components, Service HQs, DoD Agencies, and NGB.

6 <u>Country-Specific Security Cooperation Sections (CSCS).</u> CCPs include CSCS, commonly referred to as "country plans," for countries where the CCDR intends to apply significant resources. CSCSs are developed/updated annually and are synchronized with the respective Country Team's development of their Integrated Country Strategy (ICS). The ICS provides the Chief of Mission's ways and means to integrate all USG activities with the in-country security assistance and security cooperation authorities and their associated funding streams to achieve the mission's goals. CSCSs describe partner nation roles, objectives, resourcing estimates, capability and capacity shortfalls, trans-regional implications, ICS linkages, and involvement or participation by other countries and NGOs. This information will inform decisions regarding the allocation of security cooperation-related resources. Further discussion of the CSCS can be found in the JSCP.

7 <u>Contingency Plans</u>. Contingency plans are branch plans to campaign plans that are based upon potential situations for designated threats, catastrophic events, and contingent missions. They are developed pursuant to published strategic planning guidance, strategic direction, or the CCDR's direction. A contingency plan must be nested under a campaign plan with its operational ends, ways, and means looked at through the lens of the theater/functional strategic ends, ways, and means. At the same time, contingency plans inform campaign plan development by articulating required pre-crisis conditions and activities. Secretary review and approval of a contingency plan does not provide CCDR execution authority which is contingent upon a subsequent directive or order. The Secretary directs multiple CCDRs to prepare contingency plans for a designated threat or problem sets detailed or specified in strategic guidance. The Secretary establishes support relationships in planning guidance and the Chairman, via the JSCP, organizes the JPEC to support planning efforts for joint operations. The supported CCDR for a problem set will lead integrated planning and risk analysis ICW all supporting CCMDs and organizations and identify integrated planning outputs for all plans related to the problem set. Supporting CDRs will ensure their plans are integrated and synchronized across the problem set. Resource considerations must include the total global requirements of all supported and supporting

CCMDs and organizations; all associated plans related to the problem set must be planned within the constraint of apportioned forces.

a <u>Notional Time-Phased Force and Deployment Data</u>. The notional TPFDD is the planning data in a TPFDD format reflecting the time-phased deployment of force, non-unit cargo and personnel requirements associated with a plan that contains a TPFDD (level 3T or 4). A notional TPFDD may contain units (identified preferred forces or contingency sourcing) and unit movement data (type unit characteristics (TUCHA) data or tailored TUCHA) as planning assumptions in order to determine a plan's transportation feasibility. Units identified as planning assumptions in a notional TPFDD may or may not reflect the actual units that are execution sourced should the plan be implemented.

b <u>Levels of Plan Detail.</u> The JSCP assigns levels to contingency plans dependent upon the level of planning detail required. There are four levels of plans, ascending in order with a required increase of structure and content.

- <u>Level 1- Commander's Estimate</u>. This level involves the least amount of detail and focuses on producing and recommending one or more COAs to address a potential contingency. Normally, these will consist of a concise narrative statement of how the CDR intends to accomplish the mission, providing the necessary focus for plan development and will center on military capabilities in terms of forces available, response time, and significant logistics considerations.

- <u>Level 2 - Base Plan</u>. This describes the CONOPS, major forces, concepts of support, and anticipated timelines for completing the mission. It normally does not include annexes or a notional TPFFD.

- <u>Level 3 - Concept Plan (CONPLAN).</u> A CONPLAN is an OPLAN in an abbreviated format that may require considerable expansion or alteration to convert it into a complete and detailed OPLAN (Level 4), or OPORD. It includes a plan summary, a base plan with the following Annexes: A (Task Organization), B (Intelligence), C (Operations), D (Logistics), J (Command Relations), K (Communications), S (Special Technical Operations), V (Interagency Coordination), and Z (Distribution). If the development of a notional TPFDD is directed for the CONPLAN, the planning level is designated as 3T and will also include Annex W (Operational Contract Support).

- <u>Level 4 - OPLAN</u>. An OPLAN is a complete and detailed joint plan containing a full description of the CONOPS, all annexes applicable to the plan, and a notional TPFDD. The OPLAN identifies the force requirements, functional support, and resources required to execute the plan and provide closure estimates for their flow into the theater. An OPLAN requires the development of an OPORD for the plan to be implemented.

8 <u>Operation Order</u>. An OPORD is an executable version of a plan. When events or situations within the OE prompt military action planning is initiated either based upon an existing plan or developed from scratch. The result of the planning effort is an OPORD that can readily be transitioned to execution. Under crisis conditions, Joint planning and execution is adapted to the constraints of time and an OPORD may be developed in parallel with initial execution functions. Execution functions such as execution sourcing of force requirements, mobilization and deployment preparations may be concurrent to OPORD development in order to expedite plan implementation.

(3) **Transition and Conflict Termination Planning.** Termination criteria are defined as the specified standards approved by the President and/or Secretary that must be met before a joint operation can be concluded. However, at the CCMD-level, plans may transition vice terminate as subsequent plans may be required to achieve sustained successful outcomes. The process of defining transition or termination criteria requires careful dialogue between civilian and military leadership to define conditions that are politically acceptable and militarily attainable. Typically, transition involves two aspects. <u>First</u>, security transition from military operating forces to the appropriate host nation or other designated security forces (i.e., United Nations, etc.) responsible for maintenance of security conditions. <u>Sec-</u>

ondly, transition or restoration of political governance from military control to the appropriate host nation civilian or political entities responsible for governance. Development of transition or termination criteria begins at the outset of the planning process and updates as the plan evolves. The criteria should account for a wide variety of operational events that the joint force may need to accomplish, to include: disengagement, force protection, transition to post-conflict operations, reconstitution, and redeployment.

(4) **Execution**. The results of operational assessment of campaign activities combined with situational awareness may identify conditions that warrant the execution of a military operation. Execution encompasses the implementation of military plans via orders for the conduct of military activities as directed and authorized by the President or Secretary. Execution, as an operational activity, may encompass multiple simultaneous execution efforts spanning the range of military activities. Planning continues into execution as orders are modified and branch/sequel plans are developed in response to actual or anticipated changes in the OE.

(a) Plan Implementation. Military plans and orders should be prepared to facilitate implementation and seamless transition to execution. The use of common formats and collaborative systems, tools, and processes can facilitate the transition to execution by supported and supporting commands. CCMD plan implementation should:

1 Confirm assumptions. Analyze the current operating environment and establish as fact any assumptions made during plan development. Adapt or refine the plan as necessary to account for false assumptions.

2 Model the TPFDD to assess force sourcing and transportation feasibility. Validate the availability of forces and mobility resources used during plan development that are currently available.

3 Establish execution timings. Set timelines or conditions to initiate operations to allow; synchronization of execution activities.

4 Confirm authorities for execution. Request and receive the President's or Secretary's authority to conduct military operations to include legal constraints that apply.

5 Conduct *execution sourcing* from assigned forces or request Secretary allocation of additional forces (see Chapter 3, *GFM*).

6 Issuing necessary orders for execution. The Chairman issues orders implementing the directions of the President or Secretary to conduct military operations. CCDRs and Service Secretaries then issue their own orders directing the activities of subordinate CDRs that are subsequently conveyed down the chain of command.

(b) Global Force Management in Execution. During execution, CCDRs employ assigned and allocated forces. CCDRs exercise combatant command (command authority) (COCOM) of assigned forces and have authority to employ them to accomplish the missions assigned to the command. CCDRs normally exercise operational control (OPCON) of allocated forces (see Chapter 3, *GFM*).

(c) Execution Synchronization. Approval of planning does not convey authority to execute or direct operations. Authorization to implement a plan or order is directed in an EXORD or other directive or order issued by the Chairman at the direction of the Secretary. For execution of campaign plan activities, CCDRs may need to coordinate additional authorities through respective Joint Staff Regional Joint Operations Directorates dependent upon the type of activity (e.g., operation, exercise, security cooperation). Authorities for execution may include, but are not limited to, Action Memorandums, Memorandums of Understanding and Executive correspondence. In any case, President or Secretary approval is required for executing military activity. Upon authorization to conduct military operations by the President or Secretary, the supported CCDR is the coordinator of execution activities IAW the command relationships specified in the EXORD or other order. The Chairman monitors the deployment and employment of forces, coordinates to resolve or mitigate shortfalls, and issues orders implementing the decisions of the President or Secretary.

(d) <u>Time-Phased Force and Deployment Data (TPFDD)</u>. A TPFDD is collection of data and information in a database that is built and refined incrementally and modified or updated throughout an operation's planning and execution phases. The TPFDD captures the who, what, when, where and how for forces and equipment required based on the operational sequence required to support a joint operation. TPFDDs developed during planning are considered notional as the data contained therein are planning assumptions. During development of the TPFDD the supported CCDR may identify *preferred forces* to continue the various planning effort's, but these forces are planning assumptions and not considered execution-sourced units. The TPFDD may be based upon an existing plan's notional TPFDD, but is different in that the TPFDD reflects execution sourced units (vice planning assumptions) and unit-refined movement data of the execution-sourced units (i.e., embarkation data).

<u>1</u> The notional TPFDD and TPFDD contain the same data fields and both may contain a variety of movement modes reflecting force flow in database detail aligned with a planned CONOPS. ***However, the data in the notional TPFDD is based on planning assumptions while the data in the TPFDD is actual execution data.***

<u>2</u> The transition from a notional TPFDD to a TPFDD occurs over time as individual requirements are execution sourced and the movement data is populated. For those requirements that have not yet been execution sourced, the TPFDD may contain assumed units and movement data for the purpose of assessing transportation feasibility.

(5) **Assessment**. Assessment is the sustained evaluation of plans and operations, occurring at all levels in the chain of command, which measures the effectiveness or progress towards achieving objectives. It is distinct from situational awareness in that it requires deliberate measurement of aspects of the OE against an anticipated or planned benchmark. Its purpose is to identify variance in anticipated and actual outcomes and inform decision making.[8] JP 5-0 and Chapter 6 of this book discusses plan and operational assessment methodology.

(a) <u>Operation Assessment</u>. During execution, operation assessment helps commands at all levels analyze changes in the OE to better identify the problem; anticipate potential outcomes; and understand the results of various friendly, adversary, and neutral actions and how these actions affect achieving the military objectives. Intelligence assessments from the JIPOE process supports operation assessments during execution. Operation assessment compares forecasted outcomes to actual events to determine the overall effectiveness of actions planned or taken. The results of operation assessment can validate or disprove previous planning assumptions and may lead to a refinement or adaptation of a plan or order.

<u>1</u> <u>Campaigns</u>. Strategic guidance directs CCDRs to provide annual assessments to inform senior leaders of how the campaign plan's sequenced military activities contributed to the progress towards campaign objectives. The results of a campaign assessment may prompt a CCDR to consider refinement or adaptation of the current campaign plan. Campaign assessments are also a key input to the Chairman's Comprehensive Joint Assessment (CJA) contributing to a global view of campaign progress toward desired outcomes.[9,10]

<u>2</u> <u>Operations</u>. Operations are executed as branch plans to the CCMD campaign plan and as such their operation assessment (i.e., progress towards measures of effectiveness and performance) must be linked with the associated campaign plan objectives. The operation assessment of a contingency operation may prompt a more frequent data collection against contingency specific indicators and metrics that assess progress.

---

[8] *Joint Publication 5-0, Operations*

[9] *CJCSI 3141.01, Management and Review of Campaign and Contingency Plans*

[10] *CJCSM 3130.01 Series, Campaign Planning Procedures and Responsibilities*

3 Development. The development of an operation assessment plan begins at the onset of planning and should complement the planning effort. As the plan is developed, the operation assessment plan approach and framework is developed and appropriately nested with the operational design. The operation assessment plan is ultimately documented into the completed plan or order and implemented upon execution.

(b) Contingency Plan Assessment. Developed and approved contingency plans undergo continuous refinement to adapt to changes in the OE or strategic direction. Changes to the OE identified through the JIPOE process or situational awareness may prompt a CDR to assess a contingency plan at any time. The results of the assessment may prompt a CDR to recommend a plan be refined, adapted, terminated, or executed (RATE). The CDR's assessment also provides awareness of the operational and strategic risks inherent in plan execution to the supported and supporting CCDRs.

(6) **Interaction of Operational Activities**. The operational activities occur continuously and mutually influence one another. CDRs at all levels have a decision cycle that spans all four operational activities while staff members may have a specialized focus on a single operational activity. This mutual influence of operational activities is fundamentally important as it ensures planning and execution are done with a contemporary and relevant context.

3. Summary

Operational activities comprise a sustained cycle of situational awareness, planning, execution, and assessment that occur continuously to support leader decision-making cycles at all levels of command. All four operational activities continue in a complementary and iterative process.

Most operational planning activities are dependent on, or are the results of other activities and are performed and reviewed at multiple echelons of commands in overlapping timeframes. Concurrent execution may increase the tempo in which these planning activities must occur. Likewise, multiple tasks must be accomplished across a broad spectrum of activities which include: data gathering and fact finding, mission analysis, preparation and distribution of planning guidance, development and refinement of force and logistic requirements, identification of forces and sustainment requirements, transportation feasibility, and review and re-planning based on changes in assumptions or the current situation.

"Always plan ahead. It wasn't raining when Noah built the ark."
Richard Cushing

III. Planning Functions

Planning Functions Overview

Joint planning and execution encompass four operational activities, four planning functions, seven execution functions, and a number of related products. The four planning functions are: 1) **strategic guidance, 2) concept development, 3) plan development, and 4) plan assessment.**[1]

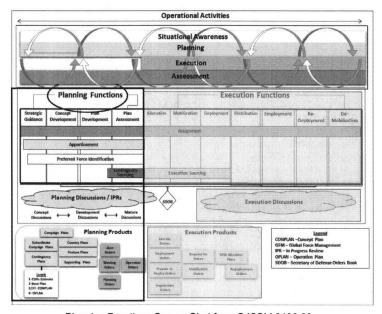

Planning Functions Screen Shot from CJCSM 3130.02

These planning functions facilitate an early understanding of the situation, problems, objectives and measures which will lead to the production of plans or orders that can be rapidly and effectively transitioned to execution and accomplish specified military objectives and to give military options to the President and SecDef as they seek to shape the environment and respond to contingencies. These planning functions are not mechanical. Planners perform JPP steps nested within these planning functions while considering the continuous operational activities that apply to each function. Effective planners also consider design, depicted as integration of the operational activities, while performing each function.

While these functions are depicted and may be performed sequentially, planning is iterative and functions may be re-visited as the planning conditions require to include planning continuing into execution. The SecDef, CJCS, CCDR, or any other Joint CDR may direct the planning staff to refine or adapt a joint plan by reentering the planning process at any of the earlier functions. The time spent accomplishing each activity and function depends on the circumstances.

[1] *CJCSM 3122.01 Joint Operation Planning and Execution System (JOPES) Volume I, Planning Policies and Procedures*

Planning Functions

Planning functions can be performed in series or in parallel as the situation dictates. During a crisis, planning functions are tailored to the time available and may be truncated, compressed, or conducted in parallel with execution functions. Instead of planning assumptions, crisis planning may be based upon the actual conditions of the OE.

Joint plans should be based upon strategic direction, reflect the current operating environment, limitations, developed and documented in standardized products and formats that are required to facilitate plan implementation and transition to execution.

Strategic Guidance - Function I *(see Chapter 5-I(a))*
The President, SecDef, and the CJCS, with appropriate consultation, formulate suitable and feasible military objectives to counter threats. The CCDR may provide input through one or more CDR's Assessments. This function is used to develop planning guidance for preparation of COAs. This process begins with an analysis of existing strategic guidance (e.g., JSCP for contingency plans or a CJCS WARNORD, PLANORD or ALERTORD for a crisis). The primary end product is a CDR's Mission Statement for contingency planning and a CDR's Assessment (OPREP-3PCA) or CDRs Estimate in a crisis.

Concept Development - Function II *(see Chapter 5-III(a))*
During contingency planning, the supported CCDR develops the CCDR's CONOPS for SecDef approval, based on SecDef, CJCS, and Service Chief planning guidance and resource apportionment provided in the JSCP and Service documents. In a crisis, concept development is based on situational awareness guidance, resource allocations from approved contingency plans, and a CJCS PLANORD, or ALERTORD. Using the CCDR's mission statement, CCMD planners develop preliminary COAs and staff estimates. COAs are then compared and the CCDR recommends a COA for SecDef approval in a CDR's *Estimate*. The CCDR also requests SecDef guidance on interagency coordination. The approved COA becomes the basis of the CONOPS containing conflict termination planning, supportability estimates, and, time permitting, an integrated time-phased database of force requirements, with estimated sustainment.

Plan Development - Function III *(see Chapter 6-I)*
This function is used in developing an OPLAN, CONPLAN or an OPORD with applicable supporting annexes and in refining preliminary feasibility analysis. This function fully integrates mobilization, deployment, employment, conflict termination, sustainment, redeployment, and demobilization activities. Detailed planning begins with SecDef approval for further planning in a non-crisis environment or a CJCS WARNORD, PLANORD or ALERTORD in a crisis situation; it ends with a SecDef-approved Plan or OPORD.

Plan Assessment – Function IV *(see Chapter 6-III)*
During this function, the CCDR refines the complete plan while supporting and subordinate CCDRs, Services and supporting agencies complete their supporting plans for his/her review and approval. CCDRs continue to develop and analyze branches and sequels as required or directed. The CCDR and the JS continue to evaluate the situation for any changes that would trigger plan revision or refinement.

   a. The JS, Services, CCMDs, and Agencies monitor current readiness and availability status to assess sourcing impacts and refine sourcing COAs should the plan be considered for near-term execution.

   b. The CCDR may conduct as many plan reviews as are required with the SecDef during Plan Assessment. These reviews could focus on branches/options and situational or assumption changes requiring major reassessment or significant plan modification/adaptation, but might also include a variety of other pertinent topics (e.g., information operations, special access programs, nuclear escalation mitigation).

Global Force Management (GFM)

The author and publisher would like to acknowledge and thank Mr. Timothy Conway for his subject matter expertise, contributions and thought—leader review to this chapter on global force management.

"Because we cannot be certain when, where, or under what conditions the next fight will occur, the Joint Force must maintain a boxer's stance -with the strength, agility, endurance, resilience, flexibility, and awareness to fight and win against any potential adversary."[1]

The global security environment presents an increasingly complex set of challenges and opportunities to which all elements of U.S. national power must be applied. To protect U.S. national interests and achieve the objectives of the NSS and NDS in this environment, the finite Joint Force will need to be used wisely.[2] The methods used to inform leaders of the options, risks and COAs requires a complex process that evaluates the ends, ways, means, and risks of using military forces to pursue strategic and operational objectives. That process is GFM and it determines which forces are employed at acceptable risk to current and future strategic and operational objectives. To build the Joint Force for the future requires a continuous recalibration of its capabilities and making those additional investments allowing us to succeed in all missions. Determining the best Joint Force of the future should be informed by near-term force needs and shortfalls, which is done through GFM assessments.

The DoD's enduring mission is to provide combat-credible military forces needed to deter war and protect the security of our nation. Should deterrence fail, the Joint Force is prepared to win. Reinforcing America's traditional tools of diplomacy, the Department provides military options to ensure the President and our diplomats negotiate from a position of strength. The experiences with operations such as Operation Urgent Response in Haiti while continuing to execute combat operations in Iraq and Afghanistan demonstrated that even while sourcing major combat operations in one part of the world, we may be called upon to react to a crisis in disparate regions of the globe. However, an earthquake in one AOR while conducting major operations in another can have a rippling impact on force sourcing for current operations and long-term security planning.

Per the NDS, long-term strategic competitions with our adversaries are the principal priorities for the Department, and require both increased and sustained investment, because of the magnitude of the threats these adversaries pose to U.S. security and prosperity today, and the potential for those threats to increase in the future. Concurrently, the Department will sustain its efforts to deter and counter rogue regimes, defeat terrorist threats to the U.S., and consolidate our gains while moving to a more resource-sustainable approach.[3] To remain dominant within this complex and uncertain security landscape, the ability to **dynamically align the force pool** must improve and keep pace with the complexity of the operational environment. The wicked problem of balancing the force against global and institutional demand requires strict purposeful design to allow for timely and informed decisions and to ultimately satisfy the broadest array of objectives.

---

[1] *2018 National Military Strategy Framework, Joseph Dunford Jr., 19th Chairman of the JCS*

[2] *National Defense Strategy (NDS)*

[3] *Ibid*

1. Purpose

This chapter provides an overview of the GFM process, which starts and ends with the SecDef. In accordance with 10 U.S.C. §§ 113, 153, 162, and 163, the SecDef directs the Services to provide ready and available forces, assigns and allocates forces/capabilities, provides planning guidance to CCMDs and provides overarching strategic guidance to CCMDs and the Chairman. The Chairman, in turn, recommends integrated solutions to employ the force to achieve the SecDef direction and develops strategic-level planning guidance including apportioned forces/capabilities to CCMDs for joint planning. CCMDs use apportioned forces as an assumption in developing plans and to coordinate force/capability planning requirements with the CJCS based on the SecDef's guidance.

The GFM processes directs the Services to provide sufficient ready and available forces to execute the NDS via the Directed Readiness Tables (DRT) (see para.3), distributes forces among the CCMD's via the assignment of forces, provides a process to temporarily adjust the distribution of forces among the CCMD's to meet dynamic challenges worldwide via the allocation process, provides apportioned forces, which is the Services' estimate of the number of forces that can reasonably be made available over a general timeline, should we be faced with executing a major operation and constantly assesses the results. The end result is a sufficient capacity of forces to execute the NDS, a risk-informed distribution of forces among the CCMDs and a starting point to begin resource-informed planning. To ensure the Joint Force remains relevant in meeting both current and future challenges, the GFM assessment processes compare supply with demands.

2. Scope

Strategic objectives are specified in strategic guidance documents, such as the NDS, DPG, and the JSCP. The Secretary also communicates strategic direction through the Chairman in the form of orders or other written or verbal communications. These strategic objectives specify the desired ends that plans and operations articulate the ways to achieve. The NDS provides a comprehensive framework to prioritize strategic objectives to shape the planning and execution of military actions to pursue the objectives and address the fundamental need to focus and apply finite resources. The resources employed in military planning and execution include interagency contract, coalition, DoD Expeditionary Civilian (DoD-EC), but are predominately military forces.

The force development processes identify, prioritize, and build the size and type of future forces necessary to pursue a strategy with acceptable risk. The Secretary through directed readiness specifies the force that must be ready and available to be employed quickly and used in creative ways.

Contingency and other plans and orders to achieve these objectives propose and direct a way to achieve the specified ends. Since campaign and contingency plans often rely on the same force pool, forces need to be prepared to execute any potential operations related to the desired objectives (ends) while also executing current operations and conducting military activities in pursuit of near-term objectives. The GFM assignment and allocation processes are the command and control (C2) mechanisms the Secretary uses to posture and distribute forces (means) which enable CCDRs to conduct operations and military activities to achieve strategic and operational objectives (ways) at acceptable risk.

The strategic environment will continue to be complex, dynamic and uncertain. The U.S. military will continue to be involved globally in executing GCPs, CCPs, other campaign and contingency plans, and ongoing operations, while being prepared to respond to domestic and overseas crises in support of NSS, NDS and NMS.[4] Success in this environment requires a coherent use of the force pool among the competing priorities in both planning and execution. This is achieved by the integrated use of the GFM processes of directed readiness, assignment, allocation, apportionment and assessment (DR4A). The goal of these processes is to provide CCDRs the forces to best support U.S. Military objectives (both

[4]SecDef directed GFM procedures are contained in the GFMIG, NDS, NMS, CPG, and CJCSM 3130.06

current and potential future) using assigned and allocated forces to accomplish missions while mitigating military risk. Directed readiness directs the Services to provide enough ready and available forces to execute the NDS. To allow feasible plans to be developed, CCMDs are provided force planning assumptions based on analysis of the force pool. The number of forces that are reasonably expected to be available, (globally, not to a specific plan or CCMD) over a general timeframe should the plan be executed, are called apportioned forces. The U.S. Military is tasked to execute the NDS objectives, which focus on major power competition. To restore readiness and build a credible deterrent force requires either building more forces or using the Services' capacity to field forces at less than the Services' maximum capacity. As the U.S. Military continues to face unpredictable fiscal challenges, the wise use of forces to meet the many global demands will become more and more important.

a. CCDRs are directed by strategic guidance and direction, and various orders, to plan and execute operations and missions. CCDRs are assigned forces that are to be used to accomplish those operations and missions; however, in the dynamic world environment, competing missions may require adjusting the distribution of assigned forces among the CCMDs and Services through allocation. Each allocation decision involves tasking a CCMD, Secretary of a Military Department, or director of DOD Agency to provide a force or individual to another CCMD. This involves risk to not only the providing Service and/or CCMD, but also to other ongoing operations, campaign and contingency plans across the Joint Planning and Execution Community (JPEC).[5]

b. <u>Integrated Deterrence</u>.[6] The NDS sets priorities to compete, deter, dissuade and, if necessary, defeat priority threats while continuing to conduct global foreign anti-violent extremist organization operations. It focuses attention on the strategic priorities. To counter and compete with multiple adversaries while continuing to resource forces globally for continued operations requires the force to be utilized more dynamically and in a more integrated deterrent manner.

The Secretary assigns and allocates forces to CCMDs. For the CCMDs to employ those forces those forces need to be ready and available. The GFM process begins with direction to the Services to build enough ready and available forces to execute the NDS and keep our military advantage into the future. This direction is contained in the GFMIG and the DRT. To comply with the DRT, the Services adjust their force development and force generation processes. Previously, the assignment and allocation processes began with the CCMDs submitting force requirements to execute their UCP assigned missions, tasked operations, and other military activities in their campaign plans.

The allocation process now begins with developing Top-Down Guidance that the Joint Staff, ICW the Services, develop a plan to align the forces against the NDS-specified strategic priorities, the GCPs, and the CCPs while maintaining a credible deterrent of ready and available forces. When the Chairman approves the Top-Down Guidance, the CCDRs develop and submit their requirements. The Joint Force Coordinator (JS JFC)** and JFPs consider the CCDRs submission as bottom-up refinement to the plan. The significant change postures the force against the strategic priorities first. Changes to the strategic posture in the Top-Down Guidance to pursue operational objectives are considered in light of

[5] *Joint Planning and Execution Community. See JP 5-0, Joint Planning.*

[6] *Global Force Management Implementation Guidance*

** *NOTE: CJCS, through the Director, J-3 (DJ-3), will serve as the Joint Force Coordinator (JFC) responsible for providing recommended sourcing solutions for all validated force and JIA requirements. In support of the DJ-3, the Joint Staff Deputy Director for Regional Operations and Force Management (J-35) assumes the responsibilities of the JFC. As such the JFC will coordinate with the Joint Staff J-3, Secretaries of the Military Departments, CCDRs, JFPs, and DoD Agencies. The Joint Force Coordinator (JFC) is referred to in current DoD GFM guidance and policy as the JFC. For clarity in this text the Joint Force Commander will be annotated by the acronym (JFC) and Joint Force Coordinator will be referred to with the acronym (JS JFC) to denote the Joint Staff Joint Force Coordinator.*

the risks to the strategic priorities. In the end, CCMDs may not get all the forces they want and assume operational risk in order to minimize strategic risk. Having a credible deterrent force and a logically postured force against the strategic priorities allows the Services to not deploy forces at or above their maximum capacity and begin to restore readiness. The C2 mechanisms the SecDef uses to posture the force are assignment and allocation. The DOD uses allocation to mitigate military risk in support of enduring and emergent requirements.

With a logically postured force against the strategic priorities and a ready and available force as a credible deterrent allows the Secretary to use the force to pursue strategic opportunities. These are typically exercises or other short duration military activities which limited forces in the credible deterrent force of ready and available forces can be used to message competitors, adversaries and allies. These strategic opportunities leverage the emergent allocation process to identify the military forces (means) to pursue a strategic opportunity (way) to further a defense objective (end) at acceptable risk.

3. Statutory Foundation and Key Strategic Guidance of GFM

a. GFM is founded in Title 10 U.S. Code.[7] Forces are assigned and allocated to Combatant Commands (CCMDs) by authority of, and under procedures prescribed by, the Secretary in accordance with (IAW) Title 10, U.S. Code (10 U.S.C.) § 162.

The statutory basis for the GFM processes are:

- **Directed Readiness** amplifies the guidance in the NDS for the Secretary to prescribe the force size and shape, force posture, and readiness to execute the NDS required by 10 U.S.C. § 113(g)(1)(B)(v). The annual DRT communicate this guidance and are posted on the JS J-8 SIPRNET web page.
- **Assignment** fulfills the 10 U.S.C. § 162(a)(1) requirement for the Secretaries of Military Departments, as directed by the Secretary, to assign forces to CCMDs or the U.S. Element of the North American Aerospace Defense Command (USELEMNORAD) for them to conduct authorized missions. The Secretary's guidance to the Secretaries of the Military Departments is the Forces For Unified Commands "Forces For" Assignment Tables. These tables are communicated annually via the "Forces For" Memorandum and posted on the J-8 Forces Div. SIPRNET web page.
- **Allocation** temporarily redistributes forces assigned to CCMDs or USELEMNORAD by transferring DoD forces and individuals between CCMDs, Services, or other DoD components to meet CCDR mission requirements or emergent needs according to 10 U.S.C. § 162(a)(3). The Secretary's direction to transfer forces is communicated in an order called the GFM Allocation Plan (GFMAP). The Secretary's Global Force Management Implementation Guidance (GFMIG) contains the allocation procedures required by 10 U.S.C. § 162(a)(3)(B). The President documents approval in the Unified Command Plan.
- **Apportionment** informs planning by providing a Service estimate of quantities of major types of forces that can reasonably be expected to be available, over a general period of time, in the event of a major operation fulfilling the Chairman's planning function according to 10 U.S.C. § 153(a). The quarterly apportionment tables on the JS J-8 SIPRNET web page communicate this estimate.

b. **Advise and Consultation**. Per 10 U.S.C. § 153 and § 163, the Chairman advises the Secretary on the use of forces, as necessary, to address transregional, multidomain, multifunctional threats and recommends the readiness, assignment, and allocation of forces. Per 10 U.S.C. § 151(c), the Chairman shall consult with the other members of the Joint Chiefs of Staff (JCS) and CCDRs when developing recommendations to the Secretary. The Chairman must also represent differing opinions to the Secretary. Differing opinions are communicated as nonconcurrence in the GFM staffing processes. As required by 10 U.S.C. § 151(d), the Chairman shall establish procedures and issue appropriate policy to ensure their advice is both timely and fully informed by the CCDRs and JCS. GFM staffing procedures specified in the GFMIG and CJCSM 3130.06 (GFM Allocation Policies and Procedures) ensure that the Chairman's advice and recommendations are not unduly delayed

[7] *Title 10 U.S.C. Armed Forces*

while awaiting the opinions of the other members of the JCS and CCDRs. IAW 10 U.S.C § 164(h), CCDRs shall provide information to the Chairman as necessary to perform his duties.

c. **Communication through the Chairman.** IAW 10 U.S.C. § 163(a), the President may direct that communications between the President and/or Secretary and CCDRs be transmitted through the Chairman. The President, in the UCP, directed that all communications between the President and/or Secretary and the CCDRs be communicated via the Chairman. Per 10 U.S.C. § 163(b), the Chairman is the spokesman for the CCMDs whereby he/she obtains requirements, evaluates and integrates information, and offers recommendations to the Secretary to those requirements both individually and collectively.

d. **Commands.** 10 U.S.C. § 167 and 167b defines unified combatant command for special operations forces and cyber operations, respectfully. All active and reserve special operations forces and cyber operations forces stationed in the U.S. shall be assigned to USSOCOM and USCYBERCOM, respectively.

e. **Baseline Documents.** The UCP, NSS, CPG, NDS, GFMIG, NMS, CJCSI 3110.01 (JSCP), *Forces For Unified Commands Memorandum ("Forces For")*, and the CJCSM 3130 Series of manuals are the baseline documents that establish the guidance, policy, doctrine and procedures in support of GFM. The SecDef's GFMIG specifies guidance for the GFM DR4A processes to manage the force pool from a global perspective. The CPG, NDS, NMS provides POTUS and SecDef guidance for prioritizing planning, execution and GFM (See Chapter 1 for Strategic Guidance statutory authorities and definitions).

f. **Unified Command Plan (UCP)**. Unified and Specified CCMDs were first described in the National Security Act (NSA) of 1947 and the statutory definition of the CCMDs has not changed since then. The NSA of 1947 and Title 10 U.S.C. provide the basis for the establishment of CCMDs. The UCP establishes the missions, responsibilities, and geographic AORs for CCDRs. The UCP is approved by the President, published by the CJCS, and addressed to the CCDRs. The unified command structure generated by the UCP is flexible, and changes as required to accommodate evolving U.S. national security needs. Communications between the President or the SecDef (or their duly deputized alternates or successors) and the CCDRs shall be transmitted through the CJCS unless otherwise directed. Title 10 U.S.C. § 161, tasks the CJCS to conduct a review of the UCP "not less often than every two years" and submit recommended changes to the President, through the SecDef.

(1) Seven CCMDs have geographic area responsibilities. These CCMDs are each assigned an AOR by the UCP and are responsible for all operations within their designated areas: U.S. Central Command, U.S. European Command, U.S. African Command, U.S. Indo-Pacific Command, U.S. Southern Command, U.S. Northern Command and U.S. Space Command. There are four CCDRs assigned worldwide transregional responsibilities not bounded by geography: U.S. Special Operations Command, U.S. Strategic Command, U.S. Transportation Command and U.S. Cyber Command.

(2) Several key strategic guidance documents provide direction for the execution of missions established in the UCP. Though not all-inclusive, they are: NSS, NDS, NMS, CPG, JSCP, and the GFMIG.

g. **Global Force Management Implementation Guidance (GFMIG).** The GFMIG integrates complementary policy and guidance on DR4A into a single authoritative GFM document in support of the DOD's strategic guidance. It provides required procedures to be prescribed by the SecDef IAW 10 U.S.C. § 162 (a)(3) to assign and allocate forces. These processes are applied within the force-management and force-planning constructs to better support resource informed planning and enable the force to be dynamically employed. The GFMIG combines the SecDef guidance on directed readiness, assignment, allocation, apportionment and assessment for all aspects of force management. The GFMIG:

- Outlines the process that directs the Joint Force readiness required to balance strategically competitive activity with wartime surge requirements.
- Provides SecDef direction for assigning forces to CCMDs for accomplishing CCDRs assigned missions.
- Prescribes the allocation process that provides access to forces and capa-

bilities when assigned mission requirements exceed the capacity and/or the capability of the assigned or allocated forces.

• Provides apportionment guidance to facilitate planning and assessment.

• Describes the force structure and capability assessment processes.

h. "**Forces For.**" The *"Forces For" Unified Commands ("Forces For") Assignment Tables* provides the SecDef's direction to the Secretaries of the Military Departments for assigning forces to CCMDs and USELEMNORAD. The assignment tables are contained in the GFMIG, and in years the GFMIG is not updated, the "Force For" Assignment Tables are published as an enclosure to the "Forces For" Memorandum. The assignment tables list major types of forces assigned to each CCMD and posted on the JS J8 web page. Forces not assigned to a CCMD remain assigned to the Secretary of the Military Department. Operational forces not assigned to a CCMD are retained under the Secretary of the Military Department and are referred to as "Service retained."

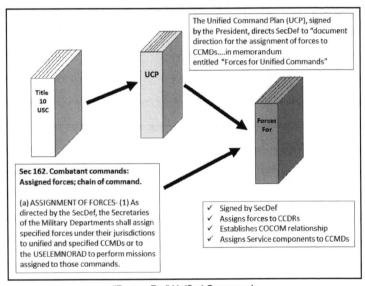

"Forces For" Unified Commands

i. <u>CJCSI 3110.01, Joint Strategic Campaign Plan (JSCP)</u>. The JSCP operationalizes the NMS and nests with the strategic direction delineated by the NSS, NDS, and the DOD's planning and resourcing guidance provided in the CPG. The JSCP also provides integrated planning guidance and direction for planners. The JSCP implements the strategic policy direction provided in the CPG and initiates the planning process for the development of campaign, campaign support, contingency, and posture plans. The JSCP contributes to the CJCS's statutory responsibility to assist the President and the Secretary in providing for the strategic direction of the Armed Forces of the U.S. to conduct planning. The goal is to provide CCMD and Service planners with meaningful, necessary guidance balanced between the details needed to conduct coordinated, sustainable steady-state activities and specific contingencies while preserving the flexibility to respond to unanticipated events by considering resourcing constraints and restraints.

I. GFM Processes

1. Foundational GFM Processes

GFM integrates complementary-directed readiness, assignment, allocation, apportionment and assessment (DR4A) information into force management and force planning constructs to support the Departments strategic direction. The GFM processes provide comprehensive insight into the strategic posture of forces and global availability of forces/capabilities for plans and operations. GFM also provides senior decision makers with a construct to assess impacts and risks associated with proposed changes to the force and how the force is used. These processes are used to align U.S. forces in support of the NDS, joint force availability requirements, and joint force assessments. Figure A depicts Secretary and Chairman GFM Products.

a. <u>Directed Readiness</u>. The Directed Readiness Tables (DRT), in support of the force planning construct, contain the Secretary's direction to the Department prescribing the capacity of forces required within specific availability windows to achieve strategic aims within a fiscal year. The Secretary issues this readiness directive to the department for budgetary planning and programming purposes. To determine the correct capacity required for a given fiscal year, the Chairman will use the Secretary's approved GCPs as the foundation of a global analysis. Based on this analysis, the Chairman will advise the Secretary on the forces required to proactively exploit identified GCP tasks, execute Secretary directed global commitments, and propose the proper amount of force capacity to withhold to generate readiness for the next contingency and maintain a credible deterrent.

b. <u>Assignment (GFMIG Section III)</u>. The Assignment Tables fulfill the 10 U.S.C. § 162(a) responsibility for the Secretary to direct the Secretaries of the Military Departments to assign forces to CCMDs or the CDRUSELEMNORAD for them to conduct authorized missions. CCDRs exercise Combatant Command authority (COCOM) over forces assigned to them. The command relationships with assigned forces are enduring until the assignment is changed. These relationships are depicted within the Automated Global Force Management Tool IAW DoDI 8260.03 GFMDI. The Secretary's guidance to the Secretaries of the Military Departments is the Forces For Unified Commands "Forces For" Assignment Tables which are communicated annually via the "Forces For" Memorandum. As with the DRT, the same strategic analysis derived from the Secretary's approved GCPs grounds the CJCS recommendation to the Secretary on the number of forces to assign to each CCMD.

c. <u>Allocation (GFMIG Section IV)</u>. Allocation temporarily redistributes forces assigned to CCMDs or USELEMNORAD by transferring DoD forces and individuals between CCMDs, Services, or other DoD components to meet CCDR mission requirements or emergent needs according to 10 U.S.C. § 162(a)(3). Forces are allocated to meet the Secretary's strategic priorities and the CCDR operational requirements. These decisions balance the risks of employing the full-Service capacity against maintaining a credible deterrent to potential enemies and enabling the Services to restore readiness. The allocation decision balances the force providers'(FP)[1] risks with the risks to current and potential future operations. When allocating a force, the Secretary specifies the command relationship (e.g., operational control (OPCON) or tactical control (TACON)) the CCDR assumes over the force

[1] *Force Providers (FPs). In accordance with the GFMIG, FPs include Secretaries of the Military Departments, CCMDs with assigned forces, the U.S. Coast Guard, DOD agencies, and OSD organizations that provide force-sourcing solutions to CCDR force requirements.*

as well as the duration, authorized missions, location, and other instructions. The Secretary's direction to transfer forces is communicated in an order called the GFM Allocation Plan (GFMAP). The Secretary's GFMIG introduces the top-down guidance the Chairman issues to shape the allocation of forces against the strategic priorities.

d. Apportionment (GFMIG Section V). Apportioned forces provide an estimate of the Military Departments'/Services'/FPs' maximum capacity to generate force elements along general timelines for planning purposes. The apportioned force types and quantities are what a CCMD can reasonably expect a Service to be able to deploy, but not necessarily an identification of the actual forces/units that will be allocated for use when a contingency plan or crisis response plan transitions to execution. Apportionment does not authorize or establish command relationships with any unit because units are not identified in the Apportionment Tables, only quantities of types of units (or force elements). Not only do CCMDs use the Apportionment Tables for operational planning, but when compared to the DRT, the Apportionment Tables provide a periodic measurement of the Department's ability to meet its readiness directive.

e. Assessment (GFMIG Section VI). The Joint Force Sufficiency Assessment (JFSA) informs the Department's assessment processes by identifying imbalances among Services' force/capability supply and current demand. It supports building a more lethal force and rebuilding warfighting readiness. JFSA will also provide assistance in determining why the Department was or was not able to meet its DRT targets. The JFSA process is broken into two parts: The GFMAP Sufficiency Assessment (GSA) and the Strategic Requirements Sufficiency Assessment (SRSA). The GSA addresses current GFM allocation related shortfalls while the SRSA focuses on the joint forces' ability to meet global demand outlined in strategic documents, both today and through the end of the Future Years Defense Program (FYDP).

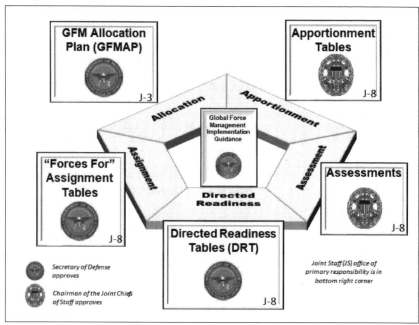

Figure A. Global Force Management Products

2. Force Pool

(Refer to Figure B, *Force Pool*)

The GFM processes are related to each other. Figure B depicts the entire DOD force pool (every military unit, Soldier, Sailor, Airman and Marine) within the "Service Institutional" and "Operational Forces" box. This force pool is further divided by assigned forces to a CCMD, unassigned forces (Service Institutional), and Service-retained forces. Most allocated forces come from operational forces assigned to a CCMD and Service retained, but in some instances, the Service may be directed to provide (allocate) forces from their Service Institutional forces (such as recruiters and schoolhouses).

Figure B also depicts the conceptual relationships of assignment, allocation, and apportionment on the DoD force pool.

The force structure shown consists of all Service forces and is divided into forces "Assigned" to CCMDs to conduct CCDR operational missions, and forces "Not Assigned" to CCMDs. Operational Forces are typically grouped into combat, combat support, and combat service support and designed to conduct operations. The Operational Forces assigned to CCMDs are labeled as "Assigned to CCMDs." CCDRs exercise COCOM authority over the forces assigned to them and have authority to employ their assigned forces to execute authorized missions. Forces "Not Assigned" to CCMDs or "Unassigned" remain assigned to the Service Secretary with an Administrative Control (ADCON) relationship. Unassigned forces consist of operational forces, labeled "Service Retained" and "Service Institutional Forces." Unassigned forces do not have a C2 relationship with any CCMD until allocated or otherwise attached. The Service Institutional Forces perform 10 U.S.C. §§ 3013,5013, and 8013 functions of the Services such as training commands, recruiting commands, and Service HQs.

The total number of operational forces is known. The number of employed forces can be determined or projected. From these two quantities, the Services can apportion forces, which is a Service estimate of the number of forces reasonably expected to be available to execute a major contingency. The quantity of forces apportioned are shown in the top box labeled "Apportioned Forces."

a. Unassigned, Service Retained and Assigned Forces may also be used by the Service to meet Service-institutional requirements. The Secretary may also direct Services to allocate institutional forces to meet operational requirements. These are why the "Projected Employed Forces" crosses the dashed line and into the "Institutional Service Forces" box. Apportioned forces are calculated by subtracting global demand from the assigned forces, and the fact that some assigned forces are employed performing Service institutional missions or are performing missions for their assigned CCDR, the employed and apportioned forces overlap. CCMD force requests are constantly changing to respond to world events. The number of forces employed and ordered to be employed by the CCMDs is considered in order to estimate the number of forces left that can reasonably be expected to be available, or apportioned.

b. 10 U.S.C. §§ 161, 162, 167 and 167b outline force assignment guidance and requirements. The President, through the UCP, instructs the Secretary to document his/her direction for assigning forces in the "Forces For." Pursuant to 10, U.S.C., § 162, the Secretaries of the Military Departments shall assign forces under their jurisdiction to unified and specified CCMDs to perform missions assigned to those commands. Such assignment defines the COCOM and shall be made as directed by the Secretary, including direction as to the command to which forces are to be assigned.

Force Pool

Force Pool

Service Institutional Forces

Functions:
* Recruiting
* Organizing
* Supplying
* Equipping
* Training
* Administering
* Mobilizing
* Demobilizing
* Maintaining
* Servicing

Operational Forces
(Combat, Combat Support, & Combat Service Support)

Apportioned Forces

Projected Available for Planning

Majority projected from

Allocated Forces

Projected Employed Forces

Service-Retained

Not Assigned to Combatant Commands "Unassigned"

Assigned to Combatant Commands

Service-Retained Forces – *AC and RC operational forces under the administrative control of respective Secretaries of the Military Departments, and not assigned to a CCMD. These forces remain under the administrative control of their respective Services and are commanded by a Service-designated CDR responsible to the Service unless allocated to a CCDR for the execution of operational missions. (GFMIG)*

Unassigned Forces – *Forces not assigned to a CCMD IAW 10 U.S.C.§ 162, and instead remain under Service control in order to carry out functions of the Secretary of a Military Department IAW 10 U.S.C. §§ 3013(b), 5013(b), and 8013(b). (GFMIG)*

Figure B. Force Pool (GFMIG).

3. GFM Principles

a. <u>Force Management Framework</u>. Figure C illustrates the conceptual force-planning construct that frames the conversation about balancing risk to strategy with risk to force inside a given fiscal year. It depicts the entire force pool and how it is being employed. It depicts the strategy-based demand signal for a ready and available credible deterrent force generated by the Secretary. The framework also captures assigned and allocated forces being employed to execute GCPs and CCDR operational missions as well as the forces undergoing reset in the Services force generation process. Finally, it depicts the institutional forces that perform Military Department statutory responsibilities. Examples of institutional forces include recruiting, training, and material commands as well as the Service HQs. The Incident Response Force (IRF) and Contingency Response Force (CRF) represent the different levels of directed readiness.

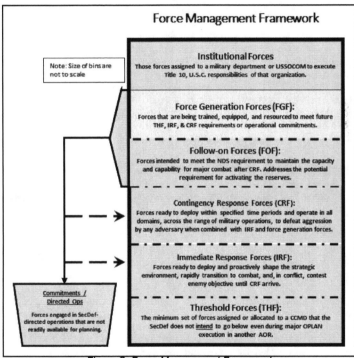

Figure C. Force Management Framework

b. <u>Global Demand</u>. In order to distribute a limited number of forces among the competing CCMD demands and preserve the ability of the force to respond to future potential contingencies, SecDef, OSD, CJCS, JCS, CCMDs, Services, Joint Staff, JS JFC, JFPs and FPs must understand the entire global demand on the force pool. By understanding the global demand, the risks of allocating or not allocating forces for a given operation can be better understood. Global demand consists of operational force (including assigned forces), Service institutional, Joint Individual Augmentee (JIA), exercise demand and future challenges IAW the GFMIG. FPs include Secretaries of the Military Departments, CCMDs with assigned forces, Commandant U.S. Coast Guard, Directors of DOD agencies, and OSD organizations that provide force-sourcing solutions to CCDR force requirements.

CCMDs, as CA, are best positioned to articulate the force requirements necessary for their respective GCPs to succeed. The Chairman's understanding of the competing demands across the GCPs enables the articulation of global risk and opportunity cost when advising the Secretary during both planning and execution. To distribute a limited number of forces among the competing GCP demands, the Chairman in consult with Services, CCMDs, JS, and JFPs will advise the Secretary on proper management of global demand using the DRT, Assignment Tables, and GFMAP. The Secretary will approve these products annually, directing the Department's acceptable level of readiness and activity. There is a separate GFMAP for each FY and the GFMAP directs forces to deploy in that FY. Each FY GFMAP has a 3-year life cycle. The GFMAP Base Order is published 10-12 months prior to the beginning of the applicable FY to support reserve notification timelines and facilitate predictability in Service deployment planning. The GFMAP and annexes are modified as necessary thereafter. There are typically about 80-90 modifications (mods) to the GFMAP. GFMAP Mods are presented to the SecDef every 3 weeks as a routine, but are also presented as required in out-of-cycle Secretary of Defense Orders Book (SDOB).

c. Operational Force Demand. Operational force demand encompasses all CCMD demand for forces in support of current operations (including shaping operations and OPLANS that require forces), activities in support of CCPs, theater security cooperation (TSC) events, activities in support of the campaign plans (including Joint Combined Exercise for Training (JCET)) and requests for federal assistance (RFA). Operational force demand includes force requirements for conventional, SOF, mobility, cyber, DOD Agency and Intelligence, Surveillance and Reconnaissance (ISR) forces and capabilities in support of the above operations and activities. Operational force demand is recorded in the Joint Capabilities Requirements Manager (JCRM), the program of record for enabling the GFM allocation process. Excluded are exercise and JIA demand.

(1) Assigned Force Demand (AFD). The demand on assigned forces is an integral part of operational force demand. CCMDs are assigned forces to perform operations and assigned missions. The CCMD to which forces are assigned may employ and deploy their assigned forces. Since the CCDR already has COCOM authority, and OPCON is inherent in COCOM, CCDRs can use their assigned forces for missions and operations they have authority to execute without additional SecDef approval. Section III (Assignment of Forces) of the GFMIG provides details on authorities granted with assigned forces.

(a) CCMDs are directed to report the requirements for their assigned forces and assign each requirement a Force Tracking Number (FTN), just as they would when requesting a force to be allocated in either the annual or emergent process. Although the SecDef does not allocate these forces back to the CCMD that has them assigned, this complete operational demand provides the JS J3, JS JFC, JFPs, FPs, and OSD this vital visibility into global demand and the number of employed forces to help in understanding the global force posture and determining risk when executing the allocation process.

(2) Service Institutional Demand. Service responsibilities addressed in Title 10 U.S.C. include recruiting, training, organizing and equipping the force. To accomplish these Title 10 U.S.C. responsibilities, Services require resources and units organized towards these missions. Normally Service units assigned to a CCMD are organized for operational missions as are Service retained units. Unassigned institutional forces are normally used by the Service to accomplish Service missions. These type units include recruiting commands, schoolhouses/training commands, experimentation, system development and Service headquarters and are normally unassigned.

(a) Occasionally units assigned to a CCMD are tasked by a Service to support Service missions. An example of this may be to operationally test a weapon system.

(b) Services currently capture their institutional demand and the risks associated with meeting that demand. Services may document their institutional demand in JCRM with FTNs, but this is voluntary. These risks to the Service are inserted into the allocation process as risk justifications that support the allocation decision process.

(3) <u>JIA Demand</u>. Included in the GFM allocation process are JIA programs and procedures that support temporary Joint Headquarters requirements for personnel in support of Presidential or SecDef-approved/directed operations. Joint and Combined Headquarters manning is documented on a Joint Manning Document (JMD). The individual billets for a JMD are each assigned a unique number in the Fourth Estate Manpower Tracking System (FMTS), the system of record to document JMDs. Each billet is classified as a coalition, contractor, unit fill, Other Government Agency (OGA), or JIA. Unit fills are force requirements for units that will be part of a joint headquarters. The unit requirements are handled via the force allocation process and reference the corresponding FTN. The billets designated as JIAs are recorded as a unique demand signal for a joint headquarters. The billets coded as JIA will be forwarded to the SecDef for an allocation decision in the GFMAP Annex D.

(a) CCDRs will also enter the requirements for all billets JMDs, including those billets which are being filled with their assigned forces.

(4) <u>Exercise Demand</u>. Exercise force demand encompasses all CCMD requirements for forces for CJCS directed and CCMD high-priority exercises. CCMDs submit exercise force requests via the Joint Training Information Management System (JTIMS), a tool that automates and supports the global Joint Training System, directly to the JS JFC and supporting JFPs with JS J3 visibility. JS JFC/JFPs coordinate exercise requests with FPs, as able. Exercise sourcing is normally a lower priority than operational force demands. Once exercise forces are sourced in JTIMS, operational force demands may reallocate the forces previously sourced to an exercise to higher priority operational force requirement. Exercise forces deploy TACON for participation in (or in Direct Support of) a CCMDs Joint exercise.

(5) <u>Future Challenges</u>. In addition to operational, institutional, JIA and exercise demand, which captures total global demand, the SecDef also considers potential use of forces/capabilities of future challenges. These future challenges are considered as potential usage of resources, but are not part of the global demand. Future challenges encompass all CCMD planning requirements for forces and capabilities required to support potential execution of a CJCS-directed or SecDef-approved plan should execution be ordered. In accordance with the CJCSM 3130.03,[2] CCMDs identify their force/capability requirements in Appendix 1 to Annex A of a plan. This list of force requirements in the JOPES[3] system is the Time Phase Force Deployment Data (TPFDD), and should execution of a plan be directed, the forces required to support the plan would become an operational demand and be submitted into the GFM process for sourcing. Forces not assigned to the supported CCMD would have to be requested and allocated by the SecDef.

d. <u>Global Force Visibility.</u> Understanding the capabilities, readiness, availability, and employment (CRAE) of the Joint force facilitates more informed decisions. CRAE information is compiled from the Services, Joint Staff, and CCMDs for many different analyses and is time and manpower-intensive. The Chairman has directed improved global force visibility capabilities and the "*Advana*" project (constellation of systems) is at the forefront of delivering the functionality to deliver globally integrated views of the Joint Force to planners and senior leaders via an IT infrastructure and tailored applications. Tailored applications in the Advana constellation of applications can be built to leverage the authoritative data in order to replicate the analytical processes currently in use and visualize the data quickly to get a current view of the disposition of the Joint Force.

4. The Global Force Management Board (GFMB)

The Global Force Management Board (GFMB) provides oversight and guidance for planning and executing all aspects of GFM and recommends near- and mid-term strategic opportunities. The GFMB is a flag officer-level body, chaired by the DJS with membership

[2] *CJCSM 3130.03, Planning and Execution Formats and Guidance*

[3] *Joint Planning and Execution System (JPES). Phase 1 of development will replace the functionality of JOPES with JPES. Subsequent phases of development are planned to integrate additional functionality to JPES.*

Global Force Management

from CCMDs, Services, JS J-Dirs, OSD (P and P&R) and select agencies to provide senior DOD decision makers the means to assess operational effects of force management decisions and provide strategic planning guidance. The GFMB does not approve assignment or allocation decisions. The GFMB usually convenes three times per year to focus on the annual sourcing process. It may also convene as required to address emergent issues.

a. <u>Purpose.</u> The purpose of the GFMB is to:

(1) Establish planning guidance for all aspects of GFM prior to developing options and recommendations.

(2) Serve as a strategic-level review panel to endorse GFM recommendations to the Secretary and ensure contentious issues are properly framed before going to the Chairman and ultimately, for SecDef decision.

b. <u>GFMB Participants</u>. The GFMB is chaired by the DJS and consists of flag/general officer (GO/FO) or SES equivalent representation from:

(1) The JS Directorates (including JS JFC).

(2) CCMDs (including JFPs).

(3) Service and Military Departments.

(4) NGB

(5) OUSD (P) & (P&R), others will attend as required.

(6) DoD Agencies. DoD agency representatives may attend as required.

5. Force Management Relationships

GFM aligns DR4A in support of the NDS, joint force availability requirements, and joint force assessments. Authorities that govern the three processes are as follows:

a. <u>Assignment</u>. 10 U.S.C. § 162(a) outlines force assignment guidance and requirements. The President, through the UCP, instructs the SecDef to document his/her direction for assigning forces in the "Forces For" Unified Commands ("Forces For") Assignment Tables. The Secretaries of the Military Departments shall assign forces under their jurisdiction to Unified and Specified CCMDs to perform missions assigned to those commands. Such assignment defines the COCOM and shall be made as directed by the SecDef, including as to the command to which forces are to be assigned. The assignment of forces is conducted annually and documented in the "Forces For" Assignment Tables of the GFMIG or, in years the GFMIG is not updated, as an attachment to the "Forces For" Unified Commands "Force For" Memorandum. These tables are the guidance from the SecDef to the Secretaries of the Military Departments to assign forces to the CCDRs. Assignment is further explained in the GFMIG within Section III.

(1) <u>Unifying Concept</u>s.

(a) <u>Force Assignment</u>. As mandated by Title 10 U.S.C., forces are assigned to CCMDs to provide those CCMDs forces to meet their UCP-established missions and assigned responsibilities.

<u>1</u> Forces permanently stationed overseas are generally assigned to the CCMD for the designated AOR, but CONUS-based force's may be assigned to GCC's outside CONUS.

<u>2</u> Special Operations Forces (SOF) are assigned to USSOCOM.

<u>3</u> Conventional forces stationed within the continental U.S. (CONUS) are Service-retained forces.

<u>4</u> Historically, other CONUS forces have been assigned as follows:

<u>a</u> West Coast conventional combat forces: Army, Navy, and Marine Corps forces to USINDOPACOM.

<u>b</u> Strategic deterrence forces to USSTRATCOM.

<u>c</u> Strategic defensive forces to U.S. Element NORAD (USELEMNORAD).

<u>d</u> Mobility forces and joint enabling capabilities to USTRANSCOM.

<u>e</u> Cyber forces to USCYBERCOM.

<u>f</u> Space forces to USSPACECOM.

(b) Service Components. Service components provide:

<u>1</u> Command relationships with Service components vary among CCMDs and are specified in the "Forces For" assignment. Service components can only be assigned CO-COM to one CCDR. However, Service-component CDRs may support multiple CCMDs in a non-COCOM, supporting CDR relationship.

<u>2</u> The Services recommend the proper headquarters to provide support, and "Forces For" is the vehicle for the Secretary to establish Service component support relationships.

<u>3</u> <u>Service Component Assignment</u>. A CCMD's Service component consists of the Service component CDR and the Service component command's forces (such as individuals, units, detachments, and organizations) that have been assigned to the CCMD.

<u>4</u> <u>Supporting Service Component Commander</u>. When a Service component is assigned to multiple CCMDs as a Supporting CDR, the Service CDR and only that portion of the CDR's assets assigned to a particular CCMD are under command authority (Support) of that particular CCDR.

Unless otherwise specified by the Secretary, the CDR tasked as a supporting CDR to additional CCMDs maintains a general support relationship for planning and coordinating regarding the CCDR's assigned mission and forces.

(2) <u>Scope</u>. The "Forces For" Assignment Tables contain the SecDef guidance to Secretaries of the Military Departments to assign specific numbers of forces to CCMDs and CDRUSELEMNORAD.

(a) CCDRs exercise COCOM over assigned forces and are directly responsible to the President and SecDef for performing assigned missions and preparing their commands to perform assigned missions. Although not a Unified CCMD, USELEMNORAD exercises COCOM over assigned U.S. forces.

(b) The assignment of forces is separate from Title 32 U.S.C. provisions that deal with the National Guard and the parts of Title 10 U.S.C. that provide for ordering National Guard and reserve forces to active duty. CCDRs will exercise COCOM over assigned RC forces when mobilized or ordered to active duty (other than training). CCDRs may employ RC forces assigned to their commands in contingency operations only when the forces have been mobilized for specific periods in accordance with the law, or when ordered to active duty with the consent of the member and validated by their parent Service. During peacetime, CCMDs will normally coordinate with Service component commands on all matters concerning assigned or attached RC forces. CCMDs exercise Training and Readiness Oversight (TRO) for assigned RC forces when not on active duty or when on active duty for training. Service Headquarters exercise TRO for Service-retained RC forces.

(c) Force reductions or new assignments are required when programmed force structure changes occur. Reassignments, from one CCMD to another, may occur as directed by the SecDef and are not necessarily tied to Presidential budget changes.

(3) <u>Assignment and Transfer of Forces</u>. A force is assigned in accordance with the guidance contained within GFMIG. Forces are allocated for execution through the GFM planning process specified in the CJCSM 3130.06 series,[4] and the GFMIG. Forces become attached when deployed via a SecDef-approved global deployment order called the GFMAP.

(a) Forces, not command relationships, are transferred between commands. When forces are transferred, the command relationship that the gaining CDR will exercise (and the losing CDR will relinquish) will be specified by the SecDef. The CCDR normally

[4] *CJCSM 3130.06, Global Force Management Allocation Policies and Procedures*

exercises OPCON over forces attached by the President or SecDef. Forces are attached when transfer of forces will be temporary. Establishing authorities for subordinate unified commands and Joint Task Forces (JTFs) may direct the assignment or attachment of their forces to those subordinate commands and delegate the command relationship as appropriate.

(b) Transient forces do not come under the chain of command of the area CDR solely by their movement across operational area boundaries, except when the CCDR is exercising TACON for the purpose of force protection.

(c) Unless otherwise specified by the SecDef, and with the exception of the USNORTHCOM AOR, a CCDR has TACON for exercise purposes whenever forces not assigned to that CCMD undertake exercises in that CCDR's AOR. TACON begins when the forces enter the AOR and is terminated at the completion of the exercise, upon departure from the AOR. In this context, TACON is directive authority over exercising forces for purposes relating to that exercise only; it does not include authority for operational employment of those forces. In accordance with the UCP, this provision for TACON normally does not apply to U.S. Transportation Command (USTRANSCOM) or U.S. Strategic Command (USSTRATCOM) assets. When USTRANSCOM, USSTRATCOM or USSPACECOM forces are deployed in a geographic CCMD's AOR, they will remain assigned to and under control of their CCDR, unless otherwise directed.

(d) Subject to the two following exceptions, in the event of a major emergency in a geographic CCMD's (GCC) AOR requiring the use of all available forces, the GCC may temporarily assume OPCON of all forces in the assigned AOR, including those of another command. The CCDR determines when such an emergency exists and, upon assuming OPCON over forces of another command, immediately advises the CJCS, the appropriate operational CDR, the Military Service Chief, and the Secretary of the Military Department of the forces concerned and the nature and estimated duration of employment of such forces. The CJCS shall notify the SecDef, who will determine whether to permit the emergency exercise of OPCON to continue.

<u>1</u> Exception 1: GCCs may not assume OPCON of all forces in the AOR for a major emergency if such use would interfere with those forces scheduled for or actually engaged in the execution of specific operational missions approved by the President or SecDef.

<u>2</u> Exception 2: CDRUSNORTHCOM's authority to assume OPCON during an emergency is limited to the portion of USNORTHCOM's AOR outside the United States. CDRUSNORTHCOM must obtain SecDef approval before assuming forces not assigned to CDRUSNORTHCOM within the United States.

b. <u>Allocation</u>. Per 10 U.S.C. § 162(a)(3), allocation is the C2 mechanism by which the Secretary temporarily adjusts the global distribution of forces among the CCDRs and CDRUSELEMNORAD. Allocation applies to all DoD forces and individuals, including DoD Expeditionary Civilian (DoD-EC). The Secretary may allocate any force or capability within the DoD regardless of current assignment or allocation. Allocation is used to fill CCMDs' force and JIA requirements that cannot be met with assigned forces by providing the most appropriate and responsive force to meet operational requirements in a globally integrated, prioritized, and risk-informed manner. The allocation process considers the readiness and availability of the force structure, the CCMDs requirements to execute specified plans and operations, and the globally integrated priorities to provide recommendations to the Secretary to distribute forces to the CCMDs.[5]

Per 10 U.S.C. §§ 153(a)(3) and 163(b)(2) the Chairman is responsible for providing globally integrated, prioritized, and risk-informed recommendations from all DoD forces to identify and recommend the most effective and efficient use of forces to achieve strategic and operational objectives. In generating the annual allocation plan, the JS J-3 considers each Services' capacity to provide ready and available forces and, with the Services, develops a recommendation to allocate those forces to CCMDs in order to posture those forces

[5] *Global Force Management Implementation Guidance*

against the Secretary's strategic priorities, which includes maintaining a credible deterrent capability of ready and available forces. A credible deterrent also provides a capacity of ready and available forces that affords the Secretary the ability to employ forces dynamically to pursue strategic opportunities.

Simply put, the allocation process adjusts the distribution of assigned forces among the CCMDs. When transferring forces, the Secretary will specify the timeframe the forces will transfer, the command relationship the gaining CDR will exercise and the losing CDR will relinquish, as well as any other necessary limitations on the use of the force. The allocation sourcing process provides fully vetted and relevant sourcing solutions to CCMDs' force requirements to the SecDef for approval, via the SDOB, in support of global and theater strategic objectives. The SecDef allocation decisions are communicated in the GFMAP.

(1) GFM Allocation Overview.[6] (See Figure D)

Top-Down Allocation Guidance. The JS J-3, in coordination with the Services and JFPs, propose the global allocation of forces to CCMDs to support the Secretary's strategic priorities while providing a credible deterrent capability. Within this context, CCMDs develop their annual submission and subsequent emergent requirements to refine the Top-Down Allocation Guidance from the bottom up. The globally integrated plan, (Top-Down Allocation Guidance), is first presented at GFMB 3, endorsed at GFMB 4, and presented to the Chairman for approval.

(a) Requirements Submission. CCMDs submit force and JIA requirements, which both include requirements for DoD-EC, for the FY in the annual submission. This submission includes demand for assigned forces, replacements for forces currently allocated (rotational forces), and any new requirements necessary for that FY. Between annual submissions, CCMDs may request additional forces and JIAs as emergent requirements. Interagency RFAs are also sponsored by a CCMD and, after OSD notification, are sourced via the allocation process.

(b) Validate Force Requirement. Upon CCMDs submission of their force requirements, the JS J-3 will validate each force request against SecDef approved validation criteria to ensure there are proper authorities for the CCMD to conduct the missions. The JS J-3 will assign the appropriate priority to each request per the NDS and Top-Down Allocation Guidance. JS J-1 conducts the same validation process for JIA. JS J-1 will advise the JS J-3 on the validity of all DoD-EC requirements contained in operational force demands

(c) FP Nomination. Once requirements are validated, the JS assigns each request to the appropriate JS JFC, and JFPs to develop a sourcing recommendation. The JS JFC and JFPs task FPs to submit sourcing nominations from their Service or assigned forces.

- The FP nomination for each request is developed from a comprehensive look across all Service forces, including forces assigned to CCMDs and Service retained forces. Services, Service Force Providers (SFPs)[7], and other FPs nominations are informed by the strategic priorities, Top-Down Allocation Guidance, and bounded by the actual ready and available forces.

- Each FP nomination will contain the FP risks inherent in providing a force to meet the requested capability. When the FP nomination recommends allocation of forces assigned to a CCMD, the FP nomination will also include the operational risks to the CCMD communicated via the CCMD's Service component.

(d) JS JFC/JFP Sourcing Recommendation. The JS JFC, and JFPs review the FP nominations, with the inherent risks, to develop a recommended sourcing solution that includes the associated operational and force providing risks justifying the recommendation. The sourcing recommendation should align with the strategic priorities and Top-Down Allocation Guidance. A recommendation may be to source, partially source, or not source a given requirement depending on the priority and risks. Contentious issues are raised,

Global Force Management

---

[6]*Global Force Management Implementation Guidance*

[7] *SFP- Perform the allocation roles and responsibilities of an FP for the Service HQ, as delegated by the Service Chief*

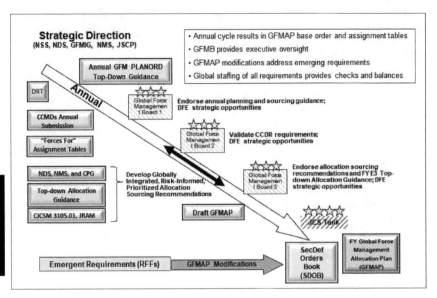

Figure D. GFM Big Picture

through a series of forums at the action officer (AO) and/or GO/FO level, as required to resolve or ensure the issue is properly framed for SecDef decision. The JS JFC and JFPs staff the joint sourcing recommendations with all CCMDs and FPs.

(e) <u>GFMAP Order Staffing</u>. The JFPs forward their sourcing recommendation in the form of a draft order to the JS JFC, who consolidates and staffs the draft order with FPs and CCMDs to ensure risks are accurately presented for SecDef decision. The draft orders are then forwarded via the SDOB.

(f) <u>Secretary of Defense Orders Book (SDOB)</u>. The JS J-3 staffs and briefs all draft orders that require SecDef approval through the JS directorates and OSD principals to the Chairman and, ultimately, to the Secretary for decision. The SDOB consists of a binder containing a book of briefs. The format of those briefs is tailored for the Secretary and has historically been modified to present the necessary information in the SecDef's desired format. The JS J-3 obtains Secretary approval via the SDOB on proposed orders authorizing the execution of military operations, directing the transfer or attachment of forces, authorizing supplemental rules of engagement that require Secretary approval, or modifying Secretary decisions in previously approved orders. The GFMIG discusses the SDOB process in greater detail. Orders and actions included in the SDOB include, but are not limited to:

- Execute Order (EXORD)
- Deployment Order (DEPORD)
- GFMAP and subsequent modifications
- Prepare-to-Deploy Order (PTDO)
- Alert Order (ALERTORD)
- NATO FORCEPREP message
- Alert and mobilization of RC forces
- Modification of CCMD authorities or previous Secretary decisions and orders.

Typically, the SDOB is briefed every three weeks; however, a special SDOB may be held whenever circumstances dictate. JS J-3 publishes SecDef-approved orders, which include the GFMAP and subsequent modifications.

(g) GFMAP/Orders. The GFMAP is a SecDef approved deployment order for all allocated forces. There is a separate GFMAP for each FY. The GFMAP Base Order is published 10-12 months prior to the beginning of the applicable FY to support reserve notification timelines and facilitate predictability in Service deployment planning. The GFMAP and annexes are modified as necessary thereafter.

1 GFMAP Annexes. The SecDef decisions to allocate forces are published within the five annexes of the CJCS GFMAP. Annexes are published with the GFMAP and subsequent modification to the GFMAP. GFMAP Annexes A-D in spreadsheet formats. They contain all the information inherent within a written order, and authorizes JFPs to order forces in subsequent GFMAP Annex Schedules. Each Annex includes force schedules as follows:

- Annex A is Conventional Forces - JS J3 as the JS JFC.
- Annex B is Special Operations Forces (SOF) - USSOCOM as the JFP.
- Annex C is Mobility Forces - USTRANSCOM as the JFP.
- Annex D is Joint Individual Augmentees - JS J3 as the JS JFC.
- Annex E is the AOR specific guidance for each CCMD.
- Annex F is Cyber-Operations Forces – USCYBERCOM as the JFP (at FOC).
- Annex G is Space Operations Forces - USSPACECOM (anticipated in FY 2024 GFMAP).

2 The GFMAP Annex serves as the DEPORD directing FPs to deploy forces at the specified dates. FPs implement the SecDef orders in the GFMAP Annexes and JFP GFMAP Annex Schedules by issuing DEPORDs, through the chain of command, to the unit or individuals deploying. The JS JFC, and JFPs monitor the FP's progress in meeting the GFMAP orders. Since most operational forces are assigned, each allocation is a decision to take a force from one CCMD and deploy it to another. Each decision has risks that must be weighted to balance the operational necessity of deploying the force with the risk to the FP. These risks also include the financial cost, the stress on the Service and the stress on the Service men and women as well as the risks to future potential contingencies. The GFMAP specifies the C2 relationship between the supported CCMD and the allocated unit. Normally, the CCDR is delegated OPCON of the force. However, the Secretary may specify other command relationships, such as TACON, as required. Forces are allocated to a CCMD for a specified period of time to execute a specific mission(s). The GFMAP authorize JFPs to order forces to deploy in subsequent JFP GFMAP Annex Schedules.

3 Each GFMAP Annex will include, at a minimum:

- Type of force or capability required (i.e., CSG, Multi-role Fighter, etc.).
- Number of units or overall AOR presence.
- Operation or mission the annual force is tasked to support (e.g., TSCP or named exercise/operation).
- Start Date and End Date of required capability.
- FTN.
- Supported CCMD.
- Type of force or capability required.
- FP.
- Dates of deployment or PTDO status.

<u>4</u> FPs implement SecDef orders in the GFMAP Annexes and JFP Annex Schedules by issuing DEPORDs, through the chain of command, to the unit or individuals deploying. Likewise, CCDRs issue orders to establish C2 relationships of assigned and allocated forces. The JS JFC and JFPs monitor the FPs' progress in meeting the deployments directed in the GFMAP.

<u>5</u> In cases that other U. S. Government (USG) departments and agencies commit capabilities to meet CCMD force requirements requested via the allocation process, the GFMAP Annex is used to communicate the sourcing solution to the supported CCMD. The Secretary does not have authority over other USG departments and agencies, and the GFMAP Annexes and Annex Schedules do not have the power of an order for these instances.

(h) <u>Annual Allocation</u>. The annual submission is essentially a consolidated request fpr forces (RFF) for all the forces required for an entire FY. The annual sourcing cycle provides a level of predictability for Services, JFPs, FPs' annual deployment scheduling, and CCMDs' operational planning (see paragraph (3(a)) for greater details).

(i) <u>Emergent Allocation</u>. Consistent with strategic direction in the NDS to be strategically predictable, but operationally unpredictable, requires continuous assessment of the global distribution of forces and the ability to adjust at the speed of relevance. An emergent RFF is a request from a CCMD, CDRUSELEMNORAD, or NATO, via the USNMR, for units and capabilities that were not anticipated at the time of the annual submission and cannot be met by the requesting headquarters, its components, or their assigned and allocated forces. CCMDs submit emergent force requirements via JCRM and an RFF record message (RMG) simultaneously (see paragraph (3(b)) for greater details).

(j) <u>Applicable to Annual and Emergent Allocation</u>. Forces are allocated to a CCMD in order to execute tasks in that CCMD's AOR, and are typically deployed for a specified period of time as outlined by Service deployment policies. Forces deploy as "units," sized from the Army Brigade Combat Team (BCT), Marine Corps Regimental Combat Team (RCT) or Marine Air Ground Task Force (MAGTF), Air Force Air and Space Expeditionary Task Force (AEF), or Navy Carrier Strike Group/Amphibious Ready Group (CSG/ARG) level or larger, down to smaller-sized capability packages (e.g., individual ships, squadrons, or mission-unique teams), and deploy in support of CCDR missions. Forces are sourced globally. The desired endstate for the annual force allocation process is a recommendation for filling validated CCMD capability requirements with appropriate joint forces to achieve an optimum level of risk in executing ongoing combat operations and other missions and tasks called for in the NDS and NMS.

(2) Secretaries of Military Departments, CCDRs and Directors of DOD agencies all report directly to the Secretary of Defense. The allocation decision process is centralized with the SecDef making the final decisions.

(a) When unassigned or Service-retained forces are allocated to a CCMD, they are normally attached via an OPCON relationship. The supported CCDR is directed to assume OPCON upon arrival in theater. The providing CCDR is tasked to relinquish OPCON of that unit during its deployment and resume OPCON following the deployment. The providing CCDR retains COCOM of the unit. When assigned forces are allocated from one CCDR to another, the gaining CCDR is usually tasked to exercise OPCON, for the time the unit will be in the gaining CCMDs AOR. Specific command relationships are specified in the CJCS GFMAP and the GFMAP Annex. Objectives of the GFM allocation process are to:

- Globally prioritize CCDR operational tasks, including war plan response posture, SCP activities, or other missions as assigned by the CCDR, taking into consideration ongoing operations, and to allocate annual forces to satisfy these tasks. Prioritization of CCDR operational tasks and the assessment of risks are essential as global demand for annual forces may exceed available supply and consequently affect the Military Department's ability to sustain rotation rates or available forces/capabilities.

- Optimize force management to reduce risk to achieve operational and strategic objectives, while balancing the Military Departments' responsibility to organize, train, and equip the force against CCDR operational tasks.

- Establish a mechanism, through the GFMB to provide joint solutions to CCDR's requirements, and enable, as the technical capacity matures, capability trade-offs among Military Departments. Remember the GFMB is a board of GO/FO/SES chaired by the DJS. The CCMDs, Services and OSD(P) and (P&R) are represented on the board. The board provides planning guidance for all aspects of GFM, endorses recommendations, and ensures contentious decisions are properly framed for CJCS recommendations and SecDef decisions. See Figure D on page 3-18 (*GFM Big Picture*) and note the GFMB role in executive oversight for the FY GFMAP.

- Provide predictability for Military Service Chiefs' annual deployment scheduling and CCMD's operational planning.

- Increase flexibility and options for senior leadership decision-makers (e.g., facilitate the Military Departments' ability to surge forces, as well as the ability, over the longer term as the capabilities to do so are fielded, to provide capability trade-off options among Military Departments, such as use of Navy Aegis Cruisers, Army Patriot assets, Navy or USAF fighters, or a mix of assets to provide defensive counter air capabilities).

(3) Annual and Emergent Allocation.

(a) Annual. This process lays out the roles, missions, and functions to support the sourcing of CCMD annual force requirements. CCMDs submit their requirements for all forces in a FY in the annual submission. These force requirements request both rotational forces (replacements for currently assigned forces) and new requirements for the FY. The Armed Forces of the U.S. provide overseas presence through a combination of allocated forces and forward-based forces and the resources (infrastructure and prepositioned equipment) necessary to sustain and maintain those forces. Forward-based forces are assigned to GCCs (e.g., USEUCOM, USINDOPACOM, and USSOUTHCOM) in the "Forces For" assignment tables.

1 Annual forces support the following activities:

a CCMD force posture against the Secretaries strategic priorities.

b CCMD OPLAN response timelines in support of CPG and JSCP requirements.

c SecDef-named operations with identified enduring requirements.

2 Allocation decisions for the annual submission are promulgated in the base order of the GFMAP.

The SecDef-approved GFMAP gives the designated JS JFC/JFP, CCMDs, and Military Departments authority for annual allocation of capabilities, forces, and units for the next fiscal year (FY+1).

3 In the event of an emerging crisis, forces allocated to other operations may be considered by the SecDef for allocation to the emerging crisis. Based on the risks, the SecDef may allocate forces previously allocated elsewhere. However, the intent of the annual allocation process is to provide some measure of predictability for CCMDs, Military Departments, Defense Agencies, and individuals.

4 Annual Force Sourcing Timeline.

a The annual allocation process is an extended effort that typically spans a year in development with the goal of publishing a given FY's annual Deployment Order, known as the GFMAP Base Order, one year in advance of execution. The primary purpose of releasing the GFMAP one year in advance is to provide better predictability to Service members, families and units, allow proper time for alert and mobilization of the Reserve Component forces, and to better synchronize with the Planning, Programming, Budgeting and Execution (PPBE) process.

<u>b</u> The annual process is facilitated by periodic (Quarterly) GFMBs that have a specific purpose and associated product to guide the development of annual force requests and sourcing recommendations. The GFMIG details the Quarterly cycle.

<u>5</u> <u>Annual Force Planning</u>. During execution, planning continues. The plan being executed is under constant review and the next step or phase of the operation is under review. During OIF and OEF, the SecDef directed that enduring force requirements be reviewed and re-validated annually. This re-validation became the basis for rotational force planning. Today, all CCMDs review their ongoing operations and submit force requirements for the upcoming FY in their annual submission. Rotational force requirements are for replacement forces to currently allocated forces. Historically, rotational force requirements are the majority of the annual submission, although the annual submission also contains new force requirements. The annual submission is, essentially, a consolidated RFF for the entire FY. CCMDs review every operation in progress and determine what forces are needed for each operation. The CCMD must also project the force requirements for the military activities in their campaign plans, to the maximum extent possible to determine the operational requirements from ongoing operations. A way to organize this task is to review the forces currently conducting the operation and validate the continuing need for each force in the coming FY for the phase of the operation that the plan will be in. Each force requirement for the current unit is refined, approved by the CCMD and submitted in the annual submission to the JS J3 to enter the annual allocation process. It is important to link current forces and their rotational force requirements because the JS JFC and JFPs must schedule units into the GFMAP Annex in order to provide uninterrupted mission coverage. Specifying the evolving missions and tasks for specific units is imperative so the Services can train, organize and equip forces to forces to be prepared to conduct those evolving missions. Once the force requirements have been submitted, newly identified, or refined, force requirements enter the emergent allocation process via an RFF. As previously mentioned, the CCMD annual submission is considered as bottom-up refinement to the Top-Down Guidance. CCMDs should articulate how each requirement is linked to a strategic or operational priority.

(b) <u>Emergent.</u> The emergent process begins with the CCMD identifying a force or individual requirement that cannot be met using available assigned forces or forces already allocated. In order to optimize the timeliness of sourcing solutions and to ensure requested Start Dates (see Table 1) are met, the CCMDs must manage expectations and submit all emergent RFFs with as much lead time as possible. Emergent RFFs fall into three categories: routine, urgent, and immediate. Immediate is the most time sensitive, followed by urgent, and routine RFFs allow for more deliberate consideration during sourcing. During time-sensitive crises, there may not be time to enter new requirements in the IT systems. Immediate requirements, when required, may be submitted using the most expeditions means available, including phone calls, secure emails or other means. Requirements relayed by expedited means should be followed up in writing and via the IT systems as soon as practicable.

<u>1</u> <u>Force Tracking Numbers (FTNs)</u>. The CCMD documents each force requirement, usually one unit per requirement. The force requirement contains information of what type of force is needed as well as the operational risk if the force is not provided. Each requirement is validated by the CCMD and assigned an FTN. The FTN is an identification number or the identification tracking number used by CCMDs and FPs to track force requirements from requesting document to unit deployment IAW the GFMIG (see Figure E).

<u>a</u> The FTN is a unique 11-character alphanumeric reference number created by the supported CCMD and assigned to a single requested force capability requirement. Adherence to the assigned eleven-character structure for each FTN position is mandatory in order to support data standardization between CCMDs and to establish a joint process. An FTN cannot be reused or duplicated for additional requirements. Each force requirement has specified data fields attached that are used by the supporting commands in developing recommended sourcing solutions and informing the SecDef of risks when making allocation decisions. Each force request is assigned a unique FTN and forwarded electronically to the JS J3 for validation.

Start Dates Defined

Start Date – *The Start Date is the date the force, including personnel and equipment, arrive in the supported CCMD's AOR to begin JRSOI, normally at the port of debarkation (POD). Start Date is requested by the CCMD, nominated by the FP, recommended by the JS JFC or JFP, and ordered in the GFMAP Annex.*

Requested Start Date – *The date the CCMD requests the force, including personnel and equipment, to arrive in the supported CCMD's AOR to begin JRSOI normally at the POD. For most Naval forces this specifies the requested numbered fleet AOR IN-CHOP (AOR entry) date. For requests for Prepare to Deploy Order (PTDO) forces it is the date the CCMD requests the force be available to deploy within the designated PTDO response time.*

Ordered Start Date – *The date the force, including personnel and equipment, is ordered to arrive in the supported CCMD's AOR to begin JRSOI, normally the POD. For most Naval forces this specifies the requested numbered fleet AOR IN-CHOP (AOR entry) date. For forces on PTDO, it is the date the unit is to be available for deployment within the designated PTDO response time.*

Table 1. Start Dates Defined

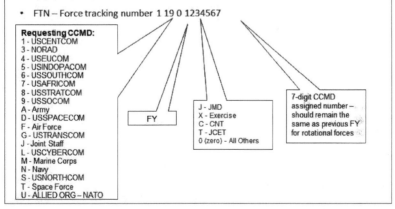

Figure E. Force Tracking Numbers

The FTN is a unique 11-character alphanumeric reference number created by the supported CCMD and assigned to a single requested force capability requirement. Adherence to the assigned eleven-character structure for each FTN position is mandatory in order to support data standardization between CCMDs and to establish a joint process. An FTN cannot be reused or duplicated for additional requirements. Each force requirement has specified data fields attached that are used by the supporting commands in developing recommended sourcing solutions and informing the SecDef of risks when making allocation decisions. Each force request is assigned a unique FTN and forwarded electronically to the JS J3 for validation.

<u>b</u> The force request is also transmitted to the JS J3 with an RFF message which requests sourcing of one or more force requirements. The JS J3 validates each force request, by FTN, provides priority based on sourcing guidance, and assigns each on to the JS JFC (conventional forces) or one of the JFPs (SOF, mobility forces and cyber operation forces) to provide a recommended sourcing solution. At the direction of the JS J3, the JS JFC coordinates with the Service HQs and the SFPs and JFPs staff the force requirement with the FPs via their assigned Service components to determine the recommended sourcing options and FP risk. The JS JFC and JFPs generate a recommended sourcing solution, drafts a modification to their respective GFMAP Annex, and staffs the draft Annex. The JS JFC consolidates the Annex Mods and staffs them with all Services, CCMDs, and applicable DoD Agency and forwards the recommendations, with the operational and FP risks to the JS J3. The JS J3 forwards the draft GFMAP Annex Mods to the Secretary for decision via the SDOB. The SDOB is physically a briefing book of all orders that require SecDef decisions. It also includes the standard briefing timelines to the JS J-Dirs, OCJCS staff, OSD, the Chairman, and ultimately the Secretary. For some contentious issues, the OPSDEPS, GFMB, or CJCS Tank may meet to review and endorse sourcing recommendations prior to the CJCS. Once the SecDef decides to allocate, the JS J3 publishes the Modification to the GFMAP Annex.

2 <u>Voice Order of the Commanding Officer (VOCO)</u>. VOCO directs action without the requirements of a written order. SecDef may approve CCDRs force requests via VOCO in order to expedite the sourcing and deployment process. During periods of rapidly developing events such as natural disasters or other emergent crises, issuing requests and/or orders using the standard RFF/RFC process may not be practical. If the situation is sufficiently time sensitive, VOCO may be used to execute the steps contained in the emergent, crisis-based force allocation process. In all cases, information will be fully documented using the standard RFF process immediately following the VOCO as soon as practicable. JS will record all VOCO approvals (date, time, and name of person relaying SecDef's order), will track the progress of force sourcing and development, and will staff a special SDOB to obtain SecDef signature for the record. Care must be taken to properly document all aspects of a VOCO authorization to ensure fiscal accountability is maintained, particularly in cases where DOD is not the Lead Federal Agency (LFA) managing a crisis or event. Additionally, the roles and responsibilities of all GFM principals, including command relationships, will remain unchanged. The VOCO process is not used merely as a convenience to supported or supporting CDRs or as a method of bypassing the checks and balances inherent in the allocation decision process. It is available only in those circumstances where time constraints render the standard process impractical. As in the standard RFF and request for capabilities (RFC) process, all VOCO requests not otherwise addressed in a Standing EXORD are presented to the SecDef via the CJCS. The SecDef's initial VOCO approval of a request implies that, when time does not allow, VOCO may be used for each step of the allocation process, provided that the allocation decision and authorities granted are fully documented using the standard procedures as soon as practicable following the VOCO authorization. Requests received by entities other than SecDef/CJCS should immediately be referred/forwarded to the JS J3 for consideration.

(4) <u>Joint Individual Augmentee (JIA) Sourcing.</u> The JIA execution sourcing allocation process mirrors the force-sourcing process with the following exceptions; the format and information requirements of the JIA request is different, and different information technology (IT) tool is used, called FMTS. The CCMD message requesting the JIA(s) is called an "Emergent JIA Request" rather than an RFF. The DJ-1 reviews and validates the JMDs and the JMD requirements. JS J3 as the JS JFC for Annex D coordinates directly with the Services for identifying recommended sourcing solutions for JIA requirements. The Service may delegate the responsibility for generating JIA sourcing recommendations to specific commands within the Service.

(a) <u>JIA Requirements.</u> A Joint Manning Document (JMD) captures all JTF HQ requirements for the contingency mission; it is not a manpower sourcing vehicle. JIA sourcing for the JMD is meant to be the last method for obtaining manpower. The optimal sourcing method for a JMD is the identification of a core unit around which to build the JTF HQ, submitting RFFs as necessary for enabling units to provide required capabilities not resident in the core unit, and finally augmenting the JTF HQ with JIAs for skills/capabilities not inherently resident in the core unit or the enabling units contributing to the JTF HQ's JMD. JIA billets can be filled by AC, RC, Coast Guard, and DoD-EC personnel.

<u>1</u> Supported CCMDs or JTF CDRs will document all Joint HQ requirements on a JMD in FMTS. This will include identification of billets filled with CCMD-assigned forces. Military Departments, Services, and Service Components will notify supported CCMDs of the JMD positions they will fill. Supported CCMDs forward requests for unfilled JMD positions to be sourced via JIA to the JS J-1. Requests to the JS J-1 will include recommended prioritization, numeric ranking priority, and estimated duration.

<u>2</u> Emergent JIA requests are those submitted between FY GFMAPs for new augmentation to existing JMDs and requests for JIA to new JTF HQ JMDs. These will be considered only as a result of SecDef or CCDR-directed mission changes.

<u>3</u> Emergent JIA requests to existing JMDs may be requested semi-annually using both FMTS and an Emergent JIA Request Message. Emergent changes to the JMD requesting core capabilities (standard units) via an RFF will be documented appropriately in FMTS and requested via the emergent allocation process. The associated FTN must be documented in FMTS against each corresponding unit billet. JIA changes that are the result of JTF mission changes directed by EXORD/PLANORD will be expected to be submitted IAW the external directives and will not be curtailed by the twice-yearly scheduled submission.

<u>4</u> JIAs will fill JTF HQ requirements and will not be used to fill permanent manning shortfalls or joint training, exercise, or TSC positions. As such, JIAs will not deploy with specialized equipment, and tactical-level deployments are not appropriate for JIA sourcing.

(5) <u>Allocation and the Inter-Agency Process</u>. Requests for interagency capabilities are not usually part of the allocation process. The interagency process is detailed in JP 3-08, *Interorganizational Coordination during Joint Operations*. There are cases in which a CCMD will request a capability via the allocation process and a non-DOD department or agency appears to offer the best solution. In these cases, the JS JFC coordinates between the agencies. The CJCS GFMAP Annex is used to relay the sourcing solution back to the JPEC and account for the requirement being filled; however, the SecDef does not have authority to direct people and capabilities of other USG Agencies.

(6) <u>Specific Requirements</u>.[8]

(a) <u>DoD-Expeditionary Civilian (DoD-EC) Requirements</u>. DoD-EC include all DoD civilians deployed to support expeditionary operations. The procedures of this section apply only to non-programmed DoD-EC. They do not apply to the sourcing of programmed Service units and force structure that contain DoD-EC personnel. See DTM 17-004, *DoD Expeditionary Civilian Workforce*, for further details on the DoD-EC process. The DoD-EC Force Pool prescribes the pool of civilians, by series and grade. FPs are directed to be prepared to deploy separate non-programmed civilian personnel from programmed Service units and force structure. The Force Pool is a cap, and establishes the maximum number of non-programmed civilian positions within each designated series and grade that an FP should be prepared to source when, and if, the requirements are ordered. The Force Pool is a reference to be used as a planning tool to improve predictability of DoD-EC deployment capacity. OUSD (P&R) will publish and maintain the DoD-EC Force Pool. OSD (CPP) reviews the Force Pool annually, and updates it if required. OSD(CPP) is required to update the force pool at least every other year. Requirements for DoD civilians may be included in JMDs as JIAs and may also be included in non-standard force requirements. CCMDs will ensure the series, grade, and position description are appropriate for the DoD-EC requirement being submitted.

(b) <u>Mobility Requirements</u>. CCMDs will submit Intra-Theater and Inter-Theater Mobility Force requirements via JCRM. Requests for mobility capabilities will be submit-

[8] *Global Force Management Implementation Guidance*

ted separate from non-mobility conventional force requirements to enable sourcing by USTRANSCOM.

(c) <u>SOF Requirements</u>. CCMDs shall submit requirements for SOF, including JCET events via JCRM. Requests for SOF will be submitted separate from non-SOF conventional force requirements to enable sourcing by USSOCOM.

(d) <u>Cyberspace Requirements</u>. CCMDs shall coordinate all cyberspace support and force requests through the USCYBERCOM Cyberspace Operations Integrated Planning Element (CO-IPE) posted at each CCMD. CCMDs shall submit all requests for support (RFS) via the appropriate USCYBERCOM RFS portal and all cyber force allocation requirements via JCRM. When developing cyberspace requirements, CCMDs will apply the terms of reference, C2, and force generation parameters specific to cyberspace forces.

(e) <u>NATO Requirements</u>. NATO develops a Combined Joint Statement of Requirements (CJSOR). The CJSOR contains force and individual requirements and identifies recommended countries to provide sourcing. The U.S. Military Representative, via the U.S. NATO National Military Representative (USNMR), coordinates with the supported CCMD to submit rotational and emergent requirements to the JS for those forces and individuals that will be under operational or tactical control or command (OPCON, TACON, OPCOM, or TACOM) of the respective NATO CDR.

(f) <u>Development of TSC Force Requirements</u>. TSC events are multi-national events in support of campaign plan objectives, many of which are authorized per 10 U.S.C. Chapter 16. Force requirements to support TSC activity are submitted in the annual submission only if the CCDR has a reasonable expectation that the TSC activity will be executed and forces will be required for the activity (i.e., usually a related allocation of resources or approval for the activity itself). CCDRs submit an RFF for emerging TSC activities when planning for a given event is refined. All operational forces requests for TSC activities are to be registered in JCRM to indicate a demand signal for DoD support. TSC events include training, SME exchanges, military-to-military engagements, and those engagement activities conducted with partner nations as part of an exercise.

(g) <u>Requests for Federal Assistance (RFA)</u>. Other U.S. Government departments and agencies may submit RFAs seeking DoD support. RFAs are normally received by the DoD Executive Secretary, coordinated with the JS, the affected CCDR, JS JFC/JFPs, and FPs and approved by the Secretary or the Secretary's designee. The affected CCMD will sponsor approved RFAs and will enter the emergent force requirements in JCRM. Emergent requirements from the Federal Emergency Management Agency (FEMA) are received by the supporting CCMD (USNORTHCOM, or USINDOPACOM), approved by the CCDR if authority has been delegated by the Secretary or forwarded to the JS, and approved by the Assistant Secretary of Defense (HD and Global Security) unless approval authority has been withheld by the Secretary, in which case the Secretary must approve. Once approved, the supporting CCMD will detail the forces required to support the RFA, in JCRM as an emergent requirement.

(h) <u>Foreign Disaster Relief (FDR) Operations</u>. Following SecDef or Deputy Secretary of Defense approval of an FDR mission and Overseas Humanitarian, Disaster, and Civic Aid (OHDACA) funding, IAW DoD Directive 5100.46, "Foreign Disaster Relief," the supported GCC may coordinate directly with FPs and CCDMD for required support beyond assigned or already allocated forces. The Secretary authorizes CCMDs to transfer requested forces in support of FDR operations to a supported GCC, who would exercise operational OPCON of those forces for up to 3O days, without additional SecDef approval. A CCDR may delegate this authority to transfer forces only to the Deputy CDR. CDRUSTRANSCOM will retain OPCON of assigned mobility forces, but may provide forces to supported GCCs in a supporting relationship. If the deployment of forces under this authority will significantly degrade the supporting CCMD's ability to respond to contingencies in its area of responsibility, the supported CCDR is to seek separate SecDef approval through the Chairman for the transfer of forces.

The Secretary also authorizes the Secretaries of the Military Departments to transfer requested forces for FDR operations to the supported GCC, who would exercise OPCON for up to 30 days, without additional SecDef approval. The Secretaries of the Military Departments may delegate this authority to transfer forces only to an Under Secretary or to a Service Chief. If the deployment of forces under this authority will significantly degrade the readiness of particular capabilities, the Secretary of the Military Department concerned must notify the Secretary and Chairman prior to providing these capabilities to the supported GCC. The deployment of forces to support FDR operations under these authorities will not affect future deployments already ordered in the GFMAP. The size and scope of forces transferred under this authority should be the minimum necessary to execute the SecDef or DepSecDef-approved tasks supporting the FDR operation.

(7) General Sourcing Requirements Guidance.

(a) Force Utilization Metrics. Deployment-to-Dwell (D2D) ratio and mobilization-to-dwell (M2D) ratios provide key measures that inform allocation decisions for AC and RC forces. Sourcing recommendations that exceed these objectives and minimum ratios will be specifically briefed to the Secretary via the SDOB.

1 An operational deployment begins when the majority of a unit or detachment, or an individual not attached to a unit or detachment, departs homeport/station/base or departs from an enroute training location to meet a SecDef-approved operational requirement. Forces deployed in support of EXORDs, OPLANs, or CONPLANs approved by the SecDef are also considered operationally deployed. An operational deployment ends when the majority of the unit or detachment, or an individual not attached to a unit or detachment, arrives back at their homeport/station/base. Forces operationally employed by SecDef orders at their home station or in PTDO status at home station are not operationally deployed.

2 AC Dwell is the period of time a unit or individual is not on an operation deployment. Dwell begins when the majority of a unit or detachment, or an individual not attached to a unit or detachment, arrives at their homeport/station/base from an operational deployment. Dwell ends when the unit or individual departs on an operational deployment. A unit is either on operational deployment or in dwell. For the RC, dwell is the period of time an individual is not mobilized. For Force Management designated deployments, dwell calculations will be determined at the first day of deployment and will not be recalculated for multiple short deployments after a return to home station. The total time the unit is deployed during this employability window shall not be in excess of force provider limits (deployment length, total time deployed).

(b) Active Component (AC) Deployment-to-Dwell (D2D). Deployment-to-dwell ratios apply to AC operational deployments to meet SecDef approved operational requirements. The planning objective for the AC D2D ratio is 1:3 (1 year deployed to 3 years in dwell). Without SecDef approval, the minimum AC D2D ratio is 1:2. MFPs will strive to achieve at least a 1:3 D2D ratio.

(c) Reserve Component (RC) Mobilization-to-Dwell (M2D). The mobilization-to-dwell (M2D) planning objective for RC units is 1 year mobilized to at least 5 years demobilized. Without SecDef approval, the minimum M2D ratio for RC forces is 1:4.

(d) Service Red Lines and Service Capacity. Service Red Lines are Service Secretary or Service Chief defined limits beyond which FP risks are unacceptable. The Secretary may, however, order a Service or JFP to exceed its Red lines.

1 Services and JFPs will brief their capacity of ready and available force capacity annually at GFMB 1. Services shall update the JS J-3 and JFPs if the capacity changes. Subsequent changes must be submitted to the JS J-3 via message traffic.

2 Service Red Lines provide guidance to SFP or Service components concerning the level of authority delegated by the Service Chief to commit to sourcing requirements. Service sourcing nominations and JS JFC and JFP sourcing recommendations will specify if the sourcing solution is within or exceeds the Service Red Line.

(e) <u>Rotation Policies</u>. Forces may not deploy longer than 365 days or remain deployed more than 30 days past the ordered deployment End Date (for deployments less than 365 days) without SecDef approval.

<u>1</u> Military Departments may conduct internal rotations that are shorter than SecDef ordered deployment lengths. JIAs will be deployed for the period of time specified in GFMAP Annex D. Service rotation lengths are vetted during the sourcing process.

<u>2</u> SecDef approval is not required for internal rotations provided that the force capabilities deployment is not greater than 365 days, or more than 30 days past the SecDef ordered deployment length for deployments shorter than 365 days.

(8) <u>Allocation Process Overview</u>. As an allocation overview (Figure F), the execution sourced forces are identified and recommended by JS JFC and JFPs, assisted by their Service components. (The JS JFC will coordinate directly with the Service Headquarters. The mobility, SOF, and cyber JFPs continue to develop sourcing recommendations from the Service via their Service component). The recommended sourcing solution is reviewed through the GFM allocation process and becomes sourced when approved by the SecDef for the execution of an approved operation of potential/imminent execution of an operation plan or other military activity. Unit reporting requirements are done in accordance with current GFM procedures per CJCSM 3130.06, *GFM Allocation Policies and Procedures.*

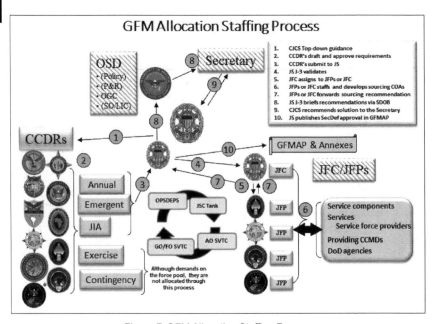

Figure F. GFM Allocation Staffing Process

(9) <u>Military Risk Assessment Informs the Chairman's Risk Assessment</u>.[9]

(a) 10 U.S.C. § 153 requires the CJCS to assess the nature and magnitude of strategic and military risks in executing the missions called for in the NMS, and to provide a report of the CJCS assessment to Congress through the SecDef. If the CJCS assesses the risk to be significant or higher, the SecDef is required to submit a plan for mitigating that risk. The SDOB contains a risk statement which includes, at a minimum, the Service or CCMD affected, the nature of the risk, who incurs the risk, duration of the risk and risk mitigation. Overall risk assessments performed in support of the annual force allocation process and emergent requirements inform the Chairman's Risk Assessment (CRA), and apply a common framework and assessment criteria. Services, CCMDs, and JS JFC/JFPs provide an overall assessment of the nature and magnitude of the risk associated with sourcing recommendations for a given fiscal year as it pertains to their ability to execute operations, activities, plans and functions with amplifying information in each of the applicable risk categories.

(b) <u>Risk Management.</u> Central to planning and execution at any level is the concept of risk. Risks are adverse outcomes that may occur during the course of a planned activity. Risk encompasses both ***probability and consequence*** of hazards to achieving objectives and to the joint force. At the strategic level, when ends, ways, and means are presented as an option or a COA, there are associated risks that should inform the civilian-military planning dialogue. Joint planning and execution involves simultaneous, concurrent planning and execution supported by a continuous operation assessment of each to identify risks for decision-makers to consider in refining, adapting, or terminating ongoing planning and/or operations. Risk management is the process of identifying, assessing, and controlling those risks arising from operational factors and making decisions that balance risk cost with mission benefits. The risk management process balances art and science. However, while there are a variety of risk analysis methods, no one solution exists that fit all circumstances. In the context of joint planning, planners consider strategic risks, and military risks:

<u>1</u> **Strategic Risk** is the potential impact upon the U.S., including the U.S. population, territory, civil society, critical infrastructure, and interests, of current and contingency events given their estimated consequences and probabilities (e.g., the security of the U.S. and its citizens).

<u>2</u> **Military Risk** is the estimated probability and consequence of the Joint Force's projected inability to achieve current or future military objectives *(risk to-mission)*, while providing and sustaining sufficient military resources *(risk-to-force)*.

(c) The GFM process incorporates an assessment framework based on the NDS when discussing overall risk associated with annual allocation submissions and plans. ***Operational, future challenges, force management, and institutional risks*** are categories used to express the overall risk associated with fiscal year requirements. When JS JFC/JFPs, FPs, and CCMDs present risk during the sourcing GFMB for fiscal year requirements, the overall risk categories of *operational, future challenges, force management, and institutional* are appropriate. CCMDs will present risks associated to operational and future challenges. In particular, supported CCMDs should focus their overall risk assessments to address the operational and future challenges and risks associated with the four major areas of the FY sourcing effort (i.e., conventional forces, SOF, ISR and mobility forces). JS JFC/JFPs and FPs will present risks associated to future challenges, force management, and institutional risk categories. CDRUSSOCOM will present risks associated with all four categories. Overall allocation risk categories are summarized in the following paragraphs.

(d) <u>Military Risk</u>. In the context of the CRA, military objectives are identified in strategic guidance and direction, including the NDS and NMS, and the sufficiency of military resources is identified in the CJA. Military risk has two complementary dimensions; ***risk-to-mission and risk-to-force.*** Both must be considered when calculating military risk, as it involves balancing a CCMD's ability to attain steady-state, current operations, and contingency plan objectives against the Services' and JFP's abilities to support CCMD missions.

---

[9] *CJCSM 3105.01, Joint Risk Analysis*

The concepts of risk-to-mission and risk-to-force can be differentiated into four risk subsets based on source of risk and time horizon. Two of the subsets measure risk-to-mission (*operational risk and future challenges risk*) and two subsets measure risk-to-force (*force management risk and institutional risk*) as depicted in Figure G and Table 2. Time horizon will remain subjective based on strategic trends, threats, the Chairman, and policy. Generally, the Joint Force considers risk in relation to three-time categories: Near-term (0-2 years), Mid-term (3-7 years), and Far-term (8-20 years). (See Military Risk Subsets graph and table.)

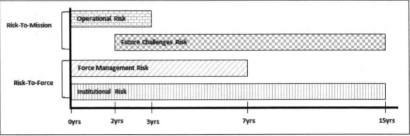

Figure G. Military Risk Subsets Graph

Military Risk Subsets

Military Risk Type	Subset	Timeframe	Assessed Against	Sources for Assessment
Risk-to-Mission	Operational Risk	0-2 Years	Current military objectives as described in current, planned, & contingency operations	Campaign Plans, Crisis Response Execution (EXORDs) CPG Objectives, GFM
Risk-to-Mission	Future Challenges Risk	0-7 Years	Futures mission objectives; capability and capacity to address emerging or anticipated threats	Joint Strategy Review, NIC Global Trends, DPG, Global Posture, GFM
Risk-to-Force	Force Management Risk	0-7 Years	Sufficient trained & ready forces to meet CCMD requirements; forces stress versus mission importance	GFM Readiness, Joint Training, BOG/DWELL Ratios, Service Strategic Plans
Risk-to-Force	Institutional Risk	0-20 Years	Organization & process effectiveness in improving national defense	UCP, DPG, Force Quality, Acquisition & Support Processes, GFM

Table 2. Military Risk Subsets Table

1 Operational Risk. (Risk-to-Mission) reflects the current force's ability to attain current military objectives called for by the current NMS, within acceptable human, material, and financial costs. Operational risk is a function of the probability and consequence of failure to achieve mission objectives while protecting the force from unacceptable losses. This risk subset considers the ability to execute campaign plans, current, planned, and contingency operations in the near term (0-2 years). The normal military planning process allocates enough time and dialogue to develop operational plans that can work in a war or crisis. These plans illuminate risks against known threats or crises. The collective assessment of these plans factors into risk assessment for the CRA, emerging crises, global force management, and other assessments, such as integrated priority lists (IPL). The SecDef's

interim progress review planning process is one of the methods used to identify risks for future plans. The TPFDD for each of these plans serves to identify and limit risk to the force. Plans without a verified TPFDD have more risk. CDRs consider the feasibility of these plans in conjunction with operational concerns to assess risk to a threat adequately.

2 Future Challenges Risk. (Risk-to-Mission) reflects the future force's ability to achieve future mission objectives over the near and mid-term (0-7 years) and considers the future force's capabilities and capacity to deter or defeat emerging or anticipated threats. Future challenges risk is a function of the probability and consequence of failure to meet future mission requirements.

3 Force Management Risk. (Risk-to-Force) reflects a Service and/or Joint Force Provider's ability to generate trained and ready forces within established rotation ratios and surge capacities to meet current campaign and contingency mission requirements; force management risk is a function of the probability and consequence of not maintaining the appropriate force generation balance ("breaking the force"). This risk subset considers the ability to execute plans today (e.g., execution) to contingency missions (e.g., potential conflict) in the near- to mid-term (0-7 years).

4 Institutional Risk. (Risk-to-Force) reflects the ability of organization, command, management, and force development processes and infrastructure to plan for, enable, and improve national defense. Institutional risk is a function of the probability and consequence of the DoD or Services failing to perform established functions. The timeframe associated with this risk subset is much broader. All three-time categories near-, mid-, and far-term—will impact institutional risk (0-20 years). It considers organization and process effectiveness, including the acquisition process, as well as Program Health, Health of the Force, and the Defense Industrial Base.

(e) Military risk. *Military risk is assessed using four probability and consequence levels (depicted in Table 3 on page 3-33.)* As with strategic risk, judgment is required to integrate different levels of probability and consequence during the Risk Characterization step. CDRs and their staffs must place risk in context through the application of costs, impacts, time, and endstates in order to inform policy-makers.

(f) Risk Levels. Provided is a separate set of categories and definitions for CCDR and FP use when presenting the impact of not sourcing or impact of sourcing risk. CCDRs and JS JFC/JFPs should use these risk categories throughout the GFM process when ad- dressing the risk associated with sourcing a specific requirement. *The definitions listed in Figure H, "Military Risk Contour and Risk Tables" on page 3-32 identify specific levels of risk in each of the overall allocation risk categories, minus operational risk, when applicable.*

The military risk matrix dipicted in Table 4 establishes standard criteria across several variables to help frame the discussion on consequences. The military risk matrix serves as a common risk framework across the Joint Force since it resides in the GFMIG. The military risk matrix and the one in the GFMIG reflect the same information with minor exceptions. Each row presents a factor for consideration with graduated consequences toward success or failure. After considering each applicable factor and assigning an expected result within the matrix, the assessor must use judgment to determine the overall expected consequence level for a situation. This tool facilitates a comprehensive picture of military risk using common metrics for the Joint Force. However, the risk analysis should not be limited to the metrics shown in the matrix; if other metrics and categories present relevant information, they should be included in the analysis to facilitate leadership making the most informed decision possible. Coupled with the strategic consequence's assessment, CDRs and staffs can reach an integrated risk assessment for strategy and military considerations.[10]

[10] Global Force Management Implementation Guidance

Military Risk Contour and Risk Tables

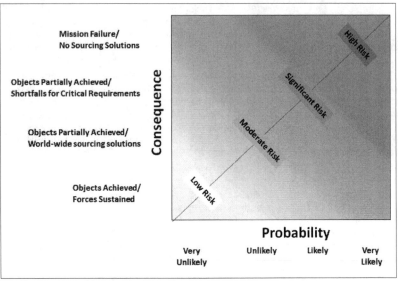

Probability/Consequence Military Risk Contour

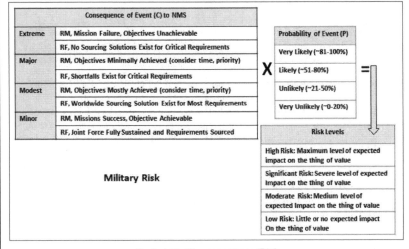

Probability/Consequences/Risk

Figure H. Military Risk Contour and Risk Tables

Low Risk. Success in achieving strategic objectives is assured, and can be achieved with planned resources and on planned timelines; unanticipated requirements can be easily managed with minimal impact on the force. FPs have the full capacity to source CCMD requirements.

Moderate Risk. Success in achieving strategic objectives is very likely, but timelines may be extended and execution may require additional resources from other plans or operations. Unanticipated requirements may necessitate adjustments to plans. FPs can source all requirements. Worldwide force allocation solutions may result in limited duration capability gaps. Intra-Service adjustments are required to source CCMD requirements.

Significant Risk. Success in achieving strategic objectives is likely, but timelines may require significant adjustments. Execution will require additional resources from other plans or operations, but some significant capability shortfalls still exist. Unanticipated requirements will necessitate adjustments to plans. FPs can source priority CCMD requirements. Worldwide force allocation solutions may result in extended duration capability gaps. Joint source solutions and force substitutions are required to source CCMD requirements.

High Risk. Success in achieving strategic objectives will require extraordinary measures. Timelines will require significant adjustments, but still may not achieve the CDR's endstate. Execution will require significant resources from other plans or operations, and some resources are severely deficient or absent altogether. Unanticipated requirements cannot be managed. FPs require full mobilization to sustain sourcing solutions to achieve strategic objectives. CCMD requirements exceed joint force capacity to substitute capability.

Table 3. Military Risk

c. <u>Apportionment</u>. Apportioned forces provide an estimate of the Military Departments/ Services/FP's maximum capacity to generate force elements along general timelines for planning purposes. Pursuant to 10 U.S.C., § 153, "The CJCS shall be responsible for preparing strategic plans, including plans which conform with resource levels projected by the SecDef to be available for the period of time for which the plans are to be effective." The CPG and JSCP provide overarching strategic guidance governing the apportionment of forces for planning, including the apportionment construct. Pursuant to the JSCP, "apportioned forces are types of combat and related support forces provided to CCMDs as a starting point for planning purposes only." Direction for generating the Apportionment Tables are contained in *Section V* of the GFMIG. These quantities of forces provide an estimate of the Military Departments' capacity to generate capabilities that can reasonably be expected to be made available, globally, along general timelines. Apportionment is a function of the number of operational forces, the readiness and availability of those forces, and the number of forces employed globally. The result is the number of apportioned forces which are published quarterly in the Apportionment Tables on the J8 Forces Division SIPRNET page. The Force Apportionment Tables do not apportion forces to a particular plan or CCMD. The Military Departments and JFPs provide their estimates of the number of forces they reasonably expect to be available and the Chairman based on guidance in the CPG approves the Force Apportionment Tables for CCMDs to utilize during planning. This estimate is a starting point to inform CCMD resource informed planning, but does not identify the actual forces or units that may be execution sourced if a plan is implemented. Force Apportionment Tables also inform the assessment of plans by providing the anticipated available force inventory. It should be noted that planning efforts toward supported and supporting plans for the same military activity are constrained to the same resources. While this resource informed planning is required for all planning, the CPG and JSCP specifically direct integrated planning within the constraints of apportionment for strategically prioritized plans. At the time of execution, competing requirements may require allocating forces among more than a single plan or operation. Planners should not use additional forces beyond those listed in the force apportionment tables without CJCS approval. To exceed the CJCS

Military Risk Matrix (GFMIG)

Military Risk Subset	Reference	CRITERIA	LOW	MODERATE	SIGNIFICANT	HIGH
Current Mission/Force	UCP CPG EXORD	Achieve Objectives (CCMD Daily Ops)	Can fully achieve all OBJs (minimal costs)	Can achieve all critical OBJs (acceptable costs)	Can achieve only most critical OBJs (substantial costs)	Potential failure: can't achieve critical OBJs (unacceptable costs)
Current & Future Mission/Force	GFM Assignment & Allocation	Meet CCDR Requirements (CCMD Daily Ops)	GFM sources ≥ 90% (some shortfalls)	GFM sources ≥ 80% (no critical shortfalls)	GFM sources ≥ 70% (critical shortfalls)	GFM sources < 70% (shortfalls cause mission failure
	CPG JSCP PLANORDs	Achieve Plan Objectives	As Planned (Minimal Costs)	Limited Delays (Acceptable Costs)	Extended Delays (Substantial Costs)	Extreme Delays (Unacceptable Costs)
	CRS/Plan Assessment (Contingency Sourcing)	Meet CCDR Requirements	Capacity to source plan requirements to achieve OBJ(s)	Shortfall cause minor plan deviations (no critical shortfalls)	Shortfalls cause major plan deviations	Shortfalls cause plan failure
	CCMD	Authorities	Full authority provided to achieve all objectives	Sufficient authority provided to achieve most objectives, no critical shortfalls	Insufficient authority provided to achieve some critical objectives	Insufficient authority for key objectives, potential mission failure
	Plan Assessment	Resources met required timelines	As Planned (Minimal Costs)	Limited Delays (Acceptable Costs)	Extended Delays (Substantial Costs)	Extreme Delays (Unacceptable Costs)
	CCMD & Services	Partnerships	Partnerships Effective	Critical Partnerships Effective	Critical Partnerships Partially Effective	Critical Partnerships ineffective, Potential Mission Failure
		Messaging	Messaging Effective	Key Messages Effective	Key Messages Partially Effective	Key Messages Ineffective, Potential Mission Failure
		DOTMLPF-P Capability vs. Threat Capability	Dominance	Superiority	Parity	Inferiority
Future Mission/Force	JFRR & DRRS	Readiness (DRRS)	Full Spectrum C1 Full Capacity	Ready for MCO C1/C2 Some capacity shortfalls	Ready for Minor Armed Conflict. Critical Capabilities C1/C2 Limited Capacity	Critical Capabilities ≤ C2 Capacity shortfalls cause mission failure
	GFM Allocation	Stress on AC Force (D2D)	D2D > 1:3	1:3 > D2D ≥1:2.5	1:2.5 > D2D ≥ 1:2	D2D < 1:2
		Stress on RC Force (M2D)	M2D > 1:5	1:5 > M2D ≥1:4	1:1:4 > M2D ≥ 1:3	M2D < 1:3
		Modernization/ Critical Maintenance	As planned (minimal cost)	Limited delays (acceptable costs)	Extended delays (substantial costs)	Extreme delays (unacceptable costs)
	JCIDS / CPR	Programmatic	Meets or exceeds schedule, IOC or FOC; incurred savings	Minor delays, milestone ≥ B Minor budget difficulty	Major delays, milestone ≥ A Over budget (Nunn/Mcurdy)	Program failure, Zeroed Out (de-funded)
	JCIDS / CJA	Force Development & Industrial Base	Meet all mission requirements	Meet priority mission requirements (no critical shortfalls)	Critical shortfalls cause major plan deviations	Failure to meet essential requirements causes mission failure
	JWC JCC	Operational Imperatives & CRCs	Achieves all operational imperatives (no capability gaps)	Achieves priority operational imperatives (no critical capability gaps)	Achieves minimal operational imperatives (minor critical capability gaps)	Operational imperatives not achieved (major critical capability gaps)

Table 4. Military Risk Matrix

approved apportioned forces in a plan requires CJCS approval. Exceeding the quantity of apportioned forces assumes additional risk because the forces required are not reasonably expected to be available. Planners must inform the plan approval authority (SecDef, Dep-SecDef, or USD(P)) if a plan exceeds apportioned forces at the IPRs and when the plan is submitted for approval.

6. GFM Assessment [11]

a. <u>Overview</u>. Each year, CCMDs' requests for forces outpace FPs ability to source, requiring DoD leadership to assume risk. The annual JFSA process captures these shortfalls to develop risk-mitigating actions. Shortfalls are the lack of forces, equipment, personnel, or capabilities and are defined as the difference between the resources demanded and those allocated, assigned, or otherwise attached to a CCMD that has affected or will adversely affect (or add risk to) the CCDR's mission accomplishment. The *Joint Force Sufficiency Assessment* looks at the issue from two vectors: 1) day-to-day shortfalls; 2) end of FYDP shortfalls to achieve NDS objectives. JFSA results inform many processes and program-matic/alternative sourcing solutions, including JCCA/JFRR and readiness assessments, Program Budget Review (PBR), GFMBs, strategic guidance development, CJCS Strategic Seminar Series, and other products as requested.

b. <u>Process</u>. The JFSA process is broken into two parts: The GFMAP Sufficiency Assessment (GSA) and the Strategic Requirements Sufficiency Assessment (SRSA). The GSA addresses current GFM allocation related shortfalls while the SRSA focuses on the joint forces' ability to meet global demand outlined in strategic documents, both today and through the end of the FYDP.

(a) <u>GFMAP Sufficiency Assessment (GSA)</u>:

<u>1</u> The Joint Staff J-8, in concert with the J-3, will provide guidance as necessary to the JFPs on capturing force sufficiency data during the annual force sourcing effort.

<u>2</u> At the beginning of each FY, JFPs will identify and submit force sufficiency data to the Joint Staff J-8 for consolidation.

<u>3</u> The Joint Staff, with assistance from the Military Departments/Services and JFPs, will assess this list to inform the PPBE strategy development processes and potentially help identify and develop potential programmatic (and other DOTMLPF) solutions to force sufficiency issues. The GFMB may provide guidance to the JFPs to plan for alternative sourcing solutions based on the annual force sufficiency assessment.

(b) While the GSA looks at shortfalls to the current year's GFMAP, the Strategic Requirements Sufficiency Assessment (SRSA) assesses the sufficiency of the Joint Force to meet the global demands identified in the current DPG and the NDS.

<u>1</u> J-8 solicits joint planning force requirement from the CCMDs on their GCPs and CCPs.

<u>2</u> J-8 requests force visibility data from the JFPs to include inventories, pre- and post-crisis rotation rates, relief in place/transfer of authority durations, and force generation using assumptions similar to those in the apportionment table generation process.

<u>3</u> J-8 conducts an assessment, comparing the demands against the force element capacity to show where gaps exist for select high priority force elements and highlighting the need for increased availability, increased capacity, and/or a decrease in demand to achieve objectives. The assessment does not assess root causes, but rather highlights areas meriting further study.

7. Overview

a. Refer to Figure I for an overview depiction of the GFM Execution Process.

[11] *Global Force Management Implementation Guidance*

GFM— the Execution Process: Overview

Global Force Management – The Execution Process

CCMDs, Services, DoD Agencies

Chairman, Joint Chiefs of Staff (J-3)
- Manages global forces
- Validates requests
- Provides strategic risk assessments
- Prioritizes competing demands
- Recommends globally integrated risk informed solutions to SecDef

Directed Readiness Assignment, and Allocation

SECDEF

CJCS Recommendation

Guidance

JFP Sourcing (for allocation)

Force Requirements

Capstone Guidance
- Title 10, U.S. Code
- Unified Command Plan
- National Security Strategy
- National Defense Strategy
- GFM Implementation Guidance
- National Military Strategy
- Joint Strategic Capabilities Plan
- Planning and Programming Guidance

Guidance

CCMDs

Force Providers

USSOCOM	USTRANSCOM
SOF JFP	Mobility JFP
USCYBERCOM	USSPACECOM
Cyberspace Ops JFP	Space Ops JFP

DoD Agencies

CCMDs
Assigned Forces

Services
Retained Forces

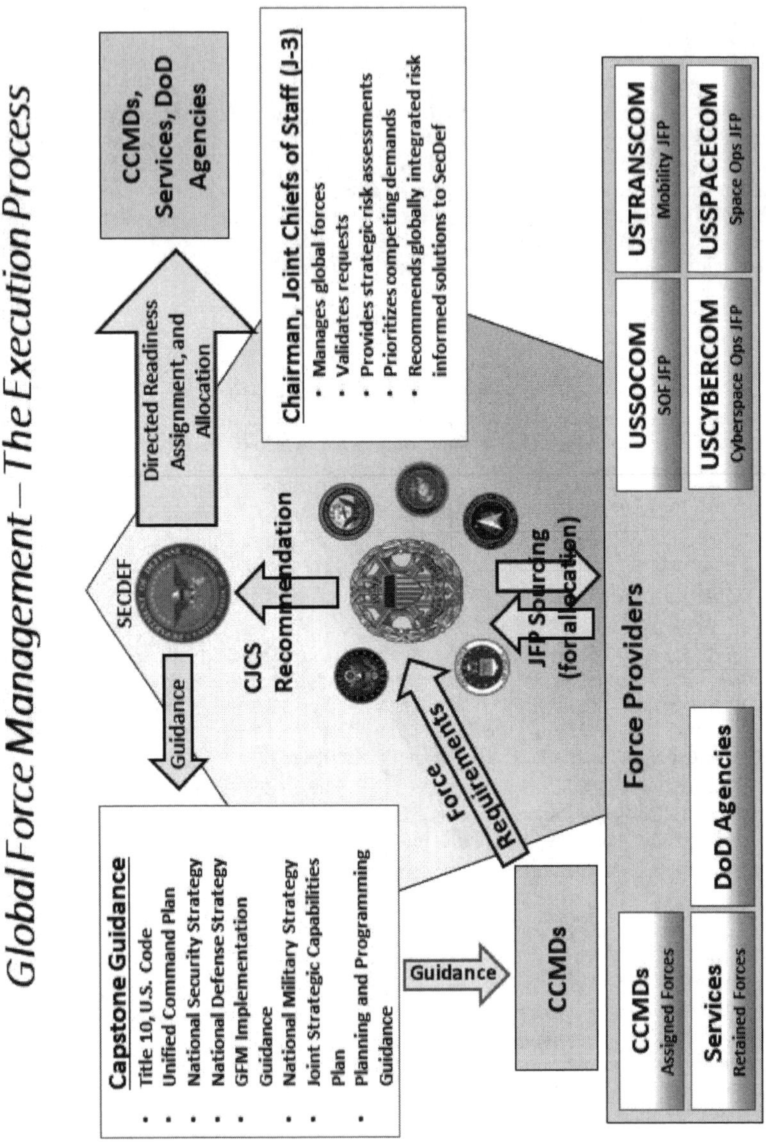

Figure I. GFM-The Execution Process Overview

II. Force Identification & Sourcing

a. <u>Overview</u>. The intent of Force Identification and Sourcing is to provide the CCMD with the most capable forces based on stated capability requirements, balanced against risks (operational, future challenges, force management, institutional) and global priorities.

b. <u>Force Providing</u>. There are three types of forces identified within the GFM process depending upon the fidelity and endstate required: **preferred force identification** (planning only), **contingency sourcing** (plan assessment) and **execution sourcing**. Within the range of these methodologies, execution and contingency are most prevalent because the force sourcing process generally results in an endstate in which the JFPs identify units to satisfy a capability requirement for either assessment or execution. The following paragraphs clarify and describe the meaning of these broad categories and related terms.

Don't confuse *preferred force identification* with apportionment. The differences are discussed below and illustrated in the following figure.

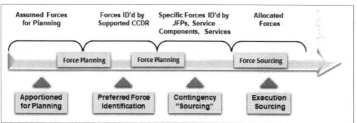

Force Sourcing

1. Preferred Forces

a. <u>Overview</u>. *Preferred forces are forces that are identified by the supported CCMD in order to continue employment, sustainment, and transportation planning and assess risk.* These forces are planning assumptions only, are not considered "sourced" units, and do not indicate that these forces will be contingency or execution sourced.

- To the degree the CCMD is able to make good assumptions with respect to preferred forces for planning, the JFPs will begin with a higher fidelity solution should the plan be designated for contingency or execution sourcing.

- CCMD Service/functional components are encouraged to work with JFPs and their components to make the best possible assumptions with respect to preferred forces for planning.

- The preferred forces identified for the plan by the CCMD should not be greater than the forces apportioned for planning unless granted permission to do so by the CJCS or designated representative.

b. Forces found in the Apportionment Tables are types of combat and related support forces provided to CCMDs as a starting point for planning purposes only. The Services provide force updates quarterly to the JS who in turn staff the inputs and publish revised Apportionment Tables on the JS J8 website. The Military Departments, USSOCOM, US-TRANSCOM, and USCYBERCOM provide their estimates of the number of forces they *reasonably* expect to be available and the Chairman approves the Force Apportionment Tables for CCMDs to utilize during planning. The Services' submittals attempt to compensate for known deployments and estimate the availability of forces for that given year, and provide an estimate of the quantities of forces (not units by name) that could be *reasonably* expected to be made available globally over a general timeline. The Apportionment Tables do not break these down to individual CCMDs or plans, but provide the total number of forces a Service can reasonably expect to be available towards the next contingency. *The takeaway is that the apportionment tables are at best a good estimate. One could reasonably argue that the day the tables are published they are already dated.*

c. While some directed plans are formally supported through the business process of contingency sourcing, discussed in the following paragraphs, the capacity of the Service and JFP planners limits the periodicity of these staffing efforts and cannot support all CCMDs plan assessment requirements. Therefore, CCMD planners, as directed by their command, may consult with the Services, Service components, the JS JFC and the JFPs and their Service components to select preferred forces.

d. Most contingency plans begin force planning utilizing the Apportionment Tables, and are later refined, as required, utilizing forces identified by the supported CCMD that place specified forces (unit level) into the plan, in lieu of apportioned forces (unit type). This allows the supported CCMD to continue employment, sustainment and transportation planning and assess risk and shortfalls with a greater fidelity. *This is preferred force identification.*

e. CCMD planners, Service and functional components are encouraged to work with the JS JFC, the JFPs and their Service components, and FPs to make the best possible assumptions with respect to identifying preferred forces. The emerging global visibility capabilities that provide capabilities, readiness, availability, and employment (CRAE) information on the Joint Force is becoming a more valuable asset to planners. As the applications improve, the speed these applications can automate analysis enables planners to devote more time to cognitive analysis of options and COAs vice gathering and analyzing preferred force information. By using these applications wisely planners can develop resource informed plans and identify integrated risks much earlier in the planning process (see previous chapter, GFM Principles).

f. The preferred forces identified for a plan by the CCMD should not be greater than the number of forces apportioned for planning unless the CJCS or his/her designee either grants permission or has provided amplifying planning guidance. To the degree the CCDR is able to make good planning assumptions when identifying preferred forces determines the feasibility of a plan, and may assist the JS JFC and/or the JFP in identifying forces should the plan be designated for contingency sourcing or transitions to execution.

g. Plan analysis achieves greater fidelity when utilizing preferred forces; however, preferred forces are still based on **planning assumptions**. Preferred forces are notional forces based on current information and are not contingency sourced nor execution sourced. Preferred forces give the plan analysis a greater fidelity, however it is still based on planning assumptions. *These units are still not considered "sourced."*

2. Contingency Sourcing

a. Overview. Contingency sourcing is a part of the plan assessment process. It entails the JS JFC and JFPs identifying forces that meet the sourcing guidance communicated in the contingency sourcing message, which is based on assumptions, and represents a snapshot depiction of sourcing feasibility for senior leaders.[1]

b. The JS J5 Joint War Plans Division (JOWPD) is responsible for collecting all contingency sourcing requests, developing a proposed Contingency Sourcing Schedule Memorandum for staffing, and presenting the recommended memorandum to the GFMB for approval. Contingency sourcing is a manpower-intensive and time-consuming requirement that competes for resources with FP execution-level sourcing responsibilities. The GFMB is the final arbiter of all contingency sourcing requested events. The Chairman, and the DJS, as delegated, tasks the JPEC to contingency source selected campaign and contingency plans.

c. The mechanics of the force-providing community in contingency sourcing closely resembles the steps they take in assisting CCMDs with identifying preferred forces. Contin-

[1]CJCSM 3130.06 Series, "Global Force Management Allocation Policies and Procedures" contains detailed step-by-step procedures for contingency sourcing.

gency sourcing is directed by the Chairman. Contingency sourcing solutions are determined by the JS JFC, for conventional forces and by the JFPs for their functional areas. CCMDs, with or without the advice of others, identify preferred forces.

d. *Contingency-sourced forces are not allocated forces.* Contingency sourcing is a sourcing feasibility assessment of a plan performed by the execution sourcing community. The term contingency "sourced" can be misleading. Contingency sourced forces are specific forces which meet the planning guidance at a specified point in time identified by the JS JFC, ICW the FPs, and the JFPs, ICW their assigned Service components. The JS J5 provides specific guidance through a list of sourcing assumptions and planning factors contained in a Contingency Sourcing Message. The JS JFC and JFPs have final approval of the sourcing solution and provide the approved solution to the supported CCDR and the CJCS prior to IPRs. The JS JFC is responsible for consolidating all JFP identified sourcing solutions and recommending global joint sourcing solutions.

e. Contingency sourcing allows greater fidelity in force planning than preferred force identification discussed previously. However, because these forces are identified based on planning assumptions and planning guidance provided for the sourcing effort, there should be *no expectation that forces sourced via contingency sourcing will be the actual forces sourced during execution sourcing.* The frequency of contingency sourcing actions is in part, dependent on the capacity of the JS JFC/JFPs and their FPs. Remember that contingency sourcing is conducted by the execution sourcing community. To the extent that the sourcing guidance, and the assumptions used to create that guidance accurately reflects actual world events if/when execution sourcing is required determines the accuracy of the sourcing assessment (contingency sourcing).

f. Variables for contingency sourced units include, but are not limited to, current disposition of forces, a specified as-of-date, categories of forces to be excluded from consideration, defined C-Day or C-Day window, substitution and mitigation options/factors, readiness reporting requirements, and training requirements.

g. *A contingency-sourced plan is not execution sourced.* The resulting sourcing, shortfalls, risks and mitigations should be interpreted as a sourcing capacity at the time of the contingency sourcing against the assumptions used to assess the plan. All else being equal if there are capacity shortfalls during the plan assessment, it is likely that there will be capacity shortfalls if the plan were to be executed. It is prudent for planners to reconsider the plan (way) and develop alternate ways to achieve the desired objectives (what) using the forces (means) assessed to be available.

h. The resulting sourcing, shortfalls, risks and mitigations of contingency sourcing are presented to the CJCS to be assessed and to the CCMD. Contingency sourcing results are typically included in the CJCS plan assessment advice to the Secretary or the delegated plan approval authority (DepSecDef or USD(P)).

3. Execution Sourcing

a. <u>Overview</u>. During execution, the supported CCDR may task their assigned forces to fill force requirements in order to perform authorized missions. CCDRs exercise COCOM over assigned forces and employ them for missions, operations, and activities they have authority to execute. These requirements constitute the assigned force demand and are documented in the JCRM (a GFM tool that manages CCMD force requirements through GFM allocation process from RFF to GFMAP) by the supported CCMD. If additional forces are required, the supported CCMD requests those forces through the GFM allocation process via the annual submission or an RFF.[2]

b. *Allocation is an execution-sourcing process* to identify forces recommended by JS JFC and JFPs (via the JFPs Service components or Services for conventional forces) and allocated by the SecDef to meet CCMD force requirements for the execution of an

---

[2] *Only execution sourcing is actual force "sourcing" defined as, "identification of actual forces or capabilities that are made available to fulfill valid CCDR requirements." CJCSM 3130.06 GFM Allocation Policies and Procedures.*

approved operation or potential/imminent execution of an operation plan or exercise. The JS J-3, in coordination with the Services and JFPs, propose the global allocation of forces to CCMDs to support the Secretary's strategic priorities while providing a credible deterrent capability. Within this context, CCMDs develop their annual submission and subsequent emergent requirements to refine the Top-Down Allocation Guidance from the bottom up. Execution sourcing of forces, including PTDO, are ordered in a DEPORD. The GFMAP and its associated annexes are the global DEPORD for all allocated forces. For execution sourcing during allocation, the JS JFC/JFP is the supported CDR for the force planning steps of identifying recommended sourcing solutions. All CCMDs and FPs are in support. The GFMAP Annex specifies the ordered FP. The GFMIG allows the FP and JS JFC/JFP some flexibility in identifying the appropriate unit for deployment.

c. Units tasked must meet minimum readiness and availability criteria as directed by the tasking authority.

d. Execution-sourced forces are considered allocated forces and are unavailable for use in other plans/operations unless reallocated by the SecDef.

e. There are four execution-force sourcing categories: **standard, joint, in-lieu-of, and ad-hoc** force sourcing solutions.

1 Standard Force Solution. A mission ready, joint capable force with associated table of organization and equipment executing its core mission. This force will also have completed core competency training associated with the RFF's requested capability(s).

2 Joint Force/Capability Solution. Joint sourcing encompasses Services providing a force/capability in place of another Service's core mission. As in a standard force solution, the capability is performing its core mission. An example of this is sourcing an Army combat heavy engineer requirement as an enabler for an Army BCT with a Naval Mobile Construction Battalion (NMCB) unit. Navy is providing a like type capability, a capability that is performing its core competency mission, in the place of another Service's core mission. Joint sourcing may also encompass a force or capability composed of elements from multiple Services merged together to develop a single force/capability meeting the requested capability. Joint sourcing solutions will increase the time required to properly train, equip, and man the force/capability prior to deployment. Additional challenges include the fact that unlike a standard force, a joint-sourcing solution may require movement of personnel and/or equipment from various locations to a single locality for consolidation and issuance of equipment. Second, once personnel and equipment are consolidated, familiarization, proper usage, and maintenance practices must also be incorporated into the training regimen to ensure that all members comprising the joint solution are well versed in required actions for sustaining operability.

3 In-lieu-of (ILO). ILO sourcing is an overarching sourcing methodology that provides alternative force sourcing solutions when standard forces sourcing options are not available. An ILO/capability is a standard force, including associated table of organization and equipment that is deployed/employed to execute missions and tasks outside its core competencies. The force can be generated by normal FPs or be a result of a change of mission (s) for forward deployed forces. An example of this is taking an existing artillery battery, providing it a complete training and equipment package and then deploying it to fill a transportation company requirement.

4 Ad-hoc. An ad-hoc capability is the consolidation of individuals and equipment from various commands/Services and forming into a deployable/employable entity properly trained, manned and equipped to meet the supported CCMD's requirements. Ad-hoc solutions will increase the time required to properly train, equip, and man the force/capability prior to deployment. Additional challenges include the fact that unlike a standard force, an ad-hoc sourcing solution will require movement of personnel and/or equipment from various locations to a single locality for consolidation and issuance of equipment. Second, once personnel and equipment are consolidated, familiarization, proper usage, and maintenance practices must also be incorporated into the training regime to ensure that all members comprising the ad-hoc solution are well versed in required actions for sustaining operability.

III. Force Planning

1. Force Planning and the GFM Process

a. <u>GFM Process during Planning</u>. The Apportionment Tables provide the number of forces reasonably expected to be available for planning. These tables should be used as a beginning assumption in planning. As the plan is refined, there may be forces identified that are required above and beyond those apportioned. Those forces should be requested, as required, to be augmented above the number apportioned for planning, or "augmentation forces." The CJCS may approve planning to continue with the revised assumption of using the identified augmentation forces. These augmentation forces are then allotted for planning. However, should the plan be executed, planners should be prepared for the risk associated with the potential of those "augmentation forces" not being available.

(1) Planners continually refine and assess the plan throughout Concept and Plan Development. To enable assessments, planners must *assume* that units are allotted to the identified plan force requirements and to enable plan assessments, planners identify preferred forces. As the plan is refined, the level of analysis used to identify preferred forces usually increases. Since contingency plans rely on a foundation of assumptions, if an event occurs that necessitates execution of a contingency plan, the planning assumptions have to be re-validated. The planners will usually verify planning assumptions against the unfolding event and re-perform planning functions from Strategic Guidance to Plan Assessment, as required to adapt it to the realities sur- rounding the event rather than transitioning directly to execution. These planning functions may be performed very deliberately or in a time-constrained environment, as time allows.

(2) As a contingency plan is either approved or nearing approval, the CJCS or, if delegated, the DJS may direct the JS JFC/JFPs to contingency source a plan to support CJCS and/or SecDefs' strategic risk assessments or IPRs. CCDRs may request contingency sourcing of specific plans. These requests are evaluated by the JS J5 and a contingency sourcing schedule is presented to the GFMB. The GFMB endorses the schedule and the CJCS orders the JS JFC/ JFPs to contingency source specific plans per the schedule (see contingency sourcing).

b. <u>GFM Process during a Crisis</u>. The same planning steps that are used to develop contingency plans are used during a crisis, but the time to conduct the planning is constrained to the time available. For planning during a crisis, preferred force identification is used the same as it was during contingency planning. Contingency sourcing is rarely used for a crisis due to the time constraints involved, but if time allows, the option exists for the CJCS to direct JS JFC/JFPs to contingency source a plan.

(1) In planning, the difference in force planning is the level of detail done with the force requirements for the plan. With contingency plans the number of planning assumptions prevents generating the detailed force requirements needed by the JS JFC/JFPs to begin execution sourcing. During crisis planning, a known event has occurred and there are fewer assumptions. *The focus of crisis planning is usually on transitioning to execution quickly.* The detailed information requirements specified to support the execution sourcing process, either emergent or annual, preclude completion until most assumptions are validated.

(2) CCDRs usually have a good understanding of the availability of their assigned forces. Availability entails the readiness of the unit, as well as the unit's time in the deployment cycle and whether it meets SecDef deployment-to-dwell (D2D) ratio requirements, and

whether the unit is already allocated to another mission. The supported CCDR generally reviews the force requirements for the contingency plan and conducts a review of assigned and previously allocated forces to determine if the mission can be done without requesting additional forces. If forces are already assigned and/or allocated that can perform the mission, the CCDR may direct those forces to perform the mission, within the constraints of the allocation authorities in the GFMAP. If additional forces are required, the CCDR will forward an RFF with all the details necessary, both electronically and by message RFF. The emergent force allocation process is the process to identify force requirements in support of planning during a crisis.

(3) Crisis planning transitions to execution when the order is given to execute a mission or operation, although planning continues throughout execution. During crisis planning, the CCDR considers using assigned and already allocated forces to respond to the situation. If the CCDR identifies additional forces that are required, a force request is submitted. Once this request is approved by the CCDR, that force request is considered a CCDR requirement. The force request is sent from the CCDR to the SecDef via the JS J3. The vehicle for the force request is a message called an RFF to the SecDef and JS J3 info the JS JFC/JFPs, FPs, OSD, and all other CCMDs as specified in CJCSM 3130.06, *GFM Allocation Policies and Procedures*. Each individual force requested is serialized with an FTN. An RFF message may contain one or more FTNs. To request JIAs for a JTF Headquarters, the message is called an Emergent JIA Request. The initial force or JIA request to perform a mission is an emergent JIA request.

(4) For operations that are longer in duration, the SecDef mandates that CCMD's validate their forces annually to determine the forces that require rotation. The process used to source annual forces is fundamentally the same as an emergent force request, but the annual process is necessarily modified to handle the large number of forces necessary to fulfill all CCMD requests for an entire FY. The annual submission is effectively the first RFF for a FY and should include all the forces for all the operations the CCMD anticipates executing.

2. GFM in Exercise Planning

Requests for forces to participate in exercises do not follow the same sourcing process as operational requests. Per reference *CJCSI 3500.01, Joint Training Policy and Guidance for the Armed Forces of the United States*, JS JFC/JFPs receive exercise force requests directly from the supported CCDRs. Supportability by JS JFC, the JFPs (and their Service-retained conventional forces) is determined and the resulting sourcing solution is provided back directly to the supported CCDR. The SecDef is not required to allocate forces for exercises, including exercises with other countries. Subsequent deployment of these exercise sourcing solutions is effected and tracked by the JS JFC/JFPs in concert with the supported CCDR. Under most circumstances, the GFMIG authorizes JS JFC/JFPs to transfer TACON of forces to support CCDR exercises and does not require a GFMAP mod to be approved by the SecDef.

IV. Mutually Supporting, Interrelated DoD Processes

1. The GFM Process

GFM DR4A processes support strategic guidance and joint force availability requirements. It provides DOD senior leadership with comprehensive insight into the global availability of forces and risk and impact of proposed force changes. The GFMB serves as a guiding body that provides complementary strategic focus and direction for the DR4A process.

a. In the following figure the mutually supporting, interrelated DOD processes are viewed through a GFM lens. The stakeholders depicted include OSD, JS J3, Services (including theater Service Components), JFPs (including their assigned Service Components and subordinate commands), CCMDs (including their assigned Service Components, JTFs, and other subordinate commands).

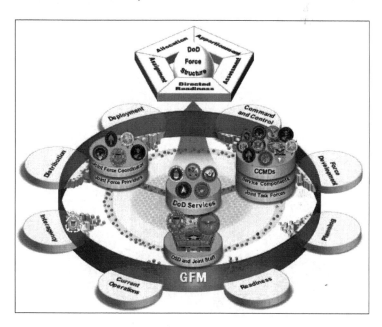

GFM Operational View (OV-1).

b. The GFM alignment (DR4A) processes, tools, and data maintain synchronization across stakeholders and integration with related processes. This enhances the ability to efficiently and effectively align the force structure to respond to the complex, dynamic global environment. The GFM DR4A processes are among the many sub-processes within Joint Planning and Execution. Each of those sub-processes considers a specific aspect of a plan or operation and are influenced by GFM. Likewise, the other sub-processes all influence and are influenced by the others. This dependency on the multitude of variables makes planning a recursive and iterative process.

c. Each stakeholder shares data and information to collaboratively determine the best use of the force structure to meet a situation. Impacts and risks of re-aligning the force structure are visible and all stakeholders collaboratively develop mitigation strategies. Planners obtain common force structure data directly from the entities responsible for building and maintaining the data.

d. Formally and rigorously specified force structure data contains unambiguously defined semantics implemented so GFM-related computer programs can readily exploit the data. Stakeholders share a common understanding of the meaning of GFM data. Changes in any of the processes depicted as overlapping and interacting with GFM influence not only GFM directly, but often influence changes in other processes. Seamless iterative interaction and integration between these related processes are necessary for the success of each of these processes as well as success of the missions these processes exist to support.

"Get there firstest with the mostest"
General Nathan Bedford Forrest

V. GFM Summary

a. The DoD's enduring mission is to provide combat-credible military forces needed to deter war and protect the security of our nation. Today's increasingly complex security environment is defined by rapid technological change, challenges from adversaries in every operating domain, and the impact on current readiness from the longest continuous stretch of armed conflict in our Nation's history. In this environment, there can be no complacency—we must make difficult choices and prioritize what is most important to field a lethal, resilient, and rapidly adapting Joint Force.[1] The future operating environment has the potential to produce more challenges than the U.S. and its military forces can respond to effectively. Therefore, resource management must enable the SecDef and other senior civilian and military leadership to balance resources, to include forces, among the strategic priorities of ongoing operations, shaping activities, and contingency plans. Foremost in the resource management effort is the GFM enterprise which aligns the DR4A processes in support of strategic guidance and a common vision and agreed-upon end result. As a cornerstone to joint planning and execution, GFM provides leaders at all levels with a clear picture of global force readiness, availability, and the risks associated with changing the alignment of the forces via the assignment and allocation processes. The GFM enterprise provides the Secretary and Chairman the decision support structure to make informed decisions on the use of forces. However, the metrics, information, and data requirements for these decisions can be situation dependent, making the GFM processes highly dynamic.

b. Doctrine and policy will continue to be challenged keeping current with the highly dynamic GFM processes. The many "independent" variables involved in GFM make data transmission and information gathering inconsistent and, at times, it is perceived as cumbersome. Compounding these issues is the fact that the number of individuals who fully understand the details of sourcing and force-providing from multiple perspectives is limited. The GFM enterprise is evolving rapidly and our senior Service schools, war colleges and universities must maintain pace with this evolution to ensure our graduates understand the principles and have a working knowledge of GFM and how GFM relates to the other processes. They will have the sword passed to them, and are expected to guide GFM into the next decade and beyond.

c. The dynamic complexity of GFM requires constant attention to provide comprehensive insights into the global availability of U.S. military forces and capabilities to quickly and accurately assess the impact and risk of force management decisions. With continued focused attention, we will leverage the hard lessons of the past, capitalize on the savvy and knowledge of our people to operate today, and lead the evolution of GFM to be ready for tomorrow's challenges.

> "When working toward the solution of a problem, it always helps if you know the answer."
>
> Rule of Accuracy

[1] National Defense Strategy

GFM Strategic Documents

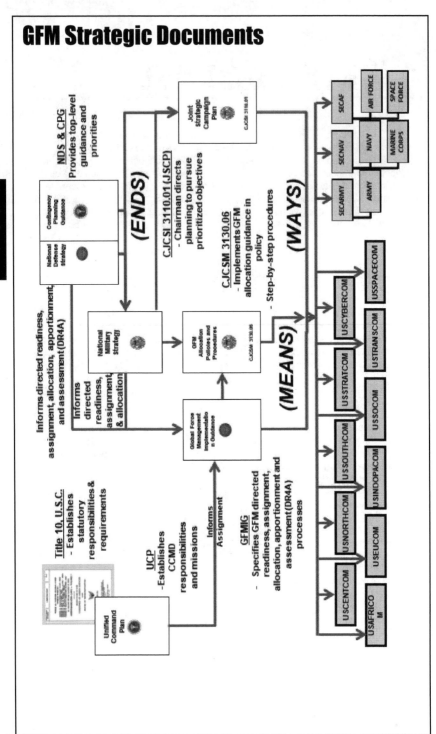

I. (JIPOE) Joint Intel Prep of the Op Environment

MISSION ANALYSIS and JOINT INTELLIGENCE PREPARATION of the OPERATIONAL ENVIRONMENT (JIPOE)

> *"Nothing is more worthy of the attention of a good general than the endeavor to penetrate the designs of the enemy."*
>
> *Niccolo Machiavelli, The Discourses on Livy, 1517*

1. JIPOE Overview

a. Joint Intelligence Preparation of the Operational Environment (JIPOE) is the analytical process used by joint intelligence organizations to produce intelligence assessments, estimates, and other intelligence products in support of the CDR's decision-making process. It is a continuous process that involves four major steps: (1) defining the total OE (OE); (2) describing the impact of the OE; (3) evaluating the adversary; and (4) determining and describing adversary potential courses of action (COAs), particularly the adversary's most likely COA and the COA most dangerous to friendly forces and mission accomplishment.

b. The process is used to analyze the physical domains (air, land, maritime and space); the information environment (which includes cyberspace), political, military, economic, social, information, and infrastructure (PMESII) systems; and all other relevant aspects of the OE, and to determine an adversary's capabilities to operate within that environment. JIPOE products are used by joint force, component, and supporting command staffs in preparing their estimates and are also applied during the analysis and selection of friendly COAs.[1]

2. JIPOE and the Intelligence Cycle

a. JIPOE is a dynamic process that both supports, and is supported by, each of the categories of intelligence operations that comprise the intelligence process.

(1) JIPOE and Intelligence Planning and Direction. The JIPOE process provides the basic data and assumptions regarding the adversary and other relevant aspects of the OE that help the CDR and staff identify intelligence requirements, information requirements, and collection requirements. By identifying known adversary capabilities, and applying those against the impact of the OE, JIPOE provides the conceptual basis for the CDR to visualize and understand how the adversary might threaten the command or interfere with mission accomplishment. This analysis forms the basis for developing the CDR's priority intelligence requirements (PIRs), which seek to answer those questions the CDR considers vital to the accomplishment of the assigned mission. Additionally, by identifying specific adversary COAs and COGs, JIPOE provides the basis for wargaming in which the staff "fights" each friendly and adversary COA. This wargaming process identifies decisions the CDR must make during execution and allows the J-2 to develop specific intelligence requirements to facilitate those decisions. JIPOE also identifies other critical information gaps regarding the adversary and other relevant aspects of the OE, which form the basis

---

[1] JP 2-01-3, Joint Intelligence Preparation of the Operational Environment

of a collection strategy that synchronizes and prioritizes collection needs and utilization of resources within the phases of the operation.[2] (see figure below)

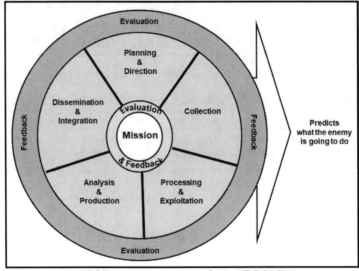

JIPOE and the Intelligence Cycle (JP 2-01.3).

   (2) <u>JIPOE and Intelligence Collection</u>. JIPOE provides the foundation for the development of an optimal intelligence collection strategy by enabling analysts to identify the time, location, and type of anticipated adversary activity corresponding to each potential adversary COA. JIPOE products include several tools that facilitate the refinement of information requirements into specific collection requirements. JIPOE templates facilitate the analysis of all identified adversary COAs and identify named areas of interest (NAIs) where specified adversary activity, associated with each COA, may occur. JIPOE matrices are also produced that describe the indicators associated with each specified adversary activity. In addition to specifying the anticipated locations and type of adversary activity, JIPOE templates and matrices also forecast the times when such activity may occur, and can therefore facilitate the sequencing of intelligence collection requirements and the identification of the most effective methods of intelligence collection.

   (3) <u>JIPOE and Processing and Exploitation</u>. The JIPOE process provides a disciplined yet dynamic time-phased methodology for optimizing the processing and exploiting of large amounts of data. The process enables JIPOE analysts to remain focused on the most critical aspects of the OE, especially the adversary. Incoming information and reports can be rapidly incorporated into existing JIPOE graphics, templates, and matrices. In this way, JIPOE products not only serve as excellent processing tools, but also provide a convenient medium for displaying the most up-to-date information, identifying critical information gaps, and supporting operational and campaign assessments.

   (4) <u>JIPOE and Analysis and Production</u>. JIPOE products provide the foundation for the J-2's intelligence estimate. In fact, the JIPOE process parallels the paragraph sequence of the intelligence estimate format.[3]

   (5) <u>JIPOE and Dissemination and Integration</u>. The J-2's intelligence estimate provides vital information that is required by the joint force staff to complete their estimates, and for

[2] See JP 2-0, Joint Intelligence, for a more in-depth discussion of the relationship between intelligence requirements and information requirements. See JP 2-01, Joint and National Intelligence Support to Military Operations, for detailed guidance on the request for information (RFI) process

[3] See JP 2-01.3, JIPOE for greater details

subordinate commanders to continue concurrent planning activities. Timely dissemination of the intelligence estimate is therefore paramount to good operation planning. If time does not permit the preparation and dissemination of a written intelligence estimate, JIPOE templates, matrices, graphics, and other data sources can and should be disseminated to other joint force staff sections, and component and supporting commands, in order to facilitate their effective integration into operation planning. JIPOE geospatial perspectives should also be provided to systems supporting the common operational picture.

(6) JIPOE and Evaluation and Feedback. Consistent with the intelligence process, the J-2 staff continuously evaluates JIPOE products to ensure that they achieve and maintain the highest possible standards of intelligence excellence as discussed in JP 2-0, *Joint Intelligence*. These standards require that intelligence products anticipate the needs of the CDR and are timely, accurate, usable, complete, objective, and relevant. If JIPOE products fail to meet these standards, the J-2 should take immediate remedial action. The failure of the J-2 staff to achieve and maintain intelligence product excellence may contribute to the joint force's failing to accomplish its mission.

b. Roles and Responsibilities. Some critical billets of the JPIOE process are:

(1) CCDR. The CCDR is responsible for ensuring the standardization of JIPOE products within the command and subordinate joint forces, and for establishing theater procedures for collection management and the production and dissemination of intelligence products.

(2) J-2. The J-2 has the primary staff responsibility for planning, coordinating, and conducting the overall JIPOE analysis and production effort at the joint force level. Through the JIPOE process, the J-2 enhances the JFC's and other staff elements' ability to visualize all relevant aspects of the OE. The J-2 uses the JIPOE process to formulate and recommend PIRs for the CDR's approval, and develops information requirements that focus the intelligence effort (collection, processing, production, and dissemination) on questions crucial to joint force planning.

(3) CCDR Joint Intelligence Operations Center (JIOC). The JIOC is the focal point for the overall JIPOE analysis and production effort within the CCMD. It is responsible for managing collection requirements related to JIPOE and Intelligence Preparation of the Battlespace (IPB) efforts, and for producing intelligence products for the CCDR and subordinate commanders that support joint operation planning and ongoing operations. The JIOC ensures that the JIPOE production effort is accomplished in conjunction with all appropriate CCMD staff elements, particularly the Geospatial Intelligence (GEOINT), Meteorological and Oceanographic (METOC), and Information Operations (IO) staff officers. The JIOC also ensures that its JIPOE analysis is fully integrated with all IPB and JIPOE products produced by subordinate commands and other organizations. With the assistance of all appropriate joint force staff elements, the JIOC identifies information gaps in existing intelligence databases and formulates collection requirements and requests for information (RFIs) to address these shortfalls. Additionally, the CCMD JIOC may be requested to support another CCDR's federated intelligence requirements, to include JIPOE requirements. As a federated partner, the JIOC must be prepared to integrate into the overall federated intelligence architecture identified by the supported CCDR. All CCMD JIOCs are eligible to participate in federated intelligence support operations.

(4) Subordinate JFC/CDR. The subordinate CDRs clearly state their objectives, CONOPS, and operation planning guidance to their staffs and ensure that the staff fully understands their intent. Based on wargaming and the joint force staff's recommendation, the CDR selects a friendly COA and issues implementing orders. The CDR also approves the list of intelligence requirements associated with that COA. The CDR then identifies those intelligence requirements most critical to the completion of the joint force's mission as PIRs.

(5) JTF Joint Intelligence Support Element (JISE) or JIOC. The intelligence organization at the JTF level is normally a JISE. However, the limited resources of a JISE will usually preclude a full JIPOE effort at the JTF level without substantial augmentation, reliance on reach-back capability, and national-level assistance. To overcome this limitation, the CCDR may authorize the establishment of a JTF-level JIOC based on the scope, duration, and mission of the unit or JTF. A JTF JIOC is normally larger than a JISE and is responsible

for complete air, space, ground, and maritime order of battle (OB) analysis; identification of adversary COGs; analysis of command and control (C2) and communications systems, targeting support; collection management; and maintenance of a 24-hour watch. Additionally, the JTF JIOC (if formed) serves as the focal point for planning, coordinating, and conducting JIPOE analysis and production at the subordinate joint force level. Most important, DIOCC forward element personnel and liaison officers from DOD intelligence organizations provide the JTF JIOC with the means to obtain national support for the JIPOE effort. The JTF JIOC conducts its JIPOE analysis in conjunction with all other appropriate joint force and component command staff elements, particularly the Geospatial Information and Services (GI&S) and METOC staff officers.

(6) <u>Subordinate Component Commands</u>. The intelligence staffs of the subordinate component commands should ensure that appropriate IPB products are prepared for each domain in which the component command operates. Subordinate component commands should evaluate the specific factors in the OE that will affect friendly, neutral, and adversary COAs in and around their operational area and impact perceptions and support within their Area of Interest (AOI). More importantly, the analysis of the OE should better define those who are potentially friendly, potentially neutral, and potentially adversarial and the actions which would determine their orientation. These component command IPB products provide a level of detail and expertise that the J-2 should not attempt to duplicate, but must draw upon in order to form an integrated or "total" picture of an adversary's joint capabilities and probable COAs. Accordingly, the component commands should coordinate their IPB effort with the J-2 and with other component commands that have overlapping IPB responsibilities. This will ensure their IPB products are coordinated and disseminated in time to support the joint force's JIPOE effort.

(7) <u>The Operations Directorate (J-3) and/or the Plans Directorate (J-5) Representative</u>. The J-3 and/or J-5 ensure that all participants in the JIPOE effort are continuously updated on planning for both current and follow-on missions as well as on any anticipated change to the operational area. The J-3 and/or J-5 representative consolidates information on our own dispositions and provides the cell a clear understanding of friendly COGs, capabilities, and vulnerabilities. The J-3 and/or J-5 will conduct wargames that test friendly COAs against the complete set of adversary COAs developed during the JIPOE process. Based on the results of these wargames, the J-3 and/or J-5 will refine and determine the probability of success of each friendly COA against each adversary COA identified during the JIPOE process, and will make a recommendation to the CDR regarding which friendly COA best accomplishes the joint mission within the CDR's guidance and intent.[4]

> "Know the enemy, know yourself -- your victory will never be endangered. Know the ground, know the weather -- your victory will then be total."
>
> Sun Tzu
> The Art of War, C. 500 B.C.

3. Joint Intelligence Preparation of the Operational Environment (JIPOE)

a. JIPOE is a continuous process which enables CDRs and their staffs to visualize the full spectrum of adversary capabilities and potential COAs across all dimensions of the OE. JIPOE is a process that assists analysts to identify facts and assumptions about the OE and the adversary. This facilitates campaign planning and the development of friendly COAs by the joint force staff. JIPOE provides the basis for intelligence direction and synchronization that supports the COA selected by the CDR. JIPOE's main focus is on providing situational aware-

---

[4] For more detailed guidance, see JP 2-01, Joint and National Intelligence Support to Military Operations.

> **Purpose of Joint Intelliegence (JP 2-0)**
> - Inform the commander
> - Identify, define, and nominate objectives
> - Support the planning and execution of operations
> - Counter adversary deception and surprise
> - Support friendly deception efforts
> - Assess the effects of operations

ness and understanding of the OE and a predicative intelligence estimate designed to help the CDR discern the adversary's probable intent and most likely and most dangerous COA.

b. The JIPOE process assists CDRs and their staffs in achieving information superiority by identifying adversary COGs, focusing intelligence collection at the right time and place, and analyzing the impact of the OE on military operations. Understanding JIPOE is critical to mission success. Intelligence must be integrated with the overall plan from beginning to end utilizing the JPP. JIPOE is a product of the intelligence staff estimate and is an integral part of the mission analysis process. JIPOE is the analytical process used by joint intelligence organizations to produce intelligence assessments, estimates, and other intelligence products in support of the joint force CDR's decision-making process. The primary purpose of the JIPOE is to support the CCDR's decision-making and planning by seeking to understand the OE and identifying, assessing, and estimating the enemy's COG, critical factors, capabilities, limitations, intentions, and enemy COAs (ECOA) that are most likely to be encountered based on the situation. Although JIPOE support to decision-making is both dynamic and continuous, it must also be "front loaded" in the sense that the majority of analysis must be completed early enough to be factored into the CDR's decision-making effort. JIPOE supports mission analysis by enabling the CDR and staff to visualize the full extent of the OE, to distinguish the known from the unknown, and to establish working assumptions regarding how adversary and friendly forces will interact within the OE. JIPOE also assists CDRs in formulating their planning guidance by identifying significant adversary capabilities and by pointing out critical OE factors, such as the locations of key geography, attitudes of indigenous populations, and potential land, air, and sea avenues of approach. JIPOE provides predictive intelligence designed to help the CDR discern the adversary's probable intent and most likely future COA. Simply stated, JIPOE helps the CDR to stay inside the adversary's decision-making cycle in order to react faster and make better decisions than the adversary.[5]

c. The intelligence directorates (J-2s) at all levels coordinate and supervise the JIPOE effort to support joint operation planning, enable commanders and other key personnel to visualize the full range of relevant aspects of the OE, identify adversary COGs, conduct assessment of friendly and enemy actions, and evaluate potential adversary and friendly COAs. The JIPOE effort must be fully coordinated, synchronized, and integrated with the separate IPB efforts of the component commands and Service intelligence centers. Additionally, JIPOE relies heavily on inputs from several related, specialized efforts, such as geospatial intelligence preparation of the environment (GPE) and medical intelligence preparation of the OE (MIPOE). All staff elements of the joint force and component commands fully participate in the JIPOE effort by providing information and data relative to their staff areas of expertise. However, CDRs and their subordinate commanders are the key players in planning and guiding the intelligence effort, and JIPOE plays a critical role in maximizing efficient intelligence operations, determining an acceptable COA, and developing a CONOPS. Therefore, commanders should integrate the JIPOE process and products into the joint force's planning, execution, and assessment efforts.

[5] JP 2-01.3, Joint Intelligence Preparation of the Operational Environment.

d. The analysts look at the OE from a systems perspective – looking at the OE major subsystems and then providing an assessment of the interrelationships between these systems. One approach is to examine the political, military, economic, social, informational, and infrastructure aspects of the OE, which are factors generally referred to as PMESII and desired effects. The PMESII construct offers a means to capture this information. Each PMESII factor is relevant and should be looked at critically when analyzing the OE. Understanding this environment has always included a perspective broader than just the adversary's military forces and other combat capabilities within the operational area. The planning, execution and assessment of joint operations require a holistic view of all *systems* (both military and non-military) that comprise the OE.

e. In addition to analyzing the conventional general military intelligence (GMI) products, JIPOE should analyze the environment from a systems perspective. Intelligence identifies and analyzes adversary and neutral systems and estimates how individual actions on one element of a system can effect other system components.

f. Using the JIPOE process, the Joint force J-2 manages the analysis and development of products that provide a systems understanding of the OE. This analysis identifies a number of nodes related to identified friendly objectives and effects—specific physical, functional, or behavioral systems, forces, information, and other components of the system. JIPOE analysts also identify links—the behavioral, physical, or functional relationship between nodes. Link and nodal analysis provide the basis for the identification of adversary COGs and decisive points for action to influence or change adversary system behavior. This methodology also provides the means by which intelligence personnel develop specific indicators of future adversary activity and support J3/5 COA development. It also enables analysts to understand how specific actions activities within the OE will influence other aspects of the OE.

g. JIPOE from a systems approach may require extensive resources (i.e., personnel with the proper expertise on the various aspects of the OE, and extensive collaboration). Although conceptually it is a sound practice, it may not always be possible to conduct a comprehensive JIPOE in this manner unless it is on a focused target set and OE, and sufficient time is allotted for this effort. It is critically important for intelligence personnel to understand the external resources available to support this effort. Like all intelligence collection and analysis, it is never complete, requiring continual update throughout planning and execution.

h. JIPOE Consists of Four Major Steps (figure on facing page). Analysts use the JIPOE process to analyze, correlate, and fuse information pertaining to all aspects of the OE. The OE consists of the air, land, sea, space and associated adversary, friendly, and neutral systems. JIPOE is conducted both prior to and during joint force operations, as well as during planning for follow-on missions. The most current information available regarding the adversary situation and the OE is continuously integrated into the JIPOE process.

i. The Staff Planners' Role in JIPOE. The joint force J-2 has primary responsibility for planning, coordinating, and conducting the overall JIPOE analysis and production at the joint force level. However, JIPOE is a staff process – *not just* a J-2 process, and should be driven by the chief of staff. To ensure you're obtaining relevant and accurate intelligence support material for the CDR, and to ensure the most efficient and productive use of intelligence resources, the staff should take an active role in meeting with the J-2 and those analysts working on your production requirements. The staff provides information and data on the OE relative to their staff areas of expertise. They now also know you and understand intelligence in context with the operation you are planning or executing.

JIPOE Consists of Four Major Steps:
- Define the OE.
- Describe the effects of the OE.
- Evaluate the adversary.
- Determine adversary COAs.

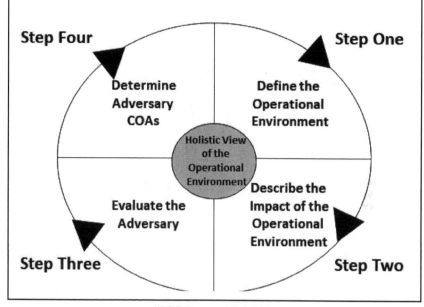

JIPOE Process (JP 2-01.3)

j. <u>Step 1 — Define the Operational Environment</u>. During Step 1, the joint force staff assists the JFC and component CDRs in determining the dimensions of the joint force's OE by identifying the significant characteristics of the OE and gathering information relating to the OE and the adversary. Successfully defining the command's OE is critical to the outcome of the JIPOE process. The joint force J-2 staff works with other joint force and component command staff elements, including the IO planning staff, to formulate an initial survey of adversary, environmental, and other characteristics that may impact the friendly joint mission. Additionally, the joint force staff must also recognize that the OE extends beyond the geographic dimensions of land, air, sea, and space. It also includes nonphysical dimensions, such as the electromagnetic spectrum, automated information systems, and public opinion. These nonphysical dimensions may extend well beyond the joint force's designated operational areas, which will also impact determining the Area of Interest, or, according to the DOD Dictionary, *"that area of concern to the CDR, including the area of influence, areas adjacent thereto, and extending into enemy territory to the objectives of current or planned operations. This area also includes areas occupied by enemy forces that could jeopardize the accomplishment of the mission."* Understanding which characteristics are significant is done in context with the adversary, weather and terrain, neutral or benign population or elements, and most importantly with the CDR's intent and the mission, if specified. The significant characteristics, once identified, will provide focus and guide the remaining steps of JIPOE. Therefore, it is essential to conduct effective analysis of the OE to ensure the "right" characteristics were identified as significant. Identifying the wrong significant characteristics or simply not addressing them jeopardizes the integrity of the operation plan (see figure on following page.)

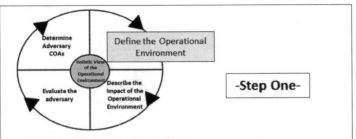

1. Identify the joint forces operational area
2. Analyze the mission's and joint force commander's intent
3. Determine the significant characteristics of the operational environment
4. Establish the limits of the joint force's area of interest
5. Determine the level of detail required and feasible within the time available
6. Determine intelligence and information gaps, shortfalls and priorities
7. Collect material and submit requests for information to support further analysis

(1) The joint force J-2 staff evaluates the available intelligence databases to determine if the necessary information is available to conduct the remainder of the JIPOE process. In nearly every situation, there will be gaps in the existing data bases. The gaps must be identified early in order for the joint force staff to initiate the appropriate intelligence collection requirements. The joint force J-2 will use the CDR's stated intent and initial PIR to establish priorities for intelligence collection, processing, production, and dissemination. The joint force J-2 staff initiates collection operations and issues RFIs to fill intelligence gaps to the level of detail required to conduct JIPOE. As additional information and intelligence is received, the J-2 staff updates all JIPOE products. If any assumptions are repudiated by new intelligence, the CDR, the J-3, and other appropriate staff elements should reexamine any evaluations and decisions that were based on those assumptions.

(2) Products from step one may include assessments of each significant characteristic, overlays of each, if applicable, and an understanding and graphical depiction of the operational area and possibly of the area of interests and entities therein which could affect our ability to accomplish our mission.

k. Step 2 — Describe the Impact of the Operational Environment. Step 2, describing the OE impacts, *focuses on the environment.* The first action in describing OE effects is to analyze the military aspects of the terrain. The acronym that aids in addressing the various aspects of the OE is OCOKA - Observation and Fields of Fire, Concealment and Cover, Obstacles, Key Terrain, and Avenues of Approach. This analysis is followed by an evaluation of how these aspects of the OE will affect operations for both friendly and adversary forces (see figure below.)

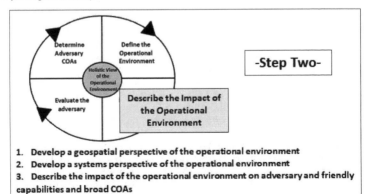

1. Develop a geospatial perspective of the operational environment
2. Develop a systems perspective of the operational environment
3. Describe the impact of the operational environment on adversary and friendly capabilities and broad COAs

(1) Products developed during this step might include overlays and matrices that depict the military effects of geography, meteorological (METOC) factors, demographics, and the electromagnetic and cyberspace environments. The primary product from JIPOE produced in Step 2 is the *Modified Combined Operations Overlay* (MCOO) and is shown in the following figure. The MCOO is "a JIPOE product used to portray the effects of each battlespace dimension on military operations. It normally depicts militarily significant aspects of the OE, such as obstacles restricting military movement, key geography, and military objectives."[6] Areas of the OE where the terrain predominantly favors one COA over others should be identified and graphically depicted. The most effective graphic technique is to construct a MCOO by depicting (in addition to the restricted and severely restricted areas already shown) such items as avenues of approach and mobility corridors, counter-mobility obstacle systems, defensible terrain, engagement areas, and key terrain.[7]

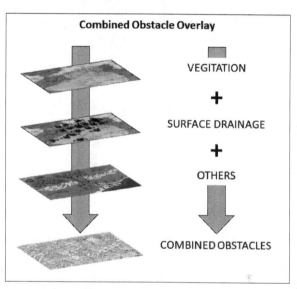

(2) A MCOO generally has standardized overlays associated with it. However, it is not a standardized product with respect to what it should portray simply because a CDR's requirements are based on their mission and intent – and they differ with each operation. Therefore, the MCOO should portray the relevant information necessary to support the CDR's understanding of the battlespace and decision-making process in context with his mission and intent. The results of terrain analysis should be disseminated to the joint force staff as soon as possible by way of the intelligence estimate (included in the order), documented analysis of the operational area, and the MCOO.

(3) Operational environments that you may be analyzing are broken down into dimensions (see respective figures on following pages), as follows:

• Land Dimension	• Cyberspace Dimension
• Maritime Dimension	• Human Dimension
• Air Dimension	• Analysis of Weather and Effects
• Space Dimension	• Other Characteristics of the OE
• Electromagnetic Dimension	

[6] *DOD Dictionary of Military and Associated Terms (DOD Dictionary)*

[7] *Refer to Joint Pub 2-01.3 JIPOE for more information concerning the types of MCOOs generated during step 2 of JIPOE*

(a) _Land Dimension_. Analysis of the land dimension of the OE concentrates on terrain features such as transportation systems (road and bridge information), surface materials, ground water, natural obstacles such as large bodies of water and mountains, the types and distribution of vegetation, and the configuration of surface drainage and weather. Observation and fields of fire, concealment and cover, obstacles, key terrain, avenues of approach, and mobility corridors are examples of what is required to be evaluated to understand the terrain effects on your plan.

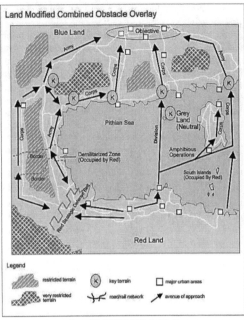

Land MCOO (JP 2-01.3).

(b) _Maritime Dimension_. The maritime dimension of the OE is the sea and littoral environment in which all naval operations take place, including sea control, power projection, and amphibious operations. Key military aspects of the maritime environment can include maneuver space and chokepoints; natural harbors and anchorages; ports, airfields, and naval bases; sea lines of communications (SLOCs), and the hydrographic and topographic characteristics of the ocean floor and littoral land masses.

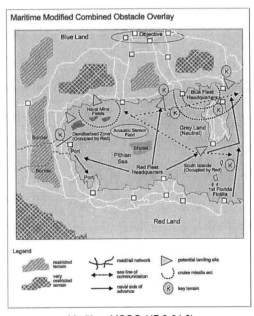

Maritime MCOO (JP 2-01.3).

(c) _Air Dimension._
The air dimension of the OE is the environment in which military air and counter-air operations take place. It is the operating medium for fixed-wing and rotary-wing aircraft, air defense systems, unmanned aerial vehicles, cruise missiles, and some theater and anti-theater ballistic missile systems. The surface and air environments located between the target areas and air operations points of origin are susceptible to METOC conditions, surface and airborne missiles, lack of emergency airfields, restrictive air avenues of approach and operating altitude restrictions, to name a few.

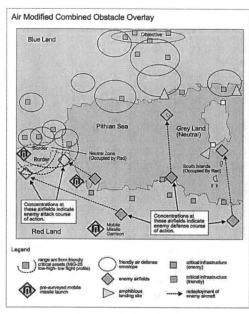

Air MCOO (JP 2-01.3).

(d) _Space Dimension._
The space dimension of the OE begins at the lowest altitude at which a space object can maintain orbit around the earth (approximately 93 miles) and extends upward to approximately 22,300 miles (geosynchronous orbit). Forces that have access to this medium are afforded a wide array of options that can be used to leverage and enhance military capabilities. However, space systems are predictable in that they are placed into the orbits that maximize their mission capabilities. Once a satellite is tracked and its orbit determined, space operations and intelligence crews can usually predict its function and future position (assuming it does not maneuver). The path a satellite makes as it passes directly over portions of the earth can be predicted and displayed on a map as a satellite ground track.

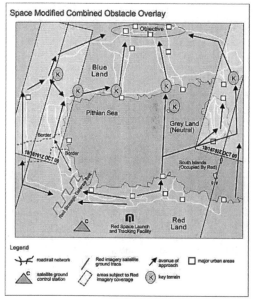

Space MCOO (JP 2-01.3).

(e) _Electromagnetic Dimension_. The electromagnetic dimension of the OE includes all militarily significant portions of the electromagnetic spectrum, to include those frequencies associated with radio, radar, laser, electro-optic, and infrared equipment. It is a combination of the civil electromagnetic infrastructure; natural phenomena; and adversary, friendly, and neutral electromagnetic OB.

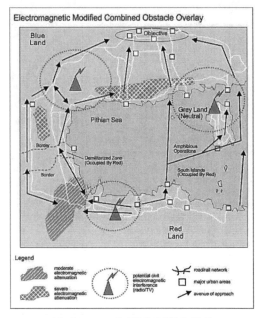

> _"Our information defines our decisions - Our decisions define our success."_
>
> _General James E. Cartwright_
> _VCJCS, April 2009_

Electromagnetic MCOO (JP 2-01.3).

(f) _Human Dimension._ The human dimension of the OE consists of various militarily significant sociological, cultural, demographic, and psychological characteristics of the friendly and adversary populace and leadership. It is the environment in which IO, such as psychological operations (PSYOP) and military deception are conducted. The analysis of the human dimension is a two-step process that: (1) identifies and assesses all human characteristics that may have an impact on the behavior of the populace as a whole, the military rank and file, and senior military and civil leaders; and (2) evaluates the effects of these human characteristics on military operations. Psychological profiles on military and political leaders may facilitate understanding an adversary's behavior, evaluating an adversary's vulnerability to deception, and assessing the relative probability of an adversary adopting various COAs.

(g) _Analysis of Weather Effects_. Weather is the state of the atmosphere regarding wind, temperature, precipitation, moisture, barometric pressure, and cloudiness. Climate is the composite or generally prevailing weather conditions of a region, averaged over a number of years. Initial studies of climatic effects may be prepared using available climatological data and/or seasonal outlooks requested from the DOD climate centers. These climate-based products are updated with outlooks and forecasts as more precise information is received concerning the actual weather conditions expected. METOC conditions affect the OE in several ways: the atmospheric and/or oceanographic environments can interact with, and thereby modify, the characteristics of each physical domain; or METOC can have a direct effect on military operations across all domains. The analysis of weather effects is a two-step process in which: (1) each military aspect of weather is analyzed; and (2) the effects of weather on military operations are evaluated. The joint force METOC officer is the source for weather information, and assists the joint force staff in determining the effects of METOC on adversary and friendly military operations. The overall effects of forecasted weather can be summarized in the form of a weather effects matrix as seen in the following figure.

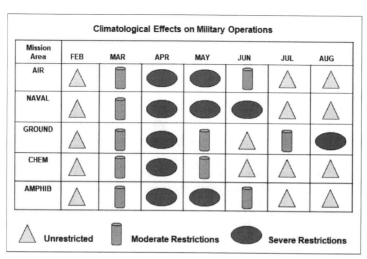

Weather Effect Matrix.

(h) <u>Others characteristics of the OE</u>. Other characteristics include all those aspects of the OE that could affect friendly or adversary COAs that fall outside the parameters of the categories previously discussed. Because the relevant characteristics will depend upon the situation associated with each mission, there can be no definitive listing of characteristics appropriate under all circumstances. For example, the characteristics of the OE that may be relevant to a sustained humanitarian relief operation will be very different from those required for a joint combat operation against an adversary. Some examples to be addressed while evaluating the battlespace environment are time, political and military constraints, environmental and health hazards, infrastructure, industry, agriculture, economics, politics, and history. The country characteristics of an adversary nation should be developed through the analytic integration of all the social, economic, and political variables listed above. Country characteristics can also provide important clues as to where a nation may use military force and to what degree.

l. <u>Step 3 — Evaluate the Adversary</u>. Step three of the JIPOE process, evaluating the adversary, identifies and evaluates the adversary's military and relevant civil COG, critical vulnerabilities (CVs), capabilities, limitations, and the doctrine and Tactics, Techniques and Procedures (TTPs) employed by adversary forces, absent any constraints that may be imposed by the OE described in step two. Failure to accurately evaluate the adversary may cause the command to be surprised by an unexpected adversary capability, or result in the unnecessary expenditure of limited resources against adversary force capabilities that do not exist (see figure on following page.)

(1) A COG can be viewed as the set of characteristics, capabilities, and sources of power from which a system derives its moral or physical strength, freedom of action, and will to act (more on COG in Chapter 5-2, *Mission Analysis*). The COG is always linked to the objective. If the objective changes, the center of gravity also could change. At the *strategic level,* a COG could be a military force, an alliance, a political or military leader, a set of critical capabilities or functions, or national will. At the *operational level* a COG often is associated with the adversary's military capabilities — such as a powerful element of the armed forces — but could include other capabilities in the OE. Since the adversary will protect the center of gravity, the COG invariably is found among strengths rather than among weaknesses or vulnerabilities. *CDRs consider not only the enemy COGs, but also identify and protect their own COGs, which is a function of the J-3.*

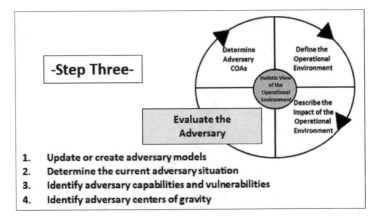

-Step Three-

Determine Adversary COAs

Define the Operational Environment

Holistic View of the Operational Environment

Describe the Impact of the Operational Environment

Evaluate the Adversary

1. Update or create adversary models
2. Determine the current adversary situation
3. Identify adversary capabilities and vulnerabilities
4. Identify adversary centers of gravity

(2) The analysis of friendly and adversary COGs is a key step within the planning process. *Joint force intelligence analysts identify adversary COGs.* The analysis is conducted after gaining an understanding of the various systems in the OE. The analysis addresses political, military, economic, social, informational, and infrastructure systems of the OE, including the adversary's leadership, fielded forces, resources, population, transportation systems, and internal and external relationships. *The goal is to determine from which elements the adversary derives freedom of action, physical strength (means), and the will to fight.* The J-2 then attempts to determine if the tentative or candidate COGs truly are critical to the adversary's strategy. This analysis is a linchpin in the planning effort. After identifying friendly and adversary COGs, CDRs and their staffs must determine how to protect or attack them, respectively. An analysis of the identified COGs in terms of critical capabilities, requirements, and vulnerabilities is vital to this process.

(3) Understanding the relationship among the COGs not only permits, but also compels, greater precision in thought and expression in operational design. Planners should analyze COGs within a framework of three *critical factors — critical capabilities (CC), critical requirements (CR), and critical vulnerabilities (CV)* — to aid in this understanding. CC are those that are considered crucial enablers for a COG to function as such, and are essential to the accomplishment of the adversary's assumed objective(s). CR are essential conditions, resources, and means for a critical capability to be fully operational. CV are those aspects or components of CR that are deficient, or vulnerable to direct or indirect attack in a manner achieving decisive or significant results. Collectively, the terms above are referred to as *critical factors* (see diagram below). In general, a JFC must possess sufficient operational reach and combat power to take advantage of an adversary's critical vulnerabilities. Similarly, a supported CDR must protect friendly critical capabilities within the operational reach of an adversary. As a best practice, the J-2 will act as a "red cell" in helping to identify the friendly forces COG and conduct COG analysis to support an understanding of what must be protected.

(4) In addition to the initial results of COG analysis, the primary products from JIPOE produced in JIPOE Step three are doctrinal templates, descriptions of the adversary's preferred tactics and options, and the identification of high-value targets (HVTs), which are targets that the enemy CDR requires for the successful completion of the mission. The loss of high-value targets would be expected to seriously degrade important enemy functions throughout the friendly CDR's area of interest.

(5) Adversary models depict how an opponent's military forces prefer to conduct operations under ideal conditions. They are based on a detailed study of the adversary's normal or "doctrinal" organization, equipment, and TTP. Adversary models are normally completed prior to deployment, and are continuously updated as required during military operations. The models consist of three major parts: graphical depictions of adversary doctrine or pat-

terns of operations (doctrinal templates), descriptions of the adversary's preferred tactics and options, and the identification of high-value targets (HVTs).

(6) Doctrinal templates illustrate the employment patterns and dispositions preferred by an adversary when not constrained by the effects of the OE. They are usually scaled graphic depictions of adversary dispositions for specific types of military (conventional or unconventional) operations such as movements to contact, anti-surface warfare operations, insurgent attacks in urban areas, combat air patrols, and aerial ambushes. JIPOE utilizes single-Service doctrinal templates that portray adversary and, sea, air, special, or space operations, and produces joint doctrinal templates that portray the relationships between all the adversary's Service components when conducting joint operations.

(7) In addition to the graphic depiction of adversary operations portrayed on the doctrinal template, an adversary model must also include a *written description of an opponent's preferred tactics.* This description should address the types of activities and supporting operations that the various adversary units portrayed on the doctrinal template are expected to perform. It also contains a listing or description of the options (branches) available to the adversary — should either the joint operation or any of the supporting operations fail — or subsequent operations (sequels) if they succeed.

(8) The adversary model must also include a list of HVTs. HVTs are those assets that the adversary CDR requires for the successful completion of the joint mission (and supporting missions) that are depicted and described on the joint doctrinal template. These targets are identified by combining operational judgment with an evaluation of the information contained in the joint doctrinal template and description. Assets are identified that are critical to the joint mission's success, that are key to each component's supporting operation, or that are crucial to the adoption of various branches or sequels to the joint operation. The joint targeting community collaborates in the identification of HVTs with the responsible producers for various intelligence product category codes.

m. Step 4 — Determine Adversary Courses of Action (COAs) (see figure below). The first three steps of the JIPOE process help to satisfy the OE awareness requirements of the CDR and subordinate CDRs by analyzing the effects of the battlespace environment, assessing adversary doctrine and capabilities, and identifying adversary COGs. The fourth step of the JIPOE process seeks to go beyond OE *awareness* to help the CDR attain *knowledge* of the OE (i.e., a detailed understanding of the adversary's probable intent and future strategy). The process for Step four provides a disciplined methodology for analyzing the set of potential adversary COAs in order to identify the COA the adversary is most likely to adopt, and the COA that would be most dangerous to the friendly force or to mission accomplishment.

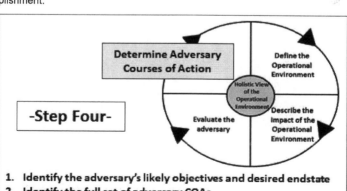

-Step Four-

Determine Adversary Courses of Action

Define the Operational Environment

Holistic View of the Operational Environment

Evaluate the adversary

Describe the Impact of the Operational Environment

1. Identify the adversary's likely objectives and desired endstate
2. Identify the full set of adversary COAs
3. Evaluate and prioritize each COA
4. Develop each COA in the amount of detail time allows
5. Identify initial collection requirements

(1) The first activity in JIPOE Step four is to identify *the adversary's likely objectives and desired endstate* by analyzing the current adversary military and political situation, strategic and operational capabilities, and the country characteristics of the adversary nation, if applicable. The JIPOE analyst should begin by identifying the adversary's overall strategic objective, which will form the basis for identifying subordinate objectives and desired endstates.

(2) During this step, a consolidated list of all potential adversary COAs is constructed. At a minimum this list will include: (1) all COAs that the adversary's doctrine considers appropriate to the current situation and accomplishment of likely objectives, (2) all adversary COAs that could significantly influence the friendly mission, even if the adversary's doctrine considers them suboptimal under current conditions, and (3) all adversary COAs indicated by recent activities or events. Each COA is generated based on what we know of the adversary and how they operate (learned from Step three of JIPOE) to determine if the adversary can in fact accomplish the COA. If not, it is eliminated. J-2 analysts study how an adversary operates compared to the environment it must operate within, which we analyzed during Step two of JIPOE. Essentially, they superimpose the doctrinal adversary mode of operation on the environment. The result of this analysis is a full set of identified adversary COAs – time permitting. Adversary COAs that meet specific criteria are then completed. Much like friendly forces determine if their COAs meet specific criteria, J-2 personnel must also weigh the identified adversary COAs against certain criteria. The criteria generally includes: (1) *suitability*, (2) *feasibility*, (3) *acceptability*, (4) *uniqueness, and* (5) *consistency with their own doctrine.*

(3) Each COA should be developed in the amount of detail that time allows. Subject to the amount of time available for analysis, each adversary COA is developed in sufficient detail to describe: (1) the type of military operation, (2) the earliest time military action could commence, (3) the location of the sectors, zones of attack, avenues of approach, and objectives that make up the COA, (4) the OPLAN, to include scheme of maneuver and force dispositions, and (5) the objective or desired endstate. Each COA should be developed in the order of its probability of adoption, and should consist of a situation template, a description of the COA, and a listing of HVTs.

(4) A full set of identified adversary COAs are evaluated and ranked according to their likely order of adoption. The purpose of the prioritized list of adversary COAs is to provide the CDRs' and their staffs with a starting point for the development of an OPLAN that takes into consideration the most likely adversary COA as well as the adversary COA most dangerous to the friendly force or mission accomplishment. The primary products produced in JIPOE Step four are the situation template and matrix, and the event template and matrix.[8]

[8] *For more information on JIPOE, refer to JP 2-01.3, Intelligence Preparation of the Operational Environment.*

II. (IPIE) Intel Preparation of the Info Environment

1. Intelligence Preparation of the Information Environment (IPIE)

a. <u>Overview</u>. IPIE continues to evolve as we realize how critical a thorough understanding of the information environment is in modern warfare, and how valuable an accurate portrayal of the information environment is in facilitating effective planning and execution of information operations. However, to be valid, IPIE must be conducted as part of the J-2's JIPOE efforts. If conducted in isolation, IPIE will fail to provide a picture of the information environment consistent with the other operating environments (i.e., land, sea, air, and space) and threat COAs generated by the J-2 staff.

Impact of the Information Environment on Military Operations

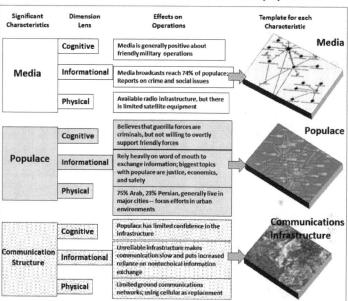

b. The information environment is where humans observe, orient, decide, and act upon information, and is therefore the principal environment of decision making. This environment is pervasive to all activities worldwide, and is a common backdrop for the air, land, maritime, and space physical domains of the CDR's OE. The actors in the information environment include military and civilian leaders, decisionmakers, individuals, and organizations. Resources include the information itself and the materials and systems employed to collect, analyze, apply, disseminate, and display information and produce information-related products such as reports, orders, and leaflets. **Cyberspace** is a global domain within the information environment consisting of the interdependent network of information technology infrastructures, including the Internet, telecommunications networks, computer systems, and embedded processors and controllers. Within cyberspace, electronics and the electromagnetic spectrum are used to store, modify, and exchange data via networked systems. Significant characteristics of the information environment can be further evaluated within physical, informational, and cognitive dimensions.

The Information Environment (Dimensions)

The **physical dimension** is composed of the command and control (C2) systems, and supporting infrastructures that enable individuals and organizations to conduct operations across the air, land, sea, and space domains. It is also the dimension where physical platforms and the communications networks that connect them reside. This includes the means of transmission, infrastructure, technologies, groups, and populations.

The **informational dimension** links the physical and cognitive dimensions. The joint force uses cyberspace capabilities, cyberspace operations, and non-cyberspace ways and means to collect, process, store, disseminate, display, and protect information and related products. The informational dimension focuses on the content and flow of information, and it is in this dimension that the CDR communicates intent and commands and controls military forces.

The **cognitive dimension** encompasses the minds of those who transmit, receive, and respond to or act on information. In this dimension, people think, perceive, visualize, understand, and decide. These activities may be affected by a CDR's psychological characteristics, personal motivations, and training. Factors such as leadership, morale, unit cohesion, emotion, state of mind, level of training, experience, situational awareness, as well as public opinion, perceptions, media, public information, and rumors may also affect cognition.

c. There is only one reality and that exists within the physical dimension. Actions within the physical dimension are converted into selected data, information, and knowledge in the information dimension, which are interpreted in the mind of individuals to develop perceptions, awareness, understanding, beliefs, and values, etc., which friendly forces would like them to.

d. IPIE strives to understand the relationship between the three dimensions as it pertains to the operation being executed or planned and those factors that can affect successful mission accomplishment within the defined OE and outside of it. Understanding the relationship between the dimensions enables one to understand the first, second and third order effects of an action that takes place in the physical domain. Thus, IO planners can more effectively plan IO initiatives to achieve desired effects, contribute to ensuring the CDR avoids scenarios that could achieve undesired effects, and also act as advisors to those who must understand the second and third order effects of actions planned to take place or that have taken place within the battlespace. Keep in mind that the OE affected may extend beyond that of the geographic area within which friendly forces operate. Needless to say, the foundation for effectively interpreting actions within the information environment is sound intelligence.

e. From an IO perspective, *the OE is the conceptual volume in which the CDR seeks to dominate the enemy.* The OE expands and contracts in relation to the CDR's ability to acquire and engage the enemy, or can change as the CDR's vision of the OE changes. It encompasses all three dimensions and is influenced by the operational dimensions of time, tempo, depth, and synchronization. The OE is not assigned by a higher CDR nor is it constrained by assigned boundaries. A command's OE is determined by the range of direct fire weapons, artillery, aviation, and electronic warfare (EW). The OE extends beyond the area of operation and may include the home station of a deploying friendly force. For IO, the OE is the volume of space in which friendly forces can influence the information environment. The information environment potentially expands the command's OE as the effects of IO elements like psychological operations and public affairs can extend well beyond the range of conventional weapon systems.

III. JIPOE Support to the Joint Planning Process

JIPOE supports joint planning by identifying significant facts and assumptions about the OE. This information includes details regarding adversary critical vulnerabilities, capabilities, decisive points, limitations, COGs, and potential COAs. JIPOE products are used by the CDR to produce the CDR's estimate of the situation and CONOPS, and by the joint force staff to produce their respective staff estimates. Various intelligence products such as DIA produced dynamic threat assessments (DTAs), baseline JIPOE products, and other locally produced assessments, will contribute to developing and enhancing comprehensive intelligence estimates. JIPOE products also help to provide the framework used by the joint force staff to develop, wargame, and compare friendly COAs and provide a foundation for the CDR's decision regarding which friendly COA to adopt. JIPOE support is crucial throughout the steps of the JPP (see following figure.)

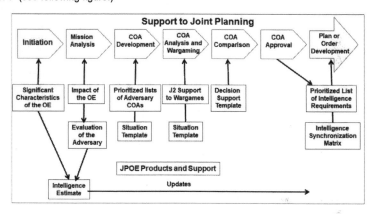

JIPOE Support to the Joint Planning Process (JPP).

a. The JIPOE effort should facilitate parallel planning by all strategic, operational, and tactical units involved in the operation. JIPOE products developed to support strategic-level planning should also be simultaneously disseminated to all appropriate operational and tactical headquarters. This is especially true during initial planning periods when headquarters at intermediate echelons may tend to filter information as it travels down to tactical units.

b. The integration of Service component IPB products with the CDRs' JIPOE effort creates a synergy in which an adversary's COAs may provide indicators as to the adversary's overall capabilities, intentions, desired endstate, and strategy. Specifically, JIPOE products facilitate operation planning by determining the following:

(1) The idiosyncrasies and decision-making patterns of the adversary strategic leadership and field CDRs.

(2) The adversary's strategy, intention, or strategic concept of operations, which should include the adversary's desired endstate, perception of friendly vulnerabilities, and adversary intentions regarding those vulnerabilities.

(3) The composition, dispositions, movements, strengths, doctrine, tactics, training, and combat effectiveness of major adversary forces that can influence friendly actions in the theater and operational areas.

(4) The adversary's principal strategic and operational objectives and lines of operation.

(5) The adversary's strategic and operational sustainment capabilities.

(6) COGs and decisive points throughout the adversary's operational and strategic depths.

(7) The adversary's ability to conduct IO and use or access data from all systems.

(8) The adversary's regional strategic vulnerabilities.

(9) The adversary's capability to conduct asymmetric attacks against friendly global critical support nodes (e.g., electric power grids, oil and gas pipelines, prepositioned supply depots).

(10) The adversary's relationship with possible allies and the ability to enlist their support.

(11) The adversary's defensive and offensive vulnerabilities in depth.

(12) The adversary's capability to operate advanced warfighting systems (e.g., smart weapons and sensors) in adverse METOC conditions.

(13) Key nodes, links, and exploitable vulnerabilities within an adversary system.

c. Plan Initiation. A preliminary or abbreviated JIPOE analysis pertaining to potential contingencies and significant characteristics of the OE should precede and inform the initiation phase of joint operation planning. During the initiation phase, DIA produces a DTA for each top priority plan and continuously updates each DTA as relevant aspects of the OE change. CCMD intelligence analysts accelerate JIPOE Step one activities by continuously monitoring the situation, alerting the CDR and staff to developments that may impact the operation planning effort, updating existing JIPOE products, and initiating new intelligence collection or production requirements. Additionally, the CDR may decide to form a JIPOE coordination cell to coordinate support and help analyze the initiating directive to determine time available until mission execution, the current status of JIPOE products and related staff estimates, and other factors relevant to the specific planning situation.

d. Mission Analysis. In order for the joint force staff to identify potential COAs, the CDR must formulate planning guidance based on an analysis of the friendly mission. This analysis helps to identify specified, implied, and essential tasks, any constraints on the application of military force, the CDR's task and purpose (restated mission), and possible follow-on missions. JIPOE supports mission analysis by enabling the CDR and joint force staff to visualize the full extent of the OE, to distinguish the known from the unknown, and to establish working assumptions regarding how adversary and friendly forces will interact within the constraints of the OE. JIPOE assists CDRs in formulating their planning guidance by identifying significant adversary capabilities and by pointing out critical factors, such as the locations of key geography, attitudes of indigenous populations, and potential land, air, and sea avenues of approach. Mission analysis and CDR guidance form the basis for the subsequent development of friendly COAs by the joint force staff. It is therefore imperative that an initial version of the impact of the OE, evaluation of the adversary and adversary COAs be briefed to the CDR at the mission analysis briefing. This is critical to enabling the CDR to provide sufficient guidance for friendly COA development.

e. COA Development. The J-2 ensures that all adversary COAs are identified, evaluated, and prioritized (JIPOE Step four) in sufficient time to be integrated into the friendly COA development effort. Additionally, the evaluation of the adversary (JIPOE Step three) is used by the J-3 and J-5 to estimate force ratios. The process of estimating force ratios may be complicated due to wide disparities between friendly and adversary unit organization, equipment capabilities, training, and morale. In such situations, the J-2, J-3, and J-5 may choose to develop local techniques and procedures for evaluating adversary units and equipment in terms of friendly force equivalents. The J-3 also depends heavily on JIPOE products prepared during the analysis of the adversary situation and the evaluation of other relevant aspects of the OE in order to formulate initial friendly force dispositions and schemes of maneuver.

Additionally, the JIPOE analysis of HVTs is used by the J-3 and J-5 to identify targets whose loss to the adversary would significantly contribute to the success of a friendly COA. These targets are refined through wargaming and are designated as HPTs. JIPOE also provides significant input to the formulation of deception plans by analyzing adversary intelligence collection capabilities and the perceptual biases of adversary decision-makers.

f. COA Analysis and Wargaming. Assumptions regarding the OE and adversary must be realistic. Avoid constructing assumptions that are deliberately designed to support premature conclusions or conceptual bias that favors one COA over another. For example, the joint force staff must guard against seizing upon one adversary COA as a "given" simply because it fits preconceived notions or is a "convenient" match for an already-favored friendly COA. Rather, the staff should plan to counter *all* adversary COAs identified during the JIPOE process. It is imperative that CDRs and their staffs recognize that the least likely adversary COA may be the one actually adopted precisely because it is the least likely, and therefore may be intended to maximize surprise. The wargame should follow a sequence of "action — reaction — counteraction" in which the J-2, JIOC, or red team personnel play the roles of adversary CDRs.

(1) Decision Support Template.
The decision support template is essentially a combined intelligence estimate and operations estimate in graphic form (see following figure.) It relates the detail contained on the event template (prepared during JIPOE Step four) to the times and locations of critical areas, events, and activities that would necessitate a command decision, such as shifting the location of the main effort or redeploying forces. Although the decision support template does not dictate decisions to the CDR, it is a useful tool for indicating points in time and space (decision points) where action by the CDR may be required.

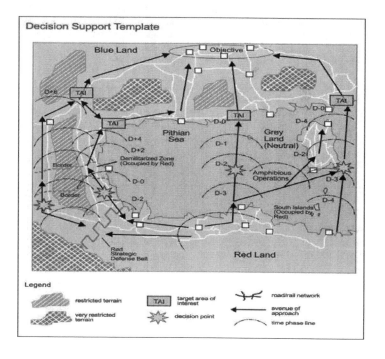

Decision Support Template

(2) The decision support template is constructed by combining the event template with data developed during the wargame. The J-2, J-3, J-4, J-5, and J-6 collaborate in the production of the decision support template, which is fully coordinated with all joint force staff elements. The decision support template displays TAIs, avenues of approach, objectives, and time phase lines derived from the JIPOE event template.

g. COA Comparison. Following wargaming, the staff compares friendly COAs to identify the one that has the highest probability of success against the full set of adversary COAs as depicted on the decision support template. Each joint force staff section uses different criteria for comparing friendly COAs, according to their own staff area of expertise. For example, the J-3 and J-5 compare friendly COAs based on the friendly force's ability to defeat each adversary COA, whereas the J-2 assesses the overall capabilities of intelligence collection and production to support each friendly COA. Additionally, each staff section must ensure that they have fully considered the CDR's initial planning guidance for COA selection.

> *"When I took a decision, or adopted an alternative, it was after studying every relevant — and many an irrelevant — factor. Geography, tribal structure, religion, social customs, language, appetites, standards — all were at my finger-ends. The enemy I knew almost like my own side. I risked myself among them a hundred times, to learn."*
>
> *Colonel T. E. Lawrence*
> *Letter to Liddell Hart, 26 June 1933*

h. COA Approval. After comparing friendly COAs, each joint force staff element presents its findings to the remainder of the staff. Together they determine which friendly COA they will recommend to the CDR. The J-3 then briefs the COAs to the CDR using graphic aids, such as the decision support template and matrix. The CDR decides upon a COA and announces the CONOPS.

i. Plan or Order Development. Using the results of wargaming associated with the selected COA, the joint force staff prepares plans and orders that implement the CDR's decision. The J-2 prioritizes intelligence requirements and synchronizes intelligence collection requirements to support the COA selected by the CDR. *Refer to JP 2-01.3, JIPOE, for a more in-depth discussion of the relationship between the JPP and JIPOE.*

> *"Nothing should be neglected in acquiring a knowledge of the geography and military statistics of their states, so as to know their material and moral capacity for attack and defense as well as the strategic advantages of the two parties."*
>
> *Jomini*
> *Precis de l'Art de la Guerre, 1838*

Joint Planning Process (Overview)

1. Joint Planning Process (JPP)

JPP is an orderly, analytical process, which consists of a set of logical steps to examine a mission; develop, analyze, and compare alternative COAs; select the best COA; and produce a plan or order. The application of operational design as explained in Chapter IV of JP 5-0 provides the conceptual basis for structuring campaigns and operations. JPP provides a proven process to organize the work of the CDR, staff, subordinate CDRs, and other partners, to develop plans that will appropriately address the problem to be solved. It focuses on defining the military mission and development and synchronization of detailed plans to accomplish that mission. CDRs and staffs can apply the thinking methodology (operational art and design) to discern the correct mission, develop creative and adaptive CONOPS to accomplish the mission, and synchronize those CONOPS so that they can be executed. Together with design, JPP facilitates interaction between the CDR, staff, and subordinate and supporting headquarters throughout planning. JPP helps CDRs and their staffs organize their planning activities, share a common understanding of the mission and CDR's intent, and develop effective plans and orders. The following figure shows the Seven Steps and Four Functions of the JPP.

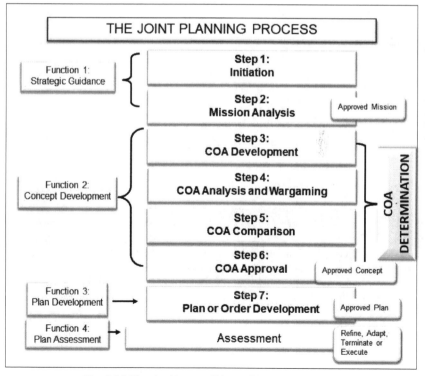

The Joint Planning Process Primary Steps and Functions

a. The seven-step JPP process aligns with the four Planning Functions discussed in Chapter 2-III, *Planning Functions*, which culminates with a published OPORD in a crisis and an OPLAN, CONPLAN, Base Plan or CDR's Estimate during contingency planning.[1]

b. The four Joint Planning Functions are: *Strategic Guidance, Concept Development, Plan Development*, and *Plan Assessment*. Each of these functions is further broken down into JPP steps, 1-7 as seen in Figure A below.

Planning Functions			
I. Strategic Guidance	**II. Concept Development**	**III. Plan Development**	**IV. Plan Assessment**
1- Planning Initiation 2- Mission Analysis	3- COA Development 4- COA Analysis 5- COA Comparison 6- COA Approval	7- Plan Development	- Refine - Adapt - Terminate - Execute

Figure A. Joint Planning Functions

The first two JPP steps (planning initiation and mission analysis) take place during the Strategic Guidance planning function. The next four JPP steps (COA Development, COA Analysis and Wargaming, COA Comparison, and COA Approval) align under the Concept Development planning function. The final JPP step (Plan or Order Development) occurs during the Plan Development planning function. While there is no JPP step associated with the Plan Assessment planning function, plans and orders are assessed with the RATE methodology in mind. Refer to Figure A above.

- Function I – Strategic Guidance consists of two steps: 1-Planning Initiation and 2-Mission Analysis.
- Function II – Concept Development consists of four steps: 3-COA Development, 4-COA Analysis and Wargaming, 5-COA Comparison and 6-COA Approval.
- Function III – Plan Development consists of 7-Plan or Order Development.
- Function IV – Plan Assessment. While there is no JPP step associated with the plan assessment planning function, plans and orders are assessed with refine, adapt, terminate, execute (RATE) methodology in mind. See Chapter 6-III, *Plan Assessment*, and JP 5-0 Chapter VI, *Operation Assessment*.

[1] *JP 5-0, Joint Planning*

c. JPP underpins planning at all levels and for missions across the full range of military operations. It applies to both supported and supporting CCDRs and to joint force component commands when the components participate in joint planning. This process is designed to facilitate interaction between the CCDR, staff, and subordinate headquarters throughout planning. JPP helps CCDRs and their staffs organize their planning activities, share a common understanding of the mission and CCDR's intent, and develop effective plans and orders as detailed in Figure B.

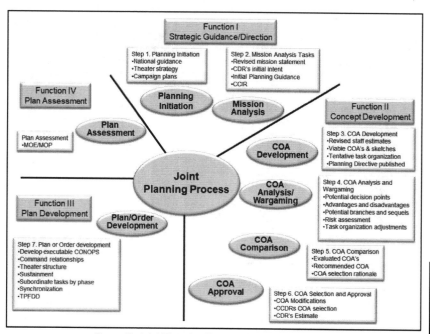

Figure B. Joint Planning Process Steps and Products

d. JPP is applicable for all planning. Like operational design, it is a logical process to approach a problem and determine a solution. It is a tool to be used by planners, but is not prescriptive. Based on the nature of the problem, other tools available to the planner, expertise in the planning team, time, and other considerations, the process can be modified as required. Similarly, some JPP steps or tasks may be performed concurrently, truncated, or modified as necessary dependent upon the situation, subject, or time constraints of the planning effort. For example, force planning, as an element of plan development, is different for campaign planning and contingency planning.[2]

e. In a crisis, the steps of JPP may be conducted simultaneously to speed the process. Supporting commands and organizations often conduct JPP simultaneously and iteratively with the supported CCMD. In these cases, once mission analysis begins it continues until the operation is complete. Moreover, steps 4-7 are repeated as often as necessary to integrate new requirements (missions) into the development of the plan.

[2]*JP 5-0, Joint Planning.*

f. Operational design and JPP are complementary tools of the overall planning process. Operational design provides an iterative process that allows for the CDR's vision and mastery of operational art to help planners answer ends–ways–means–risk questions and appropriately structure campaigns and operations. The CDR, supported by the staff, gains an understanding of the OE, defines the problem, and develops an operational approach for the campaign or operation through the application of operational design during the initiation step of JPP. CDRs communicate their operational approach to their staff, subordinates, supporting commands, agencies, multinational and NGO entities as required in their initial planning guidance so that their approach can be translated into executable plans. As JPP is applied, CDRs learn more about the operational environment and the problem and refine their initial operational approach. CDRs provide their updated approach to the staff to guide detailed planning. This iterative process between the CDR's maturing operational approach and the development of the mission and CONOPS through JPP facilitates the continuing development of possible COAs and their refinement into eventual CONOPS and executable plans.

g. The relationship between the application of operational art, operational design, and JPP continues throughout the planning and execution of the plan or order. By applying the operational design methodology in combination with the procedural rigor of JPP, the command can monitor the dynamics of the mission and OE while executing operations in accordance with the current approach and revising plans as needed. By combining these approaches, the friendly force can maintain the greatest possible flexibility and do so in a proactive vice reactive manner.

> *"If you don't know where you are going,*
> *you'll end up someplace else."*
> Yogi Berra

2. Planning Organization

At the CCMD level, a Joint Planning Group (JPG), operational planning group (OPG), or operational planning team (OPT) is typically established to direct planning efforts across the command, including implementation of plans and orders. The JPG is typically organized within the J-5 Directorate. The JPG is responsible to the J-5 and CCDR for driving the command's planning effort. Effectiveness of the JPG will be measured, in part, by the support provided to it by the principal CCMD staff officers (J-1 through J-6). The composition of the JPG is a carefully balanced consideration between group management and appropriate representation from the JTF staff and components. JPG membership will vary based on the tasks to be accomplished, time available to accomplish the tasks, and the experience level of the JPG members. Representation to the JPG should be a long-term assignment to provide continuity of focus and consistency of procedure (Figure C on the following page is a notional JPG).

3. Staff Estimates[3]

Staff Estimates are absolutely vital to establish and maintain an overall high degree of coordination and cooperation, both internally and with staffs of higher, lower, and adjacent units. The staffing process continues throughout all the operational activities, planning and execution functions. The CCDR's staff must function as a single, cohesive unit—a profes-

---

[3] *Staff estimate format. The staff estimate format contained in CJCSM 3122.01A, Appendix T, standardizes the way staff members construct estimates. The J2 (with input assistance from all staff members) will still conduct and disseminate the initial Joint Intelligence Preparation of the Operational Environment as a separate product. The Commander's Estimate format is located as Enclosure J of the same document.*

sional team. Because of the unique talents of each directorate, involvement of all is vital. Each staff estimate takes on a different focus that identifies certain assumptions, detailed aspects of the COAs, and potential deficiencies that are simply not known at any other level, but nevertheless must be considered. Such a detailed study of the COAs involves the corresponding staffs of subordinate and supporting commands. Each staff member must know their own duties and responsibilities in detail and be familiar with the duties and responsibilities of other staff members.

a. The staff's efforts must always focus on supporting the CCDR and on helping support the subordinate units. CDRs can minimize risks by increasing certainty. The staff supports the CCDR by providing better, more relevant, timely, and accurate information; making estimates and recommendations; preparing plans and orders; and monitoring execution.

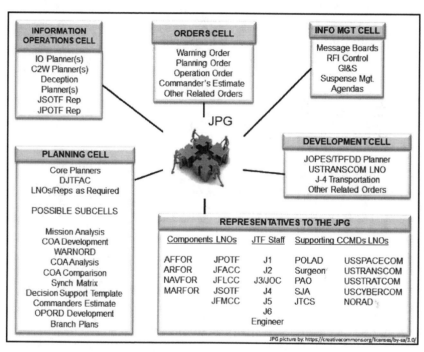

Figure C. Notional Joint Planning Group

b. **The primary product the staff produces for the CCDR, and for subordinate CCDRs, is *understanding*, or *situational awareness*.** True understanding should be the basis for information provided to CCDRs to make decisions. Formal staff processes provide two types of information associated with understanding and decision-making. All other staff activities are secondary. The first is *situational awareness information*, which creates an understanding of the situation as the basis for making an informed decision. **Simply, it is understanding oneself, the enemy, and the terrain or environment.** The second type of information, *execution information*, communicates a clearly understood vision of the operation and desired outcome after a decision is made. Examples of execution information are conclusions, recommendations, guidance, intent, concept statements, and orders.

c. Mission analysis, facts and assumptions, and the situation analysis (of the area of operations, area of interest, adversary, friendly, and support requirements, etc.) furnish the structure for the staff estimate. The estimate consists of significant facts, events, and conclusions based on analyzed data. It recommends how to best use available resources. Adequate, rapid decision-making and planning hinge on good, timely command and staff estimates. They are the basis for forming viable courses of action. Failure to make estimates can lead to errors and omissions when developing, analyzing, and comparing COAs, developing or executing plans.

d. Essential Qualities of Estimates.

(1) CCDRs control tempo by making and executing decisions faster than the adversary. Therefore, CCDRs must always strive to optimize time available. They must not allow estimates to become overly time-consuming. However, they must be comprehensive and continuous and must visualize the future.

(2) Comprehensive estimates consider both the quantifiable and the intangible aspects of military operations. They translate friendly and adversary strengths, joint weapons systems, training, morale, and leadership into combat capabilities. The estimate process requires a clear understanding of the operational environment and the ability to visualize the operational or crisis situations requiring military forces or interagency support. Estimates must provide a timely, accurate evaluation of the operation at a given time.

(3) The demand on the C2 system is continuous as opposed to cyclical. Estimates must be as thorough as time and circumstances permit. The CCDR and staff must constantly collect, process, and evaluate information. They update their estimates:

- When the CCDR and staff recognize new facts.

- When they replace assumptions with facts or find their assumptions invalid.

- When they receive changes to the mission or when changes are indicated.

(4) Estimates for the current operation can often provide a basis for estimates for future missions as well as changes to current operations. Technological advances and near-real-time information estimates ensure that estimates can be continuously updated.

e. Plan in Execution. Estimates for the plan in execution can often provide a basis for estimates for future plans, as well as changes to the plan in execution. Technological advances and near-real time information estimates ensure that estimates can be continuously updated. Estimates must visualize the future and support the CDR's visualization. They are the link between planning and execution and support continuous assessment. The CDR's vision articulated in the strategic estimate directs the endstate. Each subordinate unit CDR must also possess the ability to envision the organization's desired endstate, as well as those desired by their opposition counterpart. Estimates contribute to this vision. Failure to make staff estimates can lead to errors and omissions when developing, analyzing, and comparing COAs.

Not every situation will allow or require an extensive and lengthy planning effort. It is conceivable that a CDR could review the assigned task, receive oral briefings, make a quick decision, and direct writing of the plan to commence. This would complete the process and might be suitable if the task were simple and straightforward.

(See Chapter 5-III, *COA Development* for more on the staff estimates.)

4. Key Planning Elements and Critical Staff Involvement

CDRs at all levels need to participate in planning to the greatest extent possible from early operational design through approval of the plan or order. Understanding that CCDRs have many operations, campaigns or crisis at any given time so it's difficult as a staff to have 100% participation from the CDR especially early in the planning process. Regardless of the CDR's level of involvement, certain key planning elements require the CDR's participation and decisions. The staff will normally prepare and brief these to the CCDR and/or Chief of Staff and receive the CDR's guidance, intent, and requirements in return. These will normally include the operational approach, mission statement, CDR's planning guidance, CDR's intent, CDR's critical information requirements (CCIRs), and CONOPS.

a. <u>Operational Approach</u>. The operational approach is a CDR's initial description, to help guide further planning, of the broad actions the force must take to achieve objectives and accomplish the mission. It is the CDR's visualization of how the operation should transform current conditions into the desired conditions—the way the CDR wants the OE to look at the conclusion of operations. The operational approach is based largely on an understanding of the OE and the problem facing the JFC. Once the JFC approves the approach, it provides the basis to begin, continue, or complete detailed planning. The JFC and staff should continually review, update, and modify the approach as the OE, objectives, or problems change.

b. <u>Mission Statement</u>. The joint force's mission is what the joint force must accomplish. It is described in the mission statement, which is a sentence or short paragraph that describes the organization's essential task (or set of tasks) and purpose—a clear statement of the action to be taken and the reason for doing so. The mission statement— approved by the CDR—contains the elements of **who, what, when, where, and why** of the operation. The eventual CONOPS will specify how the joint force will accomplish the mission. The mission statement forms the basis for planning and is included in the CDR's planning guidance, the planning directive, staff estimates, CDR's estimate, and the CONOPS. The JFC should develop clear mission statements and ensure they are understood by subordinates. (See Chapter 5-II, *Mission Analysis*.)

c. <u>Commander's Planning Guidance.</u> JFCs guide the joint force's actions throughout planning and execution. However, the staff and component CDRs typically expect the JFC to issue initial guidance soon after receipt of a mission or tasks from higher authority and provide more detailed planning guidance after the JFC approves an operational approach. This guidance is an important input to subsequent mission analysis, but the completion of mission analysis is another point at which the JFC may provide updated planning guidance that affects COA development. (See Chapter 5-II, *Mission Analysis*.)

d. <u>Commander's Intent</u>. CDR's intent is the CDR's clear and concise expression of **what the force must do and the conditions the force must establish to accomplish the mission.** It includes the purpose, endstate, and associated risks. CDR's intent supports mission command and allows subordinates the greatest possible freedom of action. It provides focus to the staff and helps subordinate and supporting CDR's act to achieve the CDR's desired results without further orders once the operation begins, even when the operation does not unfold as planned. Successful CDR's demand that subordinate leaders at all echelons exercise disciplined initiative and act aggressively and independently to accomplish the mission within the CDR's intent. Subordinates emphasize timely decision making, understanding the higher CDR's intent, and clearly identifying tasks to achieve desired objectives. Well-crafted CDR's intent improves subordinates' situational awareness, which enables effective actions in fluid, chaotic situations. (See Chapter 5-II, *Mission Analysis*.)

e. <u>CCIRs.</u> CCIRs are elements of friendly and enemy information the CDR identifies as critical to timely decision-making. They focus information management and help the JFC and staff assess the OE. The CCIR list is normally a product of mission analysis, and JFCs add, delete, and update CCIRs throughout an operation. (See Chapter 4, *JIPOE* and 5-II, *Mission Analysis*.)

> "It ain't what you don't know that gets you into trouble. It's what you know for sure that just ain't so."
>
> Mark Twain

f. <u>CONOPS</u>. The CONOPS, included in paragraph 3, (Execution) of the plan or order, describes how the JFC intends to integrate, synchronize, and phase actions of the joint force components and supporting organizations to accomplish the mission. CONOPS generally include potential branches and sequels. The CONOPS is typically a detailed extension of the operational approach, but incorporates modifications based on updated information and intelligence gained during planning as well as the JFC's approved COA. The staff writes (or graphically portrays) the CONOPS in sufficient detail so that subordinate and supporting CDRs understand their mission, tasks, and other requirements and can develop their supporting plans accordingly. The CONOPS also provides the basis to develop the concept of fires, concept of intelligence operations, and theater logistics overview (TLO), which also are included in the final OPLAN or OPORD. (See Chapter 6, *Plan or Order Development*.)

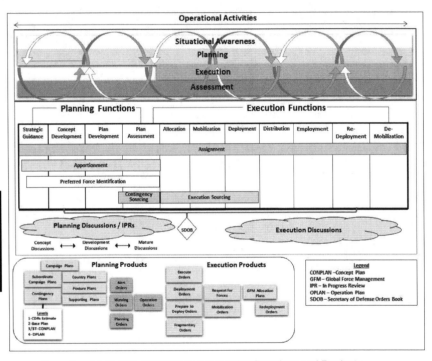

Operational Activities, Planning and Execution Functions and Products

I(a). Strategic Direction & Guidance (Function I)

> *"Strategic planning and strategic thinking are two distinct activities. Strategic planning, a process-based activity, focuses on analysis, logic, and procedures while strategic thinking, an idea-based cognitive activity, emphasizes synthesis, creativity, intuition, and innovation.*[1]

1. Understanding Strategic Direction and Guidance

a. National strategic direction is governed by the Constitution, federal law, USG policy, internationally recognized law, and the national interest as represented by national security policy. This direction provides strategic context for the employment of the instruments of national power and defines the strategic purpose that guides employment of the military as part of a global strategy.

b. The President, SecDef, and CJCS provide their orders, intent, strategy, direction, and guidance via strategic direction to the military to pursue national interests within legal and constitutional limitations. *Strategic direction is typically published in key documents, generally referred to as strategic guidance, but it may be communicated through any means available.* Strategic direction may change rapidly in response to changes in the global environment, whereas strategic guidance documents are typically updated cyclically and may not reflect the most current strategic direction.

(1) In general, the President frames the strategic context by defining national interests and goals in documents such as the NSS, Presidential Policy Directives (PPDs), executive orders, and other national strategic documents, in conjunction with the NSC and Homeland Security Council (HSC).

(2) DOD derives its strategic-level documents from guidance in the NSS. The documents outline how DOD will support NSS objectives and provide a framework for other DOD policy and planning guidance, such as the DPG, GFMIG, and the JSCP. Figure A displays strategic direction promulgated through multiple strategic guidance documents.

(3) The President approves the CPG, which is developed by the OSD. The CPG provides policy guidance and priorities to the Chairman and CCMDs for GFM and the preparation and review of campaign and contingency plans. The CJCS publishes the JSCP, which implements campaign, contingency, and posture planning guidance. The President also signs the UCP, which is developed by the OSD and the Joint Staff in coordination with the NSC. The UCP establishes CCMD missions, responsibilities, and AORs.

c. Planning usually starts with the assignment of a planning task through a directive, order, or cyclical strategic guidance depending on how a situation develops. The CDR and staff must analyze all available sources of guidance. These sources include written documents, such as the CPG and JSCP, written directives, oral instructions from higher headquarters, domestic and international laws, policies of other organizations that are interested in the situation, communication synchronization guidance, and higher headquarters' orders or estimates.

[1] Fiona Graetz, *"Strategic Thinking versus Strategic Planning: Towards Understanding the Complementarities,"* Management Decision 40, no. 5/6 (2002): 457

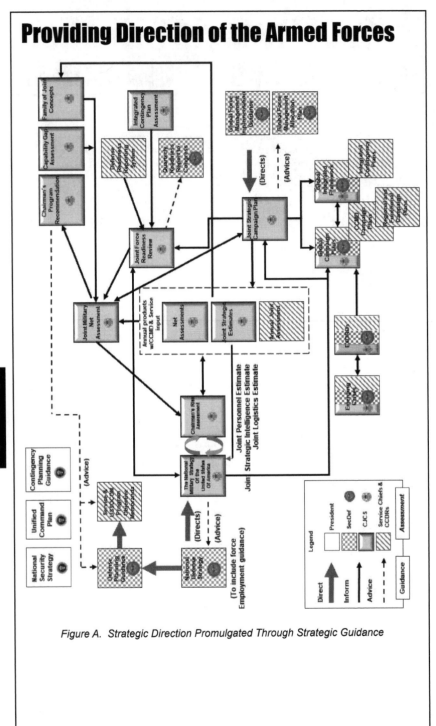

Providing Direction of the Armed Forces

Figure A. Strategic Direction Promulgated Through Strategic Guidance

d. Strategic direction from strategic guidance documents can be vague, incomplete, outdated, or conflicting. This is due to the different times at which they may have been produced, changes in personnel that result in differing opinions or policies, and the staffing process where compromises are made to achieve agreement within the documents. During planning, CDRs and staff must read the directives and synthesize the contents into a concise statement. Since strategic guidance documents can be problematic, the CCDR and staff should obtain clear, updated direction through routine and sustained civilian-military dialogue throughout the planning process. When clarification does not occur, planners and CDRs identify those areas as elements of risk.

e. Additionally, throughout the planning process, senior leaders will provide additional guidance. This can be through formal processes such as standard guidance statements (SGSs) and IPRs, or through informal processes such as e-mails, conversations, and meetings. All of this needs to be disseminated to ensure the command has a common understanding of higher CDR's intent, vision, and expectations.

f. In particular, CDRs maintain dialogue with leadership at all levels to resolve differences of interpretation of higher-level objectives and the ways and means to accomplish these objectives. Understanding the OE, defining the problem, and devising a sound approach, are rarely achieved the first time. Strategic guidance addressing complex problems can initially be vague, requiring the CDR to interpret and filter it for the staff. While CCDRs and national leaders may have a clear strategic perspective of the problem from their vantage point, operational-level CDRs and subordinate leaders often have a better understanding of specific circumstances that comprise the operational situation and may have a completely different perspective on the causes and solutions. Both perspectives are essential to a sound plan. Subordinate CDRs should be aggressive in sharing their perspective with their higher headquarters, and both should resolve differences at the earliest opportunity. While policy and strategic guidance clarify planning, it is equally true that planning informs policy formulation. A strategy or plan that cannot be realistically executed at the tactical level can be as detrimental to long-range U.S. interests as tactical actions that accomplish a task but undermine the strategic or operational objectives.

g. Strategic guidance is essential to operational art and operational design. As discussed in Chapter 1, *Strategic Organization*, the President, SecDef, and CJCS all promulgate strategic guidance. In general, this guidance provides long-term as well as intermediate or ancillary objectives. It should define what constitutes victory or success (ends) and identify available forces, resources, and authorities (means) to achieve strategic objectives. The operational approach (ways) of employing military capabilities to achieve the ends is for the supported CCDR to develop and propose, although policy or national positions may limit options available to the CDR. Connecting resources and tactical actions to strategic ends is the responsibility of the operational CDR—the CDR must be able to explain how proposed actions will result in desired effects, as well as the potential risks of such actions.

h. For situations that require the employment of military capabilities (particularly for anticipated large-scale combat), the President and SecDef may establish a set of operational objectives. However, in the absence of coherent guidance or direction, the CCDR may need to collaborate with policymakers in the development of these objectives. Achievement of these objectives should result in contributing to the strategic objective—the broadly expressed conditions that should exist after the conclusion of a campaign or operation. Based on the ongoing civilian-military dialogue, the CCDR will determine the military endstate and military objectives, which define the role of military forces. These objectives are the basis for operational design.

i. From this broad, sometimes vague, incomplete, outdated, or conflicting strategic direction, interpreted from strategic guidance, more specific national, functional, and theater strategic and supporting objectives will help focus and refine the context. These objectives will assist in guiding the military's joint planning and execution related to these objectives or a specific crisis. Integrated planning, coordination, and guidance among the Joint Staff, CCMD staffs, Service Chiefs, and USG departments and agencies translate strategic priorities into clear planning guidance, tailored force packages, operational-level objectives OPLANs, and logistical support for the joint force to accomplish its mission.[2,3]

2. Strategic Direction Focus

> "Strategy is a pattern in a stream of decisions."
> Henry Mintzberg

Strategic direction and supporting national-level activities, in concert with the efforts of CCDRs, ensure the following:

- National strategic objectives and termination criteria are clearly defined, understood, and achievable.
- Active Component (AC) is ready for combat and Reserve Component (RC) are appropriately manned, trained, and equipped in accordance with Title 10 USC responsibilities and prepared to become part of the total force upon mobilization.
- Intelligence, surveillance, and reconnaissance systems and efforts focus on the operational environment.
- Strategic guidance is current and timely.
- DOD, other intergovernmental organizations, allies, and coalition partners are fully integrated at the earliest time during planning and subsequent operations.
- All required support assets are ready.
- Multinational partners are available and integrated early in the planning process.
- Forces and associated sustaining capabilities deploy ready to support the commander's CONOPS.

3. Planning Function I: Strategic Guidance

a. <u>Strategic Guidance.</u> The purpose of the strategic guidance planning function is to initiate planning, provide the basis for mission analysis, develop a shared understanding of the environment, objectives, responsibilities and framework for subsequent concept development. This function contains the initial application of design and operational art that lays the foundation for the entire planning process. ***Planners should avoid the impulse to rush into solving the problem and producing planning products before thoroughly understanding the problem.*** Throughout this function, there must be a sustained civilian-military dialogue to ensure the alignment of military planning with strategic direction and policy. The strategic guidance function focuses on defining the objective or ends within feasible means. From this analysis, the COAs and plan (ways) are subsequently developed in follow-on planning functions. Figure B depicts the elements involved with the strategic guidance planning function.

(1) The CCDR provides input through sustained civilian-military dialogue that may include IPRs. The CCDR crafts objectives that support national strategic objectives with the guidance and consent of the SecDef; if required, the CJCS offers advice to SecDef. This

[2] JP 5-0, Joint Planning

[3] JP 3-0, Joint Operations

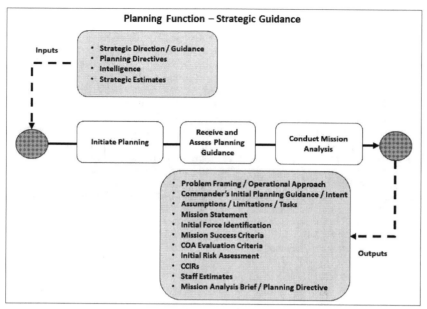

Figure B. Planning Function - Strategic Guidance

process begins with an analysis of existing strategic guidance such as the CPG and JSCP or a CJCS WARNORD, PLANORD, or ALERTORD issued in a crisis. It includes mission analysis, threat assessment, and development of assumptions, which as a minimum, will be briefed to SecDef during the strategic guidance IPR.

(2) Some of the primary end products of the strategic guidance planning function are assumptions, ID of available/acceptable resources, conclusions about the strategic and OE (nature of the problem), strategic and military objectives, and the supported CDR's mission.

(3) The CCDR will maintain dialogue with DOD leadership to ensure a common understanding of the above topics and alignment of planning to date. This step can be iterative, as the CCDR consults with the staff to identify concerns with or gaps in the guidance.

> *"However beautiful the strategy, you should occasionally look at the results."*
>
> *Winston Churchill*

b. <u>Initiate Planning</u>. Planning initiation begins when a potential or actual event is recognized that may require a military response. At the national level, the President or Secretary can direct the initiation of planning. The Presidential and SecDef direction is usually communicated via the Chairman. The Chairman also has statutory authority to direct strategic planning, and may direct CCDR planning to develop military response options. Additionally, CDRs at any level of the chain of command may initiate planning when they identify conditions that warrant planning for which planning has not been directed by higher authority. The direction to initiate planning is communicated differently under different conditions as follows:

(1) When time allows, planning is done methodically (or deliberately) to produce campaign or contingency plans. If a situation develops that warrants developing a plan for a potential threat outside of the strategic guidance development cycle, the Secretary may issue direction via a SGS or by other means, as time allows to add to or change existing planning.

(2) Planning may also be initiated in response to an emergent event or crisis. Under their separate authorities, CCMD planning may be directed by the President, Secretary, Chairman, CCDR, or any other CDR. The normal means of directing subordinates to begin planning is by an order such as a WARNORD, PLANORD, or ALERTORD. These orders convey planning and preparation authority short of execution. All plans cannot be execution sourced, however, in the event of an impending crisis there is no restriction placed on the CCMD to review available forces in order to prepare force requests should the plan transition to execution. Directed planning may be based upon an existing contingency plan for an anticipated event or without one for an unanticipated event.

(3) Upon the initiation of planning, a CDR may establish a planning team. These planning teams may have different names, but each planning team is typically comprised of functional experts tailored to the specific requirements of the planning task. Planning teams may also include liaison members from other commands to enable collaborative planning with higher, supporting, and subordinate commands. The Joint Staff establishes Integrated Planning Teams (IPTs) to focus on the GCPs. At the CCMD-level, a JPG typically directs multiple planning efforts across the command. Subordinate to the JPG may be one or more OPTs established to conduct a defined planning task or working groups established to solve a specific functional task. Specific planning efforts may be assigned to an established OPT or working group or they may be planned by the JPG directly based upon the CDR's direction.

(4) Supported CCDRs should initiate collaborative communication with supporting CCMDs, Services, and DoD Agencies upon initiation of planning in order to align their planning efforts. Early collaboration between supported and supporting commands is essential to identify key equities and issues that should be considered during problem framing and mission analysis.

(5) <u>Coordination with other non-DoD Agencies</u>. Planning may identify potential requirements for coordination with stakeholders external to DoD to include interagency and interorganizational partners. Per strategic guidance, CCDRs, ICW with the Chairman, will seek Secretary intent/guidance before beginning development of operational plans with other departments and agencies. If coordination external to DoD is authorized, interorganizational involvement in CCDR planning may be done with the oversight of USD(P) and JS J-5 through the Promote Cooperation program.

(6) <u>Classification and Access</u>. The classification and access to planning products will be established by the supported CCDR and implemented throughout the planning cycle.

(a) <u>Classification</u>. CCDRs exercise original classification authority for their plans and planning products as delegated by the Secretary per DoD Manual (DoDM) 5200.01 Volume 1, *DoD Information Security Program: Overview, Classification, and Declassification* and establish the necessary classification of individual planning products. Planners should carefully consider how the classification of the plan and specific planning products may limit collaboration with supporting or subordinate commands, Services, DoD Agencies, Interagency, NGOs, and allies/coalition partners. Planners may need collaboration platforms to accommodate multiple different levels of classification and access.

(b) <u>Access</u>. Separately, CCDRs should grant access to supporting organizations to enable synchronized planning across the JPEC. Supported CCDRs will need to grant access to the JS J-3 and J-5 in order to support sourcing feasibility, JPEC review or crisis response as required. The supported CCDRs should also grant access to supporting CCMDs, CCMD Service components, and Service HQs to enable integrated planning and collaboration. Supporting organizations will designate the specific planners that will work on the supporting plans and require access. The need to grant access to enable collaboration and support planning must be balanced with the need to limit access. For planning requiring integration of Special Access Programs (SAP) consult DoDM 5205.07 Volume 1, *DoD Special Access Program (SAP) Security Manual: General Procedures*.

c. Receive and Assess Planning Guidance. At the start of the planning function, the CDR's staff performs an analysis of existing strategic direction and preliminary planning guidance, policy, and joint doctrine to allow them to define the task.

(1) Strategic Guidance. The NSS, NDS, DSR, and NMS outline national interests and priorities. The UCP, CPG, JSCP (including supplements), force apportionment tables and GFMIG provide specific tasks and guidance for CCDRs to develop campaign and specific contingency plans. This guidance may be updated by strategic direction received through civilian-military dialogue during planning.

(a) Implementing strategic guidance are multiple DoD and CJCS policy and doctrinal issuances governing military planning and execution.

(b) Planning may also be informed by guidance and policy contained in interagency, multinational, and international law documents.

(2) Orders that Initiate Planning. Orders contain further strategic guidance that may include a description of the situation, command relationships, mission and intent, planning constraints or restraints, assumptions, COAs, and objectives.

(3) Strategic Estimate. The strategic estimate communicates the CDRs understanding of the operating environment with analysis of political, military, and economic factors and trends, and their impact on the achievement of directed objectives. While not specific to a particular planning problem, it provides a starting point for detailed staff estimates focused on a specific planning effort.

(4) Intelligence. Intelligence planners coordinate with intelligence collection managers and intelligence all-source analysis mission managers to tailor JIPOE products to support planning requirements (e.g., strategic estimates, intelligence staff estimates, and intelligence estimates) and operation assessment activities. Supported and supporting CCMDs, ICW the supporting CCMDs, intelligence CSAs, and Service intelligence centers, develop a common baseline intelligence estimate to ensure a shared understanding of the threat and OE and to facilitate integrated planning and assessment. That common baseline intelligence estimate is captured in the Intelligence Annex (Annex B) of the developing plan or order. CJCSM 3314.01 Series, Intelligence Planning, provides further guidance for intelligence planning support.

(a) For campaign plans, CCMDs use JIPOE products and Defense Intelligence Agency (DIA) Theater Intelligence Assessment (TIAs) provide a baseline understanding of the OE.

(b) For contingency plans, CCMDs use JIPOE products and DIA DTAs provide a baseline understanding of the OE.

(c) Under crisis conditions, CCMDs use the existing JIPOE, any applicable TIA or DTA, and situational awareness of the current OE.

(5) Align with Current Campaign Plans. Plan development should align with other appropriate campaign plan(s). The environment, conditions, resources, and posture established during campaign execution will influence the subsequent development/refinement of contingency plans.

(6) Review Previous Plans and Operations. Planning mission analysis may benefit from the reference of previous operations or existing plans. The Joint Center for Operational Analysis and Lessons Learned, as well as the Joint Lessons Learned Information System database (JS J-7 Future Joint Force Development Division SIPRNET website), contain specific practical lessons learned in all areas of planning and execution from previous operations. While an emergent problem may not align exactly to an existing contingency plan, a plan for a similar type of problem may provide plan elements that are useful and have already been reviewed by civilian and military leadership.

d. Conduct Mission Analysis. Mission analysis is the study and deconstruction of assigned tasks to facilitate the CDR's articulation of the enduring purpose for conducting operations. Mission analysis is conducted by the planning staff and shaped by direction from the CDR. At the strategic level, mission analysis should include a sustained civilian-military dialogue between the CCMD, JS, and OSD to ensure the alignment of national policy and military planning. CCMD planning will also connect with JPEC counterparts at supporting CCMDs, Service HQs, DoD supported command publishes the outputs of mission analysis which provides the basis for subsequent planning by supported, supporting, and subordinate commands and Services (see Chapter 5, *Mission Analysis: Key Steps*).

(1) Problem Framing. At the onset of planning, some problems may lack clear definition. By applying the tenants of design and constantly considering the persistent operational activities, the planning team focuses on the right problem. This focus shapes all future planning functions. Problem framing is used by the CDR and planning team to systematically analyze a problem and synthesize it with experience and judgement (operational art). In the context of strategic direction and an understanding of the OE, problem framing can decompose a complex or ambiguous problem, identify its root causes as well as objectives and endstate conditions that solve them. Analytical tools used for problem framing include doctrinal planning models, operational art, and operational design (see JP 5-0, *Joint Planning*).

(2) Strategic and Operational Approach. To address a national security interest, a strategic approach is developed for the combined application of one or more instruments of national power. Typically shaped within the NSC, the strategic approach provides a broad initial direction to the character of military effort to be planned. To develop this strategic approach, a menu of military options is provided to the principal decision-makers to address the problem. NSC, Secretary, and OSD feedback shapes the CCDRs development of the initial operational approach which guides the mission analysis by outlining the mission, operational concepts, tasks, and actions required to accomplish the objectives. The CDR's operational approach and mission analysis guides the preparation of planning guidance, informs the mission statement, and helps form the basis for planning.

(3) Commander's Planning Guidance. The initial CDR's planning guidance informs the planning team, including planners from supporting CCMDs, of the CDR's understanding of the environment and a framing of the problem as currently understood. Initial guidance typically provides time constraints, initial coordination requirements, authorization of movement of key capabilities within the CCDR's authority, and directs other actions as necessary. Initial planning guidance shapes the subsequent conduct of mission analysis. The CDR's understanding of the environment and problem are refined through interaction with the planning staff during mission analysis. At the conclusion of mission analysis, the CDR's planning guidance is refined to provide updated direction to the planning staff and subordinate commands for subsequent planning, and inform the planning efforts of supporting commands.

(4) Commander's Intent. The initial CDR's intent describes the purpose, desired endstate, and operational risk. Refined during planning, the CDR's intent should allow for decentralized decision making by the staff and subordinate and supporting CDRs focused on a common set of outcomes at the theater strategic level, it provides an overall vision for the campaign and identifies the major unifying efforts and where other elements of national power will play a central role. The CDR's intent must be in line with the President's, Secretary, and intent with careful consideration to not exceed the level of risk the chain of command is willing to accept.

(a) Purpose. The reason for the military action with respect to the mission of the next higher echelon. Explains why military action is being conducted.

(b) Endstate. Describing the OE and the desired conditions when an operation concludes is a useful visualization of problem framing that can inform civilian-military discussions. Developed objectives provide nearer-term, measurable goals that lead to the achievement of the desired longer-term conditions.

(c) Risk. The CDR's intent should convey where the CDR is willing to or not willing to accept risk.

(5) <u>Assumptions</u>. Assumptions are specific suppositions about an aspect of the OE that are assumed to be true in the absence of proof. Assumptions should be logical, realistic, and necessary to enable the continuation of the planning process. For planning without time constraints, limited facts are known and planners are required to derive planning assumptions in order to facilitate planning for a hypothetical scenario or potential future events. During crisis, planning usually requires fewer assumptions as plans are is based upon the actual conditions of the operating environment. Assumptions should be reviewed and evaluated throughout planning and discussed via civilian-military dialogue and during IPRs.

(a) If an assumption cannot be validated during planning, by either rejecting it or confirming it as fact, a branch plan may be considered to cover the possible conditions of an invalid assumption should the plan be executed.

(b) CCIRs should be crafted to confirm or invalidate assumptions and the potential of identifying an invalid planning assumption should be captured in a plan's risk analysis.

(6) <u>Limitations</u>. Prescribed limitations, identified during problem framing, will constrain or restrain the subsequent development of COAs. Limitations at the strategic level will be influenced by policy or political considerations. Often limitations create a degree of risk that should be examined during civilian-military dialogue and IPRs.

(7) <u>Tasks</u>. Tasks may be specified by higher authority or implied and derived from the planning guidance or mission analysis. As tasks are identified, planners need to ensure their continued alignment with military and strategic objectives.

(8) <u>Mission Statement</u>. The initial mission statement is distributed to the staff and subordinate and supporting CDRs to set the framework for concept development. The mission statement describes the essential task(s) or objectives to be accomplished and the purpose of the activity. It should contain: (1) who will execute the mission, (2) what essential actions are proposed, (3) when the action will begin, (4) where the action will occur, and (5) the purpose to be achieved.

(9) <u>Initial Force Identification</u>. During mission analysis, the planning team should consider assigned and currently allocated forces in the AOR as well as the quantity of apportioned forces and other key resources. Planners may develop a rough-order-of-magnitude list of required forces and capabilities necessary to accomplish the specified and implied tasks. While more deliberate identification of detailed force requirements continues during concept and plan development, an initial identification of readily available forces during strategic guidance effectively informs the development of an operational approach.

(a) <u>Force List</u>. The identified force requirements for a plan are initially collected and documented in a force list. Planners should consider at the onset of planning that force requirements for any plan should be documented in a format that enables feasibility analysis, plan assessment and execution sourcing, to include force allocation decisions by the Secretary, should the plan transition to execution. For planning under crisis conditions, force requirements may be documented directly as a TPFDD in order to facilitate sourcing and transportation feasibility assessments, execution sourcing and unit deployment.

(b) Assigned and allocated forces currently deployed to the CCMD AOR may be the most responsive during the early stages of an emergent crisis. Planners should consider that re-missioning of currently allocated and deployed units requires coordination and potentially SecDef approval in accordance with the GFMIG, CJCSM 3130.06, *GFM Allocation Policies and Procedures*, and procedures prescribed in JP 5-0, *Joint Planning*.

(c) Planners must take into consideration that all the force requirements from supported and supporting plans are drawing from the same pool of Joint forces. Planned force requirements may be competing with requirements for other operations happening at the time a plan is executed. Planners should also consider that deploying additional forces adds risk to other potential future operations. The Services, JS JFC, and the JFPs offer planning support and can assist in assessing sourcing feasibility and provide estimates of sourcing risks throughout the planning process. If leveraged early in the planning process, this insight and advice can significantly reduce the iterations necessary to develop feasible COAs and concepts.

(d) <u>Integrated Planning</u>. During planning, the supported CCDR develops an initial proposed distribution of the available resources among the supported and supporting plans. For integrated planning, apportioned forces should constrain the total force requirements of both supported and supporting plans. If during planning, there are points of contention among supported and supporting CCDRs, the supported CCDR should attempt to resolve force requirement conflicts with supporting CCDRs. Requirement conflicts that cannot be resolved should be raised to the JS J5. The JS J5 will facilitate bringing the issue to the OpsDeps, JCS Tank, OSD, and/or Secretary for decision.

(10) <u>Develop Mission Success Criteria</u>. Mission success criteria describe a set of standards that define mission accomplishment. At the strategic level, mission success criteria must align with strategic objectives and, for contingency plans, with termination criteria established by the Secretary or Chairman. Refined during subsequent planning, mission success criteria provide the basis for operation assessment and a frame of reference for reporting mission progress to senior leadership. JP 5-0, *Joint Planning* discusses methods of assessment in further detail.

(11) <u>Course of Action Evaluation Criteria</u>. Evaluation criteria are standards the CDR and staff will later use to evaluate the relative effectiveness of developed COAs. Developed by the planning staff during mission analysis, they are intended to provide an objective benchmark for subsequent COA evaluation.

(12) <u>Develop Initial Risk Assessment</u>. A preliminary risk assessment during mission analysis can identify obstacles or actions that may preclude mission accomplishment and determine the potential impact of each on the operation. At the strategic level, risk assessment is an essential element of civilian-military planning dialogue in order to adequately consider both military and political risks and must be reconciled with the risk tolerance levels of senior leaders.

(13) <u>Determine Initial CCIRs</u>. CCIRs are elements of information (friendly, enemy, or environment), established by the CDR, that prompt time-sensitive decisions by the CDR or senior leaders during planning or execution. They are recommended by the planning staff and approved by the CDR, but may be refined during planning and execution. CCIRs focus the efforts of the staff and subordinate commands and inform subordinate and supporting commands.

(a) Every assumption in the plan should have a CCIR assigned to ensure the CDR and staff monitor each assumption and are prepared to adapt plans when assumptions are invalidated. Historically, planners were directed to take planning assumptions proved by higher level planning direction as fact. These assumptions should be questioned, understood and ultimately validated. Assumptions, at any level, need to be validated. If assumptions provided in the planning guidance are not realistic, they should be included in the dialog up the chain of command.

(b) CCIRs should be associated with events that require decisions from senior leaders. The CCIRs should be established such that the senior leaders have sufficient time to analyze the input, make the decision, and implement the decision.

(c) CCIRs should be integrated as part of the operation assessment process as they provide the CDR a method of tracking mission progress and determining success.

(14) <u>Staff Estimates</u>. Initial staff estimates include JIPOE assessments and estimates, and functional estimates of the forces and resources available for planning. At the conclusion of mission analysis, an updated staff estimate is a key output product that informs the CDR and subsequent planning by supported, supporting, and subordinate commands. Throughout the conduct of planning, staff estimates continue to be adapted to meet planning requirements.

(15) <u>Mission Analysis Briefing</u>. At the conclusion of the mission analysis, the planning team presents the CDR with their analysis results and any identified issues. The intent of the briefing is to confirm a common understanding of the OE, mission, objectives, tasks,

and risk. It offers a forum to discuss issues that have been identified and for the CDR to provide guidance for subsequent planning.

(16) <u>Planning Directive</u>. Once the CDR approves the results of mission analysis, the staff may draft and issue a WARNORD or PLANORD to subordinate commands to enable mission planning and preparation. The order should include the mission statement, CDR's intent, CDR's planning guidance, and any other information that will assist subordinate commands with their planning. In time-sensitive situations, a WARNORD or PLANORD might not be issued, and an ALERTORD or EXORD may be the initial planning directive received by the supported or supporting CDR.

e. <u>Strategic Guidance Outcomes</u>. The strategic guidance planning function should produce the following outcomes at each level of command.

(1) <u>National Level</u>. At the national level, this planning function produces a common civilian-military understanding of the OE, strategic direction, and strategic/operational approaches. Typically, through informal dialog or an IPR, the supported CCDR provides a proposed operational approach with military ends, ways, and means that support national strategic objectives. At the national level, this discussion provides the opportunity to integrate military efforts with a whole of government approach and identify the required actions from all instruments of national power. The discussion of the operational approach should include, initial force identification, initial risk assessment, limitations, assumptions and CCDRs, as well as any President or Secretary-level decisions identified through the planning process. National level leadership should endorse the CCDR planning to date or acknowledge friction points and provide refined guidance to shape continued planning. Under crisis conditions this discussion may be significantly abbreviated to an informal, potentially virtual dialog to more rapidly develop a plan ready for execution.

(2) <u>Supported Command.</u> For the supported CDR, this planning function produces a common understanding of the OE and direction for subsequent planning by the planning team and supporting/subordinate commands. The refinement of staff estimates provides situational awareness of all facets of the OE. Problem framing defines the problem and provides a decomposition of tasks necessary to achieve objectives for subsequent development and assignment to subordinate commands. The CDR's intent, mission analysis briefing, and planning directives (when applicable) provide an enduring frame of reference to guide concept development.

(3) <u>Supporting Commands, Services, and DoD Agencies</u>. For supporting commands and organizations, this planning function produces a shared framework for supporting commands, Services, and DoD Agencies to conduct their own planning aligned with the supported CDR. While not directive of supporting commands, the supported CCDR's operational approach and outputs from the national level planning dialog inform supporting plans. Supporting planning continues in parallel based upon planning direction from the President, Secretary, or Chairman.

(4) <u>Subordinate Commands</u>. For subordinate CDRs, this planning function produces direction from higher headquarters, shaping subordinate planning. The planning directives and tasks identified during problem framing provide an understanding of the activities assigned to subordinate commands. The subordinate CDRs and their planning staffs complete their own mission analysis after the supported CCDR. The subordinate commands mission analysis is thus informed by the direction and outputs of mission analysis from their higher headquarters.

> *"Our goals can only be reached through the vehicle of a plan. There is no other route to success."*
> *Pablo Picasso*

Joint Planning Process

4. Summary

As discussed in Chapter 1, *Strategic Organization*, but worth repeating, the common thread that integrates and synchronizes the activities of the JS, CCMDs, Services, and CSAs is strategic direction. Strategic direction encompasses the processes and products (documents) by which the President, SecDef, and CJCS provide strategic vision and direction to the Joint force. Strategic direction is normally published in key documents referred to as *strategic guidance*. As seen in the Figure C these strategic guidance documents are the principle source for DOD GCPs, theater strategies, CCPs, operation plans, contingency plans, base plans, and CDRs' estimates. CCDRs, once provided the direction and guidance each prepares strategy and campaign plans in the context of national security and foreign policy goals.

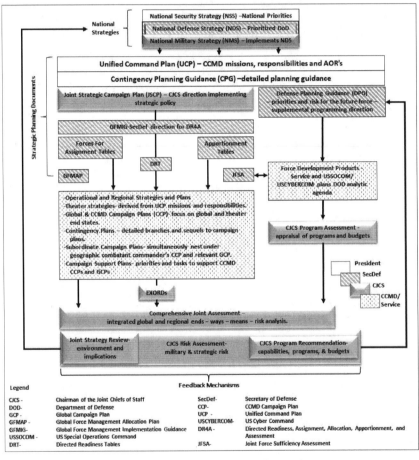

Figure C. Strategy, Planning and Resourcing

> *Strategic planning translates strategy into actionable content. Strategic thinking generates insight into the present and foresight regarding the future.*[4]

[4] T. Irene Sanders, *Strategic Thinking and the New Science: Planning in the Midst of Chaos, Complexity, and Change* (New York: Free Press, 1998), 10

I(b). Planning Initiation

1. Step 1 to JPP — Planning Initiation

a. <u>Linkage between Plan Initiation and National Strategic Endstate</u>. The first step in the Joint Planning Process (JPP) is Initiation. Prior to joint operations, planning begins when an appropriate authority recognizes a potential for military capability to be employed in response to a potential or actual crisis. The purpose of this step is to establish conditions for successful planning and on alerting the staff, forming the planning team, assessing available time for planning, Assessment Working Group (AWG) organization and deciding on a planning approach.

(1) At the strategic level planning is initiated when the President, SecDef or CJCS decides on developing military options and directs CCDRs through strategic guidance (Presidential Directives, NSS, UCP, CPG, JSCP, etc.) and related strategic guidance statements to begin planning. However, CCDRs and other CDRs may initiate planning on their own authority when they identify a planning requirement not directed by higher authority. The CJCS may also issue a WARNORD in an actual crisis. Military options normally are developed in combination with other nonmilitary options so that the President can respond with all the appropriate instruments of national power.[1]

(2) Military doctrine recognizes three levels of war: strategic, operational and tactical. These three levels overlap. Planning and execution at each level is reliant on planning and execution at other levels. Clearly delineated strategic endstates and objectives form the nucleus from which military plans at all levels evolve. Proper or improper identification of

[1] *JP 5-0, Joint Operations*

strategic endstates and objectives affect how military leaders plan and utilize military power to support attainment of those strategic measures. An incorrect interpretation of the strategic measure can lead to failure to accomplish the desired mission.

(3) Strategic endstates and objectives are approved by the President with input from their closest advisors, staff, and administration officials. These strategic measures form the foundation that subordinate agencies, departments, and military planners use to develop strategic objectives that will support the overarching desired national endstate. A clear understanding of desired political goals and endstate is imperative at the strategic level to ensure that all elements of national power are applied effectively. For the military, clear delineation of strategic endstate is essential to ensuring that military force can be effectively and efficiently applied when necessary to support strategic success.

b. The CCDR and staff will analyze the planning guidance to determine time available until mission execution, the current status of intelligence products, to include JIPOE, and staff estimates, and other factors relevant to the specific planning situation. The CCDR typically will provide their initial planning guidance, which could specify time constraints, outline initial coordination requirements, authorize movement of key capabilities within the CDR's authority, and direct other actions as necessary to provide the CDRs current understanding of the operational environment (OE), the problem, and operational approach for the campaign or operation.

c. CCDRs, subordinate CDRs, and supporting CDRs also initiate planning on their own authority when they identify a planning requirement not directed by higher authority. Additionally, analyses of the OE or developing or immediate crises may result in the President, SecDef, or CJCS directing military planning through a planning directive. CCDRs normally develop military options in combination with other non-military options so that the President can respond with all the appropriate instruments of national power. Whether or not planning begins as described here, the CDR may act within approved authorities and ROE and/or Rules for the Use of Force (RUF) in an immediate crisis.

d. For contingency planning purposes, the JSCP serves as the primary guidance to begin planning and COA development. During planning initiation, contingency planning tasks are transmitted, forces and resources are apportioned, and planning guidance is issued to the supported CCDR. CCDR's prepare contingency plans primarily in direct response to tasking in the JSCP.

(1) Strategic requirements or tasking for the planning of major contingencies may require the preparation of several alternative plans for the same requirement using different sets of forces and resources in order to preserve flexibility. For these reasons, contingency plans are based on reasonable assumptions (hypothetical situation with reasonable expectation of future action).

(2) Planning for campaign plans is different from contingency plans in that contingency planning focuses on the anticipation of future events, while campaign planning assesses the current state of the OE and identifies how the command can shape the OE to deter crisis on a daily basis and support strategic objectives.

2. Integrating Assessment

The starting point for operation assessment activities coincides with the *initiation of joint planning* and are an integral part of planning and execution of any operation. Integrating assessments into the planning cycle helps the CDR ensure the operational approach remains feasible and acceptable in the context of higher policy, guidance, and orders. This integrated approach optimizes the feedback senior leadership needs to appropriately refine, adapt, or terminate planning to be effective in the OE.[2,3]

[2] *JP 5-0, Joint Planning*

[3] *Commander's Handbook for Assessment Planning and Execution, Joint Staff, J-7*

a. Assessments are interrelated and interdependent and apply to all levels of warfare and during all military operations. Although each level of assessment may have a specific focus and a unique battle rhythm, together they form a hierarchical structure in which the conduct of one level of assessment is crucial to the success of the next. Theater strategic and operational-level assessment efforts concentrate on broader tasks, effects, objectives, and progress toward the endstate, while tactical-level assessment primarily focuses on task accomplishment. [4]

b. Assessment supports the CDR's decision-making and provide the CDR with the current state of the OE, the progress of the campaign or operation, and recommendations to account for discrepancies between the actual and predicted progress. CDRs then compare the assessment against their vision and intent and adjust operations to ensure objectives are met and the military end state is achieved. Assessment of the OE and the progress of operations are continuous.

c. As it relates to campaigns, where strategic objectives frame the CCMD's mission, assessments help CCDRs and supporting organizations refine or adapt the campaign plan and supporting plans to achieve the campaign objectives or, in coordination with SecDef and CJCS, to adapt the CPG- and/or JSCP-directed strategic objectives in response to changes in the OEs.[5]

d. Developing the *assessment plan* is a continuous process that is refined throughout all planning phases. The building of an assessment plan, including the development of collection requirements, normally begins during mission analysis after identification of the initial desired and undesired effects.

e. There is no single way to conduct an assessment. Every mission and OE has its own set of challenges, and every CDR assimilates information differently, making every assessment plan unique.

f. The following steps from ATP 5-0.3, *Multi-Service Tactics, Techniques and Procedures for Operation Assessment* are an excellent guide in the development of an effective assessment plan and assessment activities during planning, preparation and execution.[6]

Step 1 – Develop the assessment approach (planning).

Step 2 – Develop the assessment plan (planning).

Step 3 – Collect information and intelligence (preparation and execution).

Step 4 – Analyze information and intelligence (preparation and execution).

Step 5 – Communicate feedback and recommendations (preparation and execution).

Step 6 – Adapt plans or operations (planning and execution).

(See ATP 5-0.3 for a detailed discussion of each step of the assessment process)

g. Staff sections record relevant information in running estimates. Each staff section maintains a continuous assessment of current operations as a basis to determine if they are proceeding according to the CDR's intent. In their running estimates, staff sections use this new information, updated facts, and assumptions as the basis for evaluation.

h. Incorporating the assessment plan into the appropriate plans and/or orders is the recommended mechanism for providing guidance and direction to subordinate organizations or requests for key external stakeholder assistance and support. Desired and undesired effects are most effectively communicated in the main body of the base plan or order and

[4] JP 5-0, Joint Planning

[5] Commander's Handbook for Assessment Planning and Execution, Joint Staff, J-7

[6] ATP 5-0.3 Multi-Service Tactics, Techniques and Procedures for Operation Assessment

Joint Planning Process

may be repeated in the Operations Annex. The assessment plan may be included as an Appendix to the Operations Annex, or alternatively, in the Reports Annex.

See ADP 5-0, *The Operations Process*, JP 5-0 *Joint Operations* and ATP 5-0.3, *Multi-Service Tactics, Techniques and Procedures for Operation Assessment* for greater information on operations assessment.

> *"You must know the end, to know the beginning."*
> *Anonymous*

Joint Planning Process

			Operation Assessment Steps		
Step	**Operations Process Activity**	**Input**	**Personnel Involved**	**Staff Activity**	**Output**
Develop Assessment Approach	Planning	• JIPOE • Staff estimates • Operational approach development • JPP • Joint targeting • AWG	• Commander • Planners • Primary staff • Special staff • AWG personnel	• Clearly defined end states, objectives, and tasks	• Information, intelligence, and collection plans
Develop Assessment Plan	Planning	• Develop a framework • Select measures (MOE and MOP) • Identify indicators • Develop a feedback mechanism	• Operations planners • Intelligence planners • AWG personnel	• Operational approach • JIPOE • Desired end state • Feedback mechanism parameters	• Assessment plan
Collect Information and Intelligence	Execution	• Joint targeting • JIPOE • Staff estimates • IR management • ISR planning and optimization	• Intelligence analysts • Current operations • AWG personnel • Assessment cell (if established)	• Multisource intelligence reporting, and joint force resource and disposition information • Operational reports	• Estimates of OE conditions, enemy disposition, and friendly disposition
Analyze and Synthesize the Feedback	Execution	• Assessment work group • Staff estimates	• Primary staff • Special staff • AWG personnel • Assessment cell (if established)	• Intelligence assessments • Staff assessments • Analysis methods	• Estimate of joint force effects on OE (draft assessment report)
Communicate the Assessment and Recommendations	Execution	• Provide a timely recommendation to the appropriate decision-maker	• Commander • Subordinate commanders (periodically) • Primary staff • Special staff • AWG personnel • Assessment cell (if established)	• Estimate of joint force effects on OE (draft assessment report)	• Assessment report, decisions, and recommendations to higher Headquarters
Adapt Plans	Execution Planning	• Joint targeting • JPP	• Commander • Planners • Primary staff • Special staff • AWG personnel • Assessment cell (if established)	• Commander's guidance and feedback	• Changes to the operation and assessment plan

Legend:
AWG—assessment working group
IR—information requirement
ISR—intelligence, surveillance, and reconnaissance
JIPOE—joint intelligence preparation of the operational environment

OE—operational environment
JPP—joint planning process
MOE—measure of effectiveness
MOP—measure of performance

Table from ATP 5-0.3 Multi-Service Tactics, Techniques and Procedures for Operation Assessment

II. Mission Analysis Overview & Key Steps

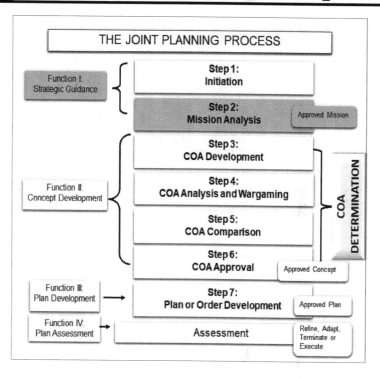

THE JOINT PLANNING PROCESS

Function I: Strategic Guidance

Step 1: Initiation

Step 2: Mission Analysis — Approved Mission

Function II: Concept Development

Step 3: COA Development

Step 4: COA Analysis and Wargaming

Step 5: COA Comparison

Step 6: COA Approval — Approved Concept

COA DETERMINATION

Function III: Plan Development → Step 7: Plan or Order Development — Approved Plan

Function IV: Plan Assessment → Assessment — Refine, Adapt, Terminate or Execute

I. Step 2 to JPP — Mission Analysis

1. Mission Analysis Overview

The mission analysis process helps to build a common understanding of the problem to be solved and boundaries within which to solve it by key stakeholders.

a. The CCDR is responsible for analyzing the mission and restating the mission for subordinate CDRs to begin their own estimate and planning efforts. Mission analysis is used to study the assigned mission and to identify all tasks necessary to accomplish it. Mission analysis is critical because it provides direction to the CCDR and the staff, enabling them to focus effectively on the problem at hand. ***There is perhaps no step more critical to the JPP and a successful plan.***

b. A primary consideration for a supported CCDR during mission analysis is the national strategic endstate and that set of national objectives and related guidance that define strategic success from the President's perspective. The endstate and national objectives will reflect the broadly expressed Political, Military, Economic, Social, Informational, Infrastructure (PMESII) and other circumstances that should exist after the conclusion of a campaign or operation. The CCDR also must consider multinational objectives associated with coalition or alliance operations.

c. The supported CCDR typically will specify a theater strategic endstate. While it will mirror many of the objectives of the national strategic endstate, the theater strategic end-state may contain other supporting objectives and conditions. This endstate normally will represent a point in time and/or circumstance beyond which the President does not require the military instrument of national power as the primary means to achieve remaining objectives of the national strategic endstate.

d. CCDRs include a discussion of the national strategic endstate and objectives in their initial planning guidance. This ensures that joint forces understand what the President wants the situation to look like at the conclusion of U.S. involvement. The CCDR and subordinate JFCs typically include the military endstate in their CDR's intent statement.

e. During mission analysis, it is essential that the tasks (specified and essential task(s)) and their purposes are clearly stated to ensure planning encompasses all requirements; limitations (restraints–can't do, or constraints–must do) on actions that the CCDR or subordinate forces may take are understood; and the correlation between the CDRs' mission and intent, and those of higher, and other CDRs is understood.

f. The joint force's mission is the task or set of tasks, together with the purpose, that clearly indicates the action to be taken and the reason for doing so. The CCDR and staff can accomplish mission analysis through a number of logical tasks. Of task and purpose, the purpose is preeminent. The CCDR can adjust the task to ensure the purpose is accomplished. This is a critical aspect of mission type orders and the ability of subordinate CDRs to re-task themselves during rapidly changing circumstances and still fulfill the CDR's intent.

g. While all of these tasks will be addressed during the plan development process, it is critical to focus on the mission essential task(s) to ensure unity of effort and maximum use of limited resources. The mission essential task(s) defines success of the assigned mission.

h. Although some Key-Steps occur before others, mission analysis typically involves substantial parallel processing (spiral development) of information by the CDR and staff, particularly in a crisis situation. A primary example is the JIPOE. JIPOE, as discussed in Chapter 4, is a continuous process that includes defining the OE, describing the effects of the operational environment, evaluating the adversary, and determining and describing adversary potential and most dangerous COA(s). This planning process must begin at the earliest stage of campaign or operations planning and must be an integral part of, not an addition to, the overall planning effort. This is also true for logistics, medical, transportation, force and deployment planning to name just a few.

2. Planner Organization

Organizing the planning team and setting goals and objectives within a specific timeline can sometimes be as time consuming as the plan itself. If you enter the planning process with the following information/guidelines defined and understood, the actual planning process will go much smoother:

• Define the seven basic questions: 1- *What needs to be accomplished*, 2- *when is the deadline*, 3- *where will this be done*, 4- *who will be responsible for it*, 5- *how will it get done*, 6- *how much time, energy, and resources are required to accomplish*, and 7- *what are the known knowns, known unknowns, and barriers*?

• Define clear organization planning responsibilities.

 - Who leads what efforts (topic, geographic, functions, makes coffee)

- Define the process for planning.
  - Planning organizations (staff sections, JPG members, contact info).
  - Product production/transition between organizations (watch the transition gaps; this is where information hides).
  - Supporting commands/organization (POC's, NIPR, SIPR, phone).
  - Subject Matter Experts (experience, knowledge, contact info).
- Information flow (with whom, how often, when, push or pull, reports).
  - Higher - keep boss informed, ask questions, don't assume the boss knows.
  - Lower - If the plan is not executable, it's useless. Ensure all effected unit planners participate from all levels, and have authority to make decisions.
  - Adjacent - Asked for all the support you need, J1, J2, J3, etc. You won't get it if you don't ask.
- Integrate other elements (POCs for each, roster, capabilities of each).
  - Coalition.
  - Interagency - may already be on the ground.
  - Host nation.
- Be sustainable in a 24/7/365 cycle.
  - Rapidly integrate augmentees.
- Focuses attention on objectives and results.
- Anticipate problems and cope with change - forecast future problems and make any necessary changes up front to avoid them.
- Guidelines for decision-making. Decisions are future oriented. If your planning team doesn't have any plans for the future, they will have few guidelines for making current decisions. Guide them. Guidelines help both CDRs and Staff keep their eyes on the endstate.
- Do not focus on the present at the expense of the future, plans forecast the future
- Do not get in a loop-de-loop just concentrating on controllable variables - Be careful not to have an aversion to the unknown.
- Do not wish away problems just because they are hard, remember the enemy also gets a vote.
- If it's improbable, it's probably doable.

Auftragstaktik (Mission Type Order): Order issued to a lower unit that includes the accomplishment of the total mission assigned to the higher headquarters, or one that assigns a broad mission (as opposed to a detailed task), without specifying how it is to be accomplished.

13* Key Steps to Mission Analysis

Key Step 1 *(See p. 5-29)*
Analyze Higher CDR's Mission and Intent

Key Step 2 *(See p. 5-30)*
Task Analysis, Determine Own Specified, Implied, and Essential Tasks

Key Step 3 *(See p. 5-32)*
Determine Known Facts, Assumptions, Current Status, or Conditions

Key Step 4 *(See p. 5-34)*
Determine Operational Limitations
- Constraints – "Must do"
- Restraints – "Can't do"

Key Step 5 *(See p. 5-35)*
Determine Conflict Termination Criteria, Own Military Endstate, Objectives and Initial Effects

Key Step 6 *(See p. 5-55)*
Determine Own and Enemy's Center(s) of Gravity, Critical Factors and Decisive Points

Key Step 7 *(See p. 5-71)*
Conduct Initial Force Structure Analysis (Apportioned Forces)

Key Step 8 *(See p. 5-73)*
Conduct Initial Risk Assessment

Key Step 9 *(See p. 5-79)*
Determine CDR's CCIR
- CFFI
- PIR

Key Step 10 *(See p. 5-86)*
Develop Tentative Mission Statement

Key Step 11 *(See p. 5-88)*
Develop and Conduct Mission Analysis Brief

Key Step 12 *(See p. 5-88)*
Prepare Initial Staff Estimates

Key Step 13 *(See p. 5-89)*
Publish CDR's Planning Guidance and Initial Intent

NOTE: JP 5-0, Joint Planning, has listed numerous mission analysis activities. JIA1-3 recognizes and addresses all activities of joint doctrine and condenses them here into the following 13 activities, or "key steps." This is done to allow a logical flow for planners to follow.

II. STEP 2 to JPP - MISSION ANALYSIS: KEY STEPS

> Just as planning is only part of the operations process, planning is only part of problem solving. In addition to planning, problem solving includes implementing the planned solution (execution), learning from the implementation of the solution (assessment), and modifying or developing a new solution as required. The object of problem solving is not just to solve near-term problems, but to do so in a way that forms the basis for long-term success. (ADRP 5-0)

1. Key Step 1: Analyze Higher Commander's Mission and Intent — "Receipt of Mission"

The decision-making process begins with the receipt or anticipation of a new mission. This can either come from an order issued by higher headquarters, or derived from an ongoing operation.

a. As soon as a new mission is received, a WARNORD is issued to the staff alerting them of the pending planning process. The staff prepares for the mission analysis immediately on receipt of a WARNORD by gathering the tools needed to do so. Once the new mission is received, the CDR and the staff must do a quick initial assessment. It is designed to optimize the CDR's use of time while preserving time for subordinate commanders to plan and complete combat preparations. This assessment:

- Determines the time available from mission receipt to mission execution.
- The staff assesses the scope of the assigned mission, intent, endstate, objectives, and other guidance from the next higher command, JSCP, etc. (purpose, method, endstate).
- Determines the time needed to plan, prepare for, and execute the mission for own and subordinate units.
- Determines the JIPOE–what we think we kown, what we need to know, what we know we know.
- Determines the staff estimates already available to assist planning.
- Determine whether the mission can be accomplished in a single operation, or will likely require a campaign due to its complexity and likely duration and intensity.

b. At the CCDR level, this would be strategic guidance issued by the President, SecDef or CJCS. The CDR must draw broad conclusions as to the character of the forthcoming military action. The CDR should not make assumptions about issues not addressed by the higher CDR and if the higher headquarters' directive is unclear, ambiguous, or confusing, the CDR should seek clarification.

c. A main concern for a CDR during mission analysis is to study not only the mission, but also the intent of the higher CDR. Within the breadth and depth of today's OE, effective decentralized control cannot occur without a shared vision. Without a CDR's intent that expresses that common vision, unity of effort is difficult to achieve. In order to turn information into decisions, and decisions into actions that are "about right," commanders must understand the higher CDR's "intent." While the CDR's intent has previously been considered inherent in the mission and concept of operations, most often you will see it explicitly detailed in the plan/order. Successfully communicating the more enduring intent allows the force to continue the mission even though circumstances have changed and the previously developed plan/concept of operations is no longer valid.

d. The higher CDR's intent is normally (five-paragraph format) found in Paragraph 3, (Execution), of the higher CDR's guidance. The intent statement of the higher echelon CDR should then be repeated in Paragraph 1, (Situation), of your own OPLAN or OPORD to ensure that the staff and supporting commanders understand it. Each subordinate CDR's intent must be framed and embedded within the context of the higher CDR's intent, and they must be nested both horizontally and vertically to achieve a common military endstate.

e. Staff officers must update their staff estimates and other critical information constantly. This information allows them to develop assumptions that are necessary to the planning process. Staff officers must be aggressive in obtaining this information. Reporting of this information must be a *push system* versus a pull system. Subordinate units must rapidly update their reports as the situation changes.

f. The critical product of this assessment is an initial allocation of available time, especially in a crisis. The CDR and the staff must balance the desire for detailed planning against the need for immediate action. The CDR must provide guidance to subordinate units as early as possible to allow subordinates the maximum time for their own planning and preparation for operations. This, in turn, requires aggressive coordination, deconfliction, integration, and assessment of plans at all levels, both vertically and horizontally.

g. As a general rule, the CDR allocates a minimum of two-thirds of available time for subordinate units to conduct their planning and preparation. This leaves one-third of the time for the CDR and his staff to do their *planning*. They use the other two-thirds for their own preparation.

h. Time, more than any other factor, determines the detail with which the staff can plan. Once time allocation is made, the CDR must determine how much time will be dedicated to each step within the JPP, or to abbreviate the process.

2. Key Step 2: Determine Own Specified, Implied, and Essential Tasks

> **Operation Iraqi Freedom (OIF) Lessons Learned at the CCMD Level:**
> • Planning organization is critical
> • Planning is evolutionary; embrace change as a constant
> • Must explain military operations in simple terms
> • Risk is different at each level

a. *Any mission consists of two elements: the task(s) to be accomplished by one's forces and their purpose.* If a mission has multiple tasks, then the priority of each task should be clearly expressed. Usually this is done by the sequence in which the tasks are presented. There might be a situation in which a CDR has been given such broad guidance that all or part of the mission would need to be deduced. Deduction should be based on an appreciation of the general situation and an understanding of the superior's objective. Consequently, deduced tasks must have a reasonable chance of accomplishment and should secure results that support the superior CDR's objective.

> **Tasks:**
> - Determine what we've been told to do (specified tasks).
> - Determine what else we need to do to accomplish our mission (implied tasks).
> - Deterimine which tasks are essential to accomplishment of the mission.
>
> **Task Analysis:**
> - State the task(s). The task is the job or function assigned to a subordinate unit or command by a higher authority. A mission can contain a single task, but it often contains two or more tasks. If there are multiple tasks, they normally will all be related to a single purpose.

b. Determine specified, implied, and essential tasks, by reviewing strategic communication guidance and other documents used during Function I, Strategic Guidance and Initiation (see Chapter 5-I), in order to develop a concise mission statement. Specified and implied strategic tasks are derived from specific Presidential, SecDef guidance, national (or multinational) planning guidance documents such as the CPG, JSCP, the UCP, or from CCDR initiatives. The national military objectives form the basis of the campaign's mission statement.

(1) Specified Task — A task that is **specifically assigned** to an organization by its higher headquarters. Tasks *listed* in the mission received from higher headquarters are specified or stated (assigned) tasks. They are what the higher CDR wants accomplished. The CDR's specified tasks may be found in Paragraph 3b, (Execution-Tasks) section of the order, but could also be contained elsewhere, for example in the JSCP or other national guidance.

> **Specified Tasks (examples):**
> - Assure allies, coalition partners and friends in the African Region are committed to uphold treaty obligations (USAFRICOM Security Cooperation Strategy).
> - Deter Mass Migration (USSOUTHCOM TCP, Haiti).
> - Prepare CONPLAN w/TPFDD for the defense of Country Blue in case of external aggression and in support of U.S. interests (JSCP).

(2) Implied Task — After identifying the specified tasks, the CDR identifies **additional major tasks necessary to accomplish the assigned mission**. Though not facts, these additional major tasks are implied tasks, which are sometimes *deduced* from detailed analysis of the order of the higher CDR, known enemy situation, and the CDR's knowledge of the physical environment. *Therefore, the implied tasks subsequently included in the CDR's restated mission should be limited to those considered critical to the accomplishment of the assigned mission.* Implied tasks do not include routine or standing operating procedures (SOPs) that are performed to accomplish any type of mission by friendly forces. Hence, tasks that are inherent responsibilities of the CDR (providing protection of the flank of his own unit, reconnaissance, deception, etc.) are not considered implied tasks. The exceptions are only those routine tasks that cannot successfully be carried out without support or coordination of other friendly commanders. An example of an implied task is: if the JTF CDR was given a specified task to seize a seaport facility, the implied task might be the requirement to establish maritime superiority within the AO before the assault.

(3) Essential Tasks — Essential tasks are determined from the list of both specified and implied tasks. **They are those tasks that must be executed to achieve the conditions that define mission success.** Depending on the scope of the mission's purpose, some of

Joint Planning Process

the specified and implied tasks might need to be synthesized and re-written as an essential task. *Only essential tasks should be included in the mission statement.*

Implied Tasks:

• Additional major tasks necessary to accomplish the assigned mission.

• Not routine or SOPs, except if the routine task cannot successfully be carried out without support or coordination with another CDR.

• Example:

- Specified Task: Seize the Port of Red.

- Implied Task: Establish maritime superiority within the AO prior to D-day.

Implied Tasks (example):

• Develop/Expand the FID program.

• Secure lines of communication in region to ensure unfettered flow of forces and equipment.

• Build capacity to conduct stability operations, HA/DR, CN.

Essential Tasks (examples):

• Defend Country Blue.

• Secure international support for the conduct of military operations.

• Maintain/Restore regional stability.

• Build usable, relevant and enduring capabilities to improve GOH's ability to provide for its own security needs (Operation Unified Response).

---

3. <u>Key Step 3</u>: Determine Known Facts, Assumptions, Current Status, or Conditions

The staff assembles both facts and assumptions to support the planning process and initial planning guidance. What does the organization know about the current situation and status?

a. <u>Fact</u>.

A *fact* is ***a statement of information known to be true*** (such as verified locations of friendly and adversary force dispositions).[1]

Facts (examples):

• Government of Haiti (GoH) does not have adequate emergency management capabilities.

• GoH has little or no control over its borders.

• The U.S. has an embassy in Red.

• Red has a mutual defense pact with Country Orange.

• Blue airfields would require extensive improvements to support modern operations.

---

b. <u>Assumptions</u>.

An assumption is ***used in the absence of facts*** that the CDR *needs* to continue planning. *It is a supposition on the current situation or a presupposition on the future course of events*, either or both assumed to be true in the absence of positive proof, necessary to enable the CDR in the process of planning to complete an estimate of the situation and

[1] *JP 5-0, Joint Planning*

make a decision on the COA. An assumption encompasses the issues over which a CDR normally does not have control, both friendly and adversary.

- If you make an assumption, you must direct resources towards turning it into a fact (intelligence collection, RFI etc.) and/or develop a branch plan.

- Assumptions that address gaps in knowledge are critical for the planning process to continue.

- Subordinate commanders must treat assumptions given by the higher headquarters as *facts*. If the CDR or staff does not concur with the higher CDR's planning assumptions, they should be challenged before continuing with the planning process. All assumptions should be continually reviewed to ensure validity.

- When dealing with an assumption, changes to the plan may need to be developed should the assumption prove to be incorrect.

- Because of their influence on planning, the fewest possible assumptions are included in a plan.

- A *valid assumption* has three characteristics: it is *logical, realistic, and essential* for the planning to continue. Assumptions should be continually re-validated.

- Assumptions are made for both friendly and adversary situations. The planner should assume that the adversary would use every capability at his disposal (i.e., nuclear, biological, and chemical (NBC), asymmetric approach, etc.) and operate in the most efficient manner possible.

- Planners should never assume an adversary has less capability than anticipated, nor assume that key friendly forces have more capability than has been demonstrated.

Assumptions (examples):

- LOCs outside the theater will remain open.

- Annual pre-planned U.S. force movements (annual and temporarily deployed) continue at historic levels.

- The USG will make significant, long-term resource commitment to reconstruction in Haiti.

- NATO countries will provide basing and over-flight.

- Red supported terrorists will conduct operations in Blue in attempt to destabilize the government.

Joint Planning Process

(1) Assumptions are used in the planning process at each command echelon. Usually, commanders and their staffs should make assumptions that fall within the scope of their OE. We often see that the higher the command echelon, the more assumptions will be made. Assumptions enable the CDR and the staff to continue planning despite a lack of concrete information. They are artificial devices to fill gaps in actual knowledge, but they play a crucial role in planning. A poor assumption may partially or completely invalidate the entire plan—to account for a possible wrong assumption, planners should consider developing branches to the basic plan. Assumptions should be kept at a minimum.

"Extended DOD participation in Haitian recovery will impact Global Force Posture/Capability."
(Operation Unified Response)

---

(2) ***Assumptions are not rigid.*** Their validation will influence intelligence collection. They must be continuously checked, revalidated, and adjusted until they are proven as facts or are overcome by events. Ask yourself three simple questions while considering an assumption:

- Is it *logical*?
- Is it *realistic*?
- Is it *essential for planning*?

(3) Probably one of the most important considerations of an assumption is that if you cannot validate the assumption during the planning process by either rejecting it or turning it into a fact, then you *MUST* consider a branch plan to cover the possible repercussions of an invalid assumption during execution.

4. <u>Key Step 4</u>: Determine Operational Limitations (Limiting Factors): Constraints/Restraints

Operational limitations are actions required or prohibited by higher authority and other restrictions that limit the CDR's freedom of action, such as diplomatic agreements, political and economic conditions in affected countries, host nation issues and support agreements.

a. *A constraint is a requirement placed on the command by a higher command that dictates an action, thus restricting freedom of action (must do).* he superior's directive normally indicates circumstances and limitations under which one's own forces will initiate and/or continue their actions. Therefore, the higher CDR may impose some constraints on the CDR's freedom of action with respect to the actions to be conducted. These constraints will affect the selection of COAs and the planning process. Examples include tasks by the higher command that specify: "Be prepared to . . . "; "Not earlier than . . . "; "Not later than . . . "; "Use coalition forces . . . " Time is often a constraint, because it affects the time available for planning or execution of certain tasks.

b. *A restraint is a requirement placed on the command by a higher command that prohibits an action, thus restricting freedom of action (can't do).*

Constraint example: General Eisenhower was required to liberate Paris instead of bypassing it during the 1944 campaign in France.

Constraint example: Simultaneous Humanitarian Assistance Operation beginning the same day as the air campaign on the kick-off of Operation Enduring Freedom.

Restraint example: General MacArthur was prohibited from striking Chinese targets north of the Yalu River during the Korean War.

Restraint example: President Musharraf's request that the American plan for Afghanistan not involve the Indian government nor their military.

---

c. Some operational limitations are commonly expressed as *Rules of Engagement* (ROE). Operational limitations may restrict or bind COA selection or may even impede implementation of the chosen COA. These ROE or operational limitations become more complex in multinational or coalition operations. CDRs must examine the operational limitations imposed on them, understand their impacts, and develop options that minimize these impacts in order to promote maximum freedom of action during execution.[2]

Note: Constraints and restraints collectively comprise "operational limitations" on the CDR's freedom of action. Remember restraints and constraints do not include doctrinal considerations. Do not include self-imposed limitations during this portion of the process.

[2] *JP 5-0, Joint Planning*

5. <u>Key Step Five</u>: Determine Conflict Termination Criteria, Own Military Endstates, Objectives and Initial Effects

> *"One should not take the first step in war without considering the last."*
>
> Carl von Clausewitz, On War, edited and translated by Michael Howard

a. <u>Termination of Joint Operations</u>.

(1) Because the very nature of conflict termination begins shaping the futures of contesting nations or groups, it is imperative that we fundamentally understand what conflict termination criteria are. ***What military conditions must be produced in the theater to achieve the strategic goal and how those military conditions serve to leverage the transition from war to peace*** is a fundamental aspect of conflict termination. It is important to recognize that these conditions defined during mission analysis may change as the operation planning, and then the operation unfolds. Nonetheless, the process of explicitly and clearly defining terminal conditions is an important one since it requires careful dialogue between civilian (strategic) and military (operational) leadership which may, in turn, offer some greater assurance that the defined endstate is both politically acceptable and militarily attainable.

Conflict Termination Criteria:

• In many cases, military power can be applied to establish a specified condition that serves as a transition point for the application of other instruments to augment or replace military efforts.

• Other means of power will provide the driving force for achievement of the objectives.

• The planner must determine what that point is and define the conditions that exist at that point in time and space.

   • This list of conditions comprises the termination criteria.

   • Those conditions are event-driven rather than time-driven to define the point at which the President no longer requires the military instrument of power.

• We re-write this list into paragraph format that contains the CDR's vision.

• We call this the military endstate.

---

(2) Based on the President's strategic objectives that comprise a desired national strategic endstate, the supported JFC can develop and propose *conflict termination criteria* which are the specified standards approved by the President or the SecDef that must be met before a joint operation can be concluded. Military operations seek to end war on favorable terms. Knowing when to end a war and how to preserve the objectives achieved are vital components of campaign design and relate to theater-strategic planning. Before hostilities begin, the theater CDR must have a clear sense of what the endstate is. The CDR also needs to know whether (and how) ending the conflict at the point recommended will contribute to the overall strategic goals.[3] Conflict termination can be considered the link or leverage between endstate and post hostilities.[4] The design and implementation of leverage and the ability to know how and when to terminate operations are part of the

---

[3] *Joint Publication 1, Doctrine for the Armed Forces of the United States, JP 3-0, Joint Operations*

[4] *Should Deterrence Fail: War Termination in Campaign Planning, James W. Reed, Parameters, Summer 1993*

overall implementation of operational design. Not only should termination of operations be considered from the outset of planning, but it should also be a coordinated effort with appropriate other government agencies (OGA), international governmental organizations (IGO), non-governmental organizations (NGO), and multi-national partners.

> **Termination Criteria (example):**
> • Blue borders are secure.
> • A stable security environment exists in Blue, Red, and Orange.
> • Red no longer poses a threat to regional countries.
> • Non-DOD agencies and/or international agencies effectively lead and conduct reconstruction and humanitarian assistance operations.
> • U.S. military forces return to shaping and security cooperation activities.

(3) If the termination criteria have been properly set and met, the necessary leverage should exist to prevent the enemy from renewing hostilities and to dissuade other adversaries from interfering. Moreover, the national strategic endstate for which the United States fought should be secured by the leverage that U.S. and multinational forces have gained and can maintain.

(4) As discussed, in order to have acceptable termination criteria, it is of utmost importance to have an achievable national strategic endstate based on clear national strategic objectives.[5] One of the first considerations in Operational Design that a CDR must address is:

(5) This consideration requires clarity first on strategic objectives (President, SecDef) before operational military objectives (CCMD, JFC) are defined. This consideration also requires a measure of clarity on those "conditions" which are required to achieve the strategic objectives. Those conditions are the conflict termination criteria which is the bridge over which armed conflict transitions into a post-conflict phase and between the political objectives and the desired strategic endstate.[6]

> "What conditions are required to achieve the objectives?" (Ends)

(6) The CDR will be in a position to provide the President, SecDef and CJCS with critical information on enemy intent, objectives, strategy, and chances of success. CDRs consider the nature and type of conflict, the objectives of military force, the plans and operations that will most affect the enemy's judgment of cost and risk, and the impact of military operations on alliance and coalition warfare.[7] Military strategic advice to political authorities regarding termination criteria should be reviewed for military feasibility, adequacy, and acceptability as well as estimates of time, costs, and the military forces required to reach the criteria.[8]

(a) The CDR will then formulate conflict termination criteria that will set the enduring conditions to achieving the military and strategic endstates. Those properly conceived termination criteria are key to ensuring that the achieved military objectives endure (Figure A).

(7) To facilitate development of effective conflict termination criteria, it must be understood that U.S. forces must follow through in not only the "dominate" phase, but also the

[5] *Joint Publication 1, Doctrine for the Armed Forces of the United States*

[6] *Joint Operational Warfare, Dr. Milan Vego, 20 September 2007, IX-180*

[7] *FM 100-5, Operations, FM 3-0, FM 3-0, Operations*

[8] *Joint Publication 1, Doctrine for the Armed Forces of the United States*

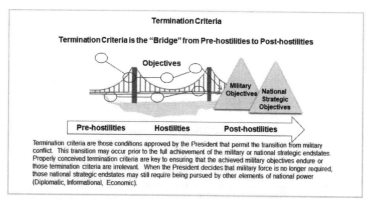

Figure A. Termination Criteria Pre-hostilities-Hostilities-Post-Hostilities.

"stabilize" and "enable civil authority" phases to achieve the leverage sufficient to impose a lasting solution.[9]

(8) Further, development of a military endstate is complementary to and supports attaining the specified conflict termination criteria and national strategic endstate. As stated, the termination of operations must be considered from the outset of planning in a coordinated effort with relevant agencies, organizations, and multinational partners. This ability to understand how and when to terminate operations is instrumental to operational design.[10]

(9) There are three approaches to obtaining national strategic objectives by military force:[11]

(a) Imposed settlement: destroy critical functions and assets and/or adversary military capabilities.

(b) Negotiated settlement: political, diplomatic and military. Negotiating power springs from two sources: military success and military potential. In American history, more than half of the wars have ended in negotiated settlements. However, some wars may not have a formal ending, because the defeated side is unwilling to negotiate or because there is no one with whom to negotiate. Such conflicts are often transformed into insurgency and counterinsurgency, for example the U.S. experience in Iraq and the French experience in Indochina.

(c) Indirect Approach: Erode adversary's power, influence, and will; undermine the credibility and legitimacy of their political authority; and undermine their influence and control over, and support by, the indigenous population. This is in relation to the irregular challenges posed by non-state actors and is irregular warfare.[12]

(10) In its strategic context, military victory is measured in the attainment of the national strategic endstate and associated conflict termination criteria. Conflict termination criteria for a negotiated settlement will differ significantly than those of an imposed settlement.

> "The success of Desert Storm militarily was astounding, but Saddam Hussein reminded the world for the next twelve years that the political victory was incomplete. In April 2003, Operation Iraqi Freedom (OIF) completed the task of ousting Saddam Hussein, but a political victory in Iraq remains unsecured today."
>
> The Theater Commander: Planning for Conflict Termination; Colonel John W. Guthrie, U.S. Army War College, Carlisle Barracks, Carlisle, PA.

[9] JP 5-0, Joint Planning, JP 3-0, Joint Operations

[10] Ibid

[11] Joint Operational Warfare, Dr. Milan Vego, 20 September 2007, p. IX-178

[12] Joint Publication 1, Doctrine for the Armed Forces of the United States

(11) At the operational level, then, the military contribution should serve to increase (or at least not decrease) the leverage available to national decision-makers during the terminal phases of a conflict.

(12) One could view conflict termination not as the end of hostilities, but as the transition to a new post-conflict phase characterized by both civil and military problems. This consideration implies an especially important role for various civil affairs functions. It also implies a requirement to plan the interagency transfer of certain responsibilities to national, international, or nongovernmental agencies. Effective battle hand-off requires planning and coordination before the fact.[13]

> *"Those who can win a war well can rarely make a good peace, and those who could make a good peace would never have won the war."*
> My Early Life: A Roving Commission, Winston Churchill

(13) Conflict Termination Summary. Knowing when to terminate military operations and how to preserve achieved advantages is essential to achieving the national strategic endstate. When and under what circumstances to suspend or *terminate* military operations is a *political* decision. Even so, it is essential that the CJCS and the supported JFC advise the President and SecDef during the decision-making process. The supported JFC should ensure that political leaders understand the implications, both immediate and long term, of a suspension of hostilities at any point in the conflict. Once established, the national strategic objectives enable the supported CDR to develop the military endstate, recommended conflict termination criteria, and supporting military objectives.

(a) Planners must plan for conflict termination from the outset of the planning process and update these plans as the campaign evolves. To maintain the proper perspective, they must know what constitutes an acceptable political-military endstate (i.e., what military conditions must exist to justify a cessation of combat operations?). In examining the proposed national strategic endstate, the CCDR and the staff must consider whether it has reasonable assurance of ending the fundamental problem or underlying conditions that instigated the conflict in the first place.

(b) When addressing conflict termination, campaign planners must consider a wide variety of operational issues to include disengagement, force protection, transition to post-conflict operations, and reconstitution and redeployment. Planners must also anticipate the nature of post-conflict operations, where the focus will likely shift to Stability, Security, Transition and Reconstruction (SSTR); for example, peace operations, foreign humanitarian assistance, or enforcement of exclusion zones.

(c) In formulating the theater campaign plan, the CCDR and staff should consider the following:

1 Conflict termination. End of the joint operation, is a key aspect of the campaign planning process.

2 Emphasize backward planning. Decision-makers should not take the first step toward hostilities or war without considering the last step.

3 Define the conditions of the termination phase. The military objectives must support the political aims — the campaign's conflict termination process is a part of a larger *implicit bargaining process*, even while hostilities continue. The military contribution can significantly affect the political leverage available to influence that process.

4 Consider how efforts to eliminate or degrade an opponent's command and control. (C2) may affect, positively or negatively, efforts to achieve the termination objectives. Will opponents be able to affect a cease-fire or otherwise control the actions of their forces?

5 Interagency coordination. Plays a major role in the termination phase. View conflict termination not just as the end of a joint operation and disengagement by joint forces,

---

[13] *Should Deterrence Fail: War Termination in Campaign Planning, James W. Reed, Parameters, Summer 1993*

but as the transition to a new post-hostilities phase characterized by both civil and military problems.[14]

(d) **Since war is fought for political aims, it is only successful when such aims are ultimately achieved.** Success on the battlefield does not always lead to a successful strategic endstate. Making sure that it does requires the close collaboration of political and military leaders. As noted, a period of post-hostilities activities may exist in the period from the immediate end of the hostilities to the accomplishment of the national strategic goals and objectives, and it is imperative that this be recognized and planned for.

> *"The object of war is a better state of peace — even if only from your own point of view. Hence it is essential to conduct war with constant regard to the peace you devise."*
>
> B.H. Liddell Hart, Strategy (2nd revised edition) New York: Penguin Group, 1991

b. Endstate. Once the termination criteria are understood, operational design continues with the determination of the *strategic and military endstates and objectives.* **The endstate gets to "why" we are developing a campaign plan and seeks to answer the question: "How does the United States strategic leadership want the OE (i.e., the region and/ or potential adversary) to behave at the conclusion of the campaign?"** Objectives normally answer the question of "what" needs to be done to achieve the endstate. As you might expect, the distinction between endstates and objectives can be very vague. In designing and planning, we must recognize and define two mutually supporting endstates in a single campaign – a national strategic endstate and a theater-strategic/military endstate.

> *"The political object is the goal, war is the means of reaching it, and means can never be considered in isolation from their purpose."*
>
> Carl Von Clausewitz, On War: edited and translated by Michael Howard and Peter Paret, 1976, p.87

(1) The *national strategic endstate* describes the President's political, informational, economic, and military vision for the region or theater when operations conclude. National strategic endstates are derived from President/SecDef guidance that is often vague. More often than not, senior military leaders will assist the President/SecDef in developing that endstate. Below is an example of a national strategic endstate:

> Secure, stable, democratic, inclusive, non-hostile and prosperous Haiti.
> (USSOUTHCOM TCP)

> An economically viable and stable Country X, without the capability to coerce its neighbors.

(2) The *theater strategic or military endstate* is a subset of the national strategic endstate discussed above and generally describes the military conditions that must be met to satisfy the objectives of the national strategic endstate. We develop our military endstate after analyzing the tasks required by strategic direction. Strategic objectives clarify and expand upon endstate by *clearly defining* the decisive goals that must be achieved in order to ensure we achieve U.S. policy, for example:

> • Loss of life and human suffering of Haitians minimized.
> • Scope of crisis no longer exceeds capability of HN, UN and international efforts.
> • U.S. military personnel are redeployed.

[14] *JP 5-0, Joint Planning*

> - Country X deterred from coercing its neighbors.
> - X ceases support to regional terrorism.
> - X's WMD inventory, production and delivery means reduced.

Getting the answers to the above decisive goals is what makes mission analysis different at this level when compared to the tactical level – you will not find the clear and definitive guidance in one location that you may be used to. There is no "higher order" to cut and paste from. Instead the NSS, NDS, NMS, PPD, SecDef and Presidential speeches, and verbal guidance all provide input to help define an endstate and corresponding objectives. With so many sources of guidance, consistency is normally an issue to overcome. Though not directive in nature, guidance contained in various United States interagency and even international directives, such as United Nations Security Council Resolutions (UNSCRs), will also impact campaign endstates and objectives.

(3) This endstate normally will represent a point in time or circumstance beyond which the President does not require the military instrument of national power to achieve remaining objectives of the national strategic endstate. While the military endstate typically will mirror many of the conditions of the national strategic endstate, it may contain other contributory or supporting conditions. Aside from its obvious role in accomplishing both the national and military strategic objectives, clearly defining the military endstate conditions promotes unified action, facilitates synchronization, and helps clarify (and may reduce) the risk associated with the joint campaign or operation. CDRs should include the military endstate in their planning guidance and CDR's intent statement.[15]

(4) The CJCS or the supported CCDR may recommend a military endstate, but the President or SecDef should formally approve it. A clearly defined military endstate complements and *supports attaining the specified termination criteria and objectives associated with other instruments of national power.* The military endstate helps affected CCDRs modify their theater strategic estimates and begin mission analysis even without a pre-existing OPLAN. The CCDR must work closely with the civilian leadership to ensure a clearly defined military endstate is established and that the military objectives support the political ones, and will lead to the desired endstate. This backward planning cycle ensures a military victory with lasting political success.

(5) The CCDR also should anticipate that military capability likely would be required in some capacity in support of other instruments of national power, potentially before, during, and after any required large-scale combat. CDRs and their staffs must understand that many factors can affect national strategic objectives, possibly causing the endstate to change even as military operations unfold. A clearly defined endstate is just as necessary for situations across the range of military operations that might not require large-scale combat. While there may not be an armed adversary to confront in some situations, the JFC still must think in terms of ends, ways, and means that will lead to success and endstate attainment.

> *"War termination is the bridge between the political objectives and the desired strategic end state. Hence, defining the desired strategic end state allows the planners to develop the most effective path to its accomplishment. Yet despite its undoubtedly critical role in planning, the political leadership pays little, if any, attention to it. All too often, the operational commander does not receive a desired strategic end state at all. For example, prior to the invasion of Panama (Operation Just Cause) in 1989, the U.S. national leadership never issued a clearly defined desired strategic endstate, although it provided the political objectives to the operational commander."*
>
> Mario A Garza, Conflict Termination: Every Conflict Must End (Newport, RI: Naval War College, 13 June 1997) p.13.

[15] JP 5-0, Joint Planning

(6) Remember that a primary consideration for a supported CCDR during mission analysis is the *national strategic endstate* — that set of national objectives and related guidance that define strategic success from the President's perspective. This national strategic endstate will reflect the "broadly" expressed political, military, economic, social, informational, and other circumstances that should exist after the conclusion of a campaign or operation. Below is an example of a broad national strategic endstate:

> *"Violent extremist ideology and terrorist attacks eliminated as a threat to the way of life of free and open societies. A global environment that is inhospitable to violent extremism, wherein countries have the capacity to govern their own territories, including both the physical and virtual domains of their jurisdictions. Partner countries have in place laws, information sharing, and other arrangements, that allow them to defeat terrorists as they emerge, at the local and regional levels."*
>
> Chairman's National Military Strategic Plan for the War on Terrorism

(7) Often, the military endstate is achieved before the national strategic endstate. While it will mirror many of the objectives of the national strategic endstate, the theater strategic or military endstate may contain other supporting objectives and conditions. An example of a theater strategic or military endstate:

> *"A defeated Country X where WMD delivery, production, and storage, as well as conventional force projection capabilities, are destroyed, and its remaining military is reorganized to adequately defend its borders."*

(8) CDRs include a discussion of the national strategic endstate in their initial planning guidance. This ensures that joint forces understand what the President wants the situation to look like at the conclusion of U.S. involvement. The CCDR and subordinate JFCs typically include the *military endstate* in their intent statement.

> *"My intent is to persuade country X through a show of coalition force to stop intimidating its neighbors and cooperate with diplomatic efforts to abandon its WMD programs. If X continues its belligerence and expansion of WMD programs, we will use force to reduce X's ability to threaten its neighbors, and restore the regional military balance of power. Before U.S. and coalition forces redeploy, X's military will be reduced by half, its modern equipment destroyed, its capability to project force across its borders eliminated, and its WMD stores, production capacity, and delivery systems eliminated."*

c. Objectives. **An objective is a clearly defined, decisive, and attainable goal toward which every military operation is directed.** Objectives and their supporting effects provide the basis for identifying tasks to be accomplished.

> *"In whatever position you find yourself, determine first your objective."*
> Ferdinand Foch

(1) Objectives Prescribe Friendly Goals. They constitute the aim of military operations and are necessarily linked to political objectives (simply defined as: what we want to accomplish). Military objectives are one of the most important considerations in campaign and operational design. They specify what must be accomplished and provide the basis for describing campaign effects.

(2) A clear and concise endstate allows planners to examine objectives that support the endstate. Objectives describe what must be achieved to reach the endstate. These are usually expressed in military, political, economic, and informational terms and help define

and clarify what military planners must do to support the achievement of the national strategic endstate. Objectives developed at the national-strategic and theater-strategic levels are the defined, decisive, and attainable goals towards which all operations-not just military operations-and activities are directed within the JOA.

> **Examples of National Strategic Objectives: (NDS)**
> • Defending Homeland from attack.
> • Ensuring common domains remain open and free.
> • Deterring adversaries from aggression against our vital interests.
> • Enabling U.S. interagency counterparts to advance U.S. influence and interests.

(3) Objective statements do not suggest or infer the ways and means for accomplishment. Passive voice is a convention that can assist a CDR in distinguishing an objective from an effect and state objectives without inferring potential ways and means, thereby broadening the range of possible actions to achieve the objectives.

(4) An objective statement should have the following attributes:

- Establishes a single goal: a desired result, providing one concise "end" toward which operations are directed.
- Is specific: identifies the key system, node or link to be affected.
- Link to higher-level objectives directly or indirectly.
- Unambiguous as possible.
- Uses passive voice.
- Prescriptive in nature.
- Does not infer causality: no words (nouns or verbs) that suggest ways and/or means.

> *"Provide direct, urgent humanitarian relief support to the government and people of Haiti."*
> *(Operation Unified Response)*

(5) Objectives are written as concise descriptive statements: *Country Red is no longer a threat to regional peace. Country Blue's IRBM capability is eliminated.*

> **Determine the Valid Objective:**
> • Eliminate terrorists in Brown.
> - Uses active voice, re-write as follows: "Terrorists are eliminated in Brown."
> • Pre-invasion borders are restored after offensive operations.
> - Uses a causality to explain how it will achieve the effect, re-write as follows: "Restoration of pre-invasion boarders."
> • A free and democratic Blue under the duly-elected government.
> - This objective is correct.

(6) At the outset of *OIF*, the SecDef set eight mission objectives for the operation. The eight mission objectives for *OIF* were:

- End the regime of Saddam Hussein.
- Eliminate Iraq's weapons of mass destruction.
- Capture or drive out terrorists.
- Collect intelligence on terrorist networks.

- Collect intelligence on Iraq's weapons of mass destruction activity.
- Secure Iraq's oil fields.
- Deliver humanitarian relief and end sanctions.
- Help Iraq achieve representative self-government and ensure its territorial integrity.

(7) The stated U.S. military objectives for *OEF* were:

- The destruction of terrorist training camps and infrastructure within Afghanistan.
- The capture of al Qaeda leaders.
- The cessation of terrorist activities in Afghanistan.
- Make clear to Taliban leaders that the harboring of terrorists is unacceptable.
- Acquire intelligence on al Qaeda and Taliban resources.
- Develop relations with groups opposed to the Taliban.
- Prevent the use of Afghanistan as a safe haven for terrorists.
- Destroy the Taliban military allowing opposition forces to succeed in their struggle.
- Military force would help facilitate the delivering of humanitarian supplies to the Afghan people.

(8) Strategic military objectives define the role of military forces in the larger context of national strategic objectives. This focus on strategic military objectives is one of the most important considerations in operational design. The nature of the political aim, taken in balance with the sources of national strength and vulnerabilities, must be compared with the strengths and vulnerabilities of the adversary and/or other factors in the OE to arrive at reasonably attainable strategic military objectives. Strategic objectives must dominate the planning process at every juncture.[16]

> *"During Desert Storm the U.S. Central Command's sweeping envelopment maneuver was brilliantly effective, not only because it neutralized the Republican Guard forces, the Iraqi army's center of gravity, it also placed a significant allied force in position to threaten Baghdad, thus creating added incentive for Iraq to agree to an early cease-fire. An operational decision had affected an opponent's strategic calculus by creating additional allied leverage."*
>
> *Should Deterrence Fail: War Termination in Campaign Planning, James W. Reed, Parameters, Summer 1993*

(9) Objectives may change over time. You may see a change in operational level objectives that still support strategic or national level objectives. You may even see changes in strategic objectives that of course will affect both operational and tactical level objectives. The dialogue boxes and figures[17] on the following pages is an historical example of how objectives can change throughout a conflict due to operational necessity and strategic guidance.

[16] *JP 5-0, Joint Planning*

[17] *Warfare Studies Institute, Joint Air Operations Planning Course, Joint Air Estimate Planning Handbook, Version 5, 30 January 2005, pg 15-16*

The Korean War clearly demonstrates the linkage between the political and military objectives. The political objectives changed three times during the conflict, mandating major revisions of and limitations to the campaign plans.

> **CHANGING OBJECTIVES**
> *The Korean War—A case of changing political and national objectives while engaged in combat.*

> **OBJECTIVE: FREE THE REPUBLIC OF KOREA**
> *The Democratic People's Republic of Korea invaded the Republic of Korea on 24 June 1950. President Harry S. Truman heeded the request of the United Nations Security Council that all members "furnish such assistance to the Republic of Korea as may be necessary to repel armed attack and restore international peace and security in the area."* [18] *This translated into guidance to United Nations Command Far East Command, then commanded by Gen Douglas MacArthur, "to drive forward to the 38th parallel, thus clearing the Republic of Korea of invasion forces."* [19] *MacArthur accomplished this by first heavily reinforcing the remaining pocket of South Korean resistance around Pusan with United States military forces. Using these forces he pushed northward and executed the highly successful amphibious landing of two divisions behind enemy lines at Inchon. Airpower was used to wage a comprehensive interdiction campaign against the enemy's overextended supply routes. United Nations forces achieved the original objective by October 1950.*

> **OBJECTIVE: CHANGE 1, FREE ALL OF KOREA**
> *In view of the success at Inchon and the rapid progress of United Nations forces northward, the original objective was expanded. "We regarded," said Secretary of Defense Marshall, "that there was no . . . legal prohibition against passing the 38th parallel."* [20] *The feeling was that the safety of the Republic of Korea would remain in jeopardy as long as remnants of the North Korean Army survived in North Korea.*[21] *This was expressed in a UN resolution on 7 October 1950 requiring "all necessary steps be taken to ensure conditions of stability throughout Korea."*[22] *MacArthur then extended the counteroffensive into North Korea. However, the enemy's logistic tail extended northward into the People's Republic of China. Because the United Nations and United States did not want to draw China into the war, targets in China were off-limits. For this reason, use of airpower was limited largely to close air support. United Nations forces advanced to near the Chinese border. The second objective was achieved, temporarily at least, by November 1950.*

[18] *Robert Frank Futrell, Ideas, Concepts, Doctrine, Vol. 1 (Maxwell AFB, Ala.: Air University Press, 1989), 293*

[19] *Ibid., 297*

[20] *Ibid*

[21] *Ibid*

[22] *Ibid*

OBJECTIVE: CHANGE 2, SEEK CEASE-FIRE, RESOLVE BY NEGOTIATION

On 26 November 1950, the Chinese Communists launched a massive counterattack that shattered the United Nations forces, forcing a retreat from North Korea. MacArthur realized he was in a no-win situation and requested permission to attack targets in China. The Joint Chiefs of Staff's (JCS) guidance was neither to win nor to quit; they could only order him to hold. They vaguely explained that, if necessary, he should defend himself in successive lines and that successful resistance at some point in Korea would be "highly desirable," but that Asia was "not the place to fight a major war." [23] On 14 December, at the request of the United States, the United Nations adopted a resolution proposing immediate steps be taken to end the fighting in Korea and to settle existing issues by peaceful means. "On 9 January 1951, the Joint Chiefs of Staff informed MacArthur that while the war would be limited to Korea, he should inflict as much damage upon the enemy as possible." [24] Limiting the conflict to Korea negated our ability to use naval and airpower to strategically strike enemy centers of gravity located within China. On 11 April 1951, Truman explained the military objective of Korea was to "repel attack. . . to restore peace. . . to avoid the spread of the conflict." [25] The political objectives and the military reality placed MacArthur in a difficult situation. MacArthur proved unwilling to accept these limited objectives and was openly critical of the Truman administration. Truman relieved him of command.

The massive Chinese attacks mounted in January and April of 1951 failed because of poor logistical support. United Nations forces sought to exact heavy casualties upon the enemy rather than to defend specific geographical objectives. As the Chinese and North Koreans pressed forward, their lines of communication were extended and came under heavy air interdiction attack. By May 1951, United Nations forces had driven forward on all fronts. With communist forces becoming exhausted, negotiations for a cease-fire began on 10 July 1951. The quest for the third objective finally ended on 27 July 1953 with implementation of a cease-fire that is still in force.

Changing Objectives.

"Sometimes military and political objectives become decisive points in the operational design because they are essential not only to reach the endstate, but critical to affecting the enemy's center(s) of gravity or protecting friendly center(s) of gravity. Above all, keep in mind that while objectives can be rephrased as decisive points, decisive points are not synonymous with objectives. Objectives always refer to endstate. Decisive points always refer to operational level center(s) of gravity."

Operational Design: A Methodology for Planners,
Dr. Keith Dickson, Professor of Military Studies, JFSC

[23] William Manchester, American Caesar (New York: Dell Publishing, 1983), 617

[24] Robert Frank Futrell, Ideas, Concepts, Doctrine, Vol. 1 (Maxwell AFB, Ala.: Air University Press, 1989), 302

[25] Ibid

d. Effects. One of the most confusing aspects encountered in recent joint doctrinal changes has been the infusion of *effects* language and processes into the planning process. The issue is made even more difficult in view of the conceptual overlap between *effects theory* with the established principles of operational art. There have been numerous effects-related models promulgated by a number of agencies. These include Effects-Based Operations (EBO) and Effects-Based Planning (EBP). Neither of these terms are codified in U.S. joint doctrine, though one will often see them used in some staffs. Presently, Joint doctrine simply uses the term "effects."

(1) Effects help CDR's and their staffs understand and measure conditions for achieving objectives. ***An effect is a physical and/or behavioral state of a system that results from an action, a set of actions, or another effect.*** Just as termination conditions support/define achieving the endstate, a *desired effect* can also be thought of as a condition that can support achieving an associated objective, while an *undesired effect* is a condition that can inhibit progress toward an objective.

> **Effects are derived from objectives.** They help bridge the gap between objectives and tasks by describing the conditions that need to be established or avoided within the JOA to achieve the desired endstate.

(2) The use of effects planning is not new; good CDRs and staffs have always thought and planned this way. Effects should not be over-engineered into a list of equations, data bases and checklists. The use of effects during planning is reflected in the steps of JPP as a way to clarify the relationship between objectives and tasks and help the CDR and staffs determine conditions for achieving objectives.

(3) Before one can appreciate the commonality between operational art and effects-based concepts, it is important to understand the fundamental tenets of operational art. This text cannot detail all aspects of operational art, and those unfamiliar with operational art must first gain an awareness of this theoretical framework before launching into operational-level planning. This lack of knowledge of operational art is often the source of much of the current disconnect found in many effects-based concepts. The most critical of these is the nature of the objective and its relationship to the COG. Please review the discussion on the COG under Key-Step 6 and in JP 5-0, *Joint Planning*.

(a) Operational design is defined by joint doctrine as, "The conception and construction of the framework that underpins a campaign or major operation plan and its subsequent execution." Central to operational design is for the planning staff to understand the *desired* endstate and requisite objectives necessary to achieve the desired endstate.

(b) An objective is a clearly defined, decisive, and attainable goal toward which every military operation is directed. Objectives and their supporting effects provide the basis for identifying tasks to be accomplished. Joint planning integrates military actions and capabilities with those of other instruments of national power in time, space, and purpose in unified action to achieve the JFC's objectives.

(4) As observed in Key-Step 6, an enemy's COG is inextricably linked to its objective (just as the friendly COG is linked to its own objectives). In fact, this is a critical aspect to determining an enemy's COG, a fact that is generally unclear in effects-based concepts. An enemy's interconnected PMESII system is meaningless if one does not first (correctly) assess an enemy's probable endstate and objectives. One must remember that the *raison d'être* for a COG is to accomplish an objective.

(5) Destroy or defeat an enemy's COG and you have severely inhibited an enemy's ability to achieve their objective (unless another source of power assumes the role of COG). In a theoretical construct, one should see an enemy COG as something that stands between the friendly COG and the friendly objective(s). Thus, an operation is invariably focused upon an enemy's COG in order to achieve *friendly* objectives.

(6) <u>Effects vs. Objectives</u>.

(a) The next point that often serves to confuse joint planners is the role of effects in the COG-Objective construct. Joint doctrine defines an effect as:

> **Effect:** (1) The physical or behavioral state of a system that results from an action, a set of actions, or another effect. (2) The result, outcome, or consequence of an action. (3) A change to a condition, behavior, or degree of freedom. JP 3-0, *Joint Campaign and Operations.*

(b) Thus, an effect is an outcome from an action, and in its simplest terms it is the condition that one hopes for (if a desired effect) upon the accomplishment of a task or action assigned to a subordinate command. In operational art terms, an effect may be considered as the consequence from attacking/controlling a decisive point or an enemy's critical vulnerability/ requirement and/or capability enroute to the defeat of an enemy's COG. A desired effect should become apparent as the planning staff completes its COG analysis and is often found in the CDR's planning guidance and/or intent. *An effect, however, is not necessarily an end onto itself.* Rather, an effect's purpose should be to ensure a friendly COG's progress to its respective objective(s). Figure B offers a graphic depiction of the concept.

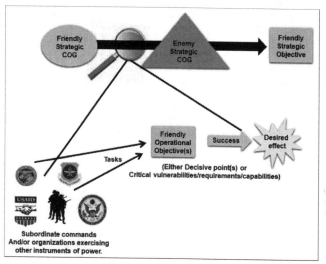

Figure B. Effects vs. Objectives (NWC 4111H).

(c) Figure B illustrates the relationship between objectives, effects, and COGs. Upon determination of one's own objectives and COG, the planning staff assesses an enemy's probable objective(s) and its related COG. The friendly operations are then focused upon defeating the enemy COG enroute to the friendly objective. Since a direct attack upon the enemy COG is frequently impractical (or too costly), friendly COAs will often focus on attacking through a combination of Critical Vulnerabilities (CV), Critical Requirements (CR), Critical Capabilities (CC), and decisive points (DP) that are integral to the enemy COG. If, as depicted in the Figure B, the enemy COG being attacked is at the strategic level of war, these attacks/operations become tasks for subordinate commands, functions and/or agencies against the applicable CVs, CRs, CCs, and DPs. The successful accomplishment of these subordinate operations produced desired effects (either facilitating the friendly COG and/or degrading the enemy COG).

These effects, coupled most likely with other friendly-induced effects from other complementary operations, are intended to combine/to defeat/degrade the enemy COG and allow for the accomplishment of the friendly strategic objective.

(7) The language used to craft effects is as critical as the language used to describe objectives and tasks. Effects' planning focuses all elements of national and international power to achieving national/coalition and theater strategic objectives.

(a) ***Although joint doctrine does not prescribe a specific convention for writing a desired effect statement, there are four primary considerations.***

- Each desired effect should link directly to one or more objectives.
- The effect should be measurable.
- The statement should not specify ways and means for accomplishment.
- The effect should be distinguishable from the objective it supports as a condition for success, not as another objective or a task.

(b) The effect should be distinguishable from the objective it supports as a condition for success, not as another objective or a task. The same considerations apply to writing an undesired effect statement.

(8) During mission analysis, the CDR considers how to achieve national and theater-strategic objectives, knowing that these likely will involve the efforts of other U.S. agencies and multinational partners.

(9) At the operational and tactical levels, subordinate commanders use the effects and broad tasks prescribed by the JFCs as the context in which to develop their supporting mission plans. Normally, the effects established by the JFCs have been developed in collaboration with subordinate commands and supporting agencies. At a minimum, the development of effects should have the full participation by the JFCs, as well as empowered representatives from interagency organizations.

(10) The use of effects in planning helps commanders and staffs use other elements of operational design more effectively by clarifying the relationships between COG, Lines of Operation (LOO), Lines of Effect (LOE), DPs, and Termination Criteria. The JFC and planners continue to develop and refine effects throughout JPP planning steps. Monitoring progress toward attaining desired effects and avoiding undesired effects continues throughout execution.[26]

(11) With a common set of desired and undesired effects, the CDR can issue guidance and intent to their staff and components, and work with other stakeholders to accomplish fused, synchronized, and appropriate actions on PMESII systems within the OE (beyond military on military) to attain the desired effects and achieve objectives.

Operational Design and Effects:

- Developed by CCMD based on an assessment of the operational environment and objectives.
- Integrate, to extent possible, all "stakeholder" input – JTF, Embassy(s) and other non-military stakeholders.
- Achieved by the cumulative outcome of the DIME tasks/activities executed over time - multiple phases.
- A JTF CDR may develop specific effects within his JOA to facilitate attainment of the campaign effects and objectives. These must be nested within those developed by the higher command.

---

(12) Key to the effects is full participation of all of the players – military and other elements of national power – in a fully inclusive process of assessing, planning and eventually directing actions.

[26] JP 5-0, Joint Planning

e. <u>Effects in the Planning Process-The Bottom Line</u>. While the CDR and unit SOP will dictate how the staff will apply effects-based processes, for the purposes of this text, one may find the following considerations of use:

(1) <u>Operational Design</u>.

(a) Apply the constructs of operational art. Ensure clarity between objectives and desired effects.

(b) When assessing possible enemy COGs, ensure the JIPOE is inclusive of the full PMESII (as appropriate to the level of operation) in its analysis. Recognize the duplication of the concept of nodes and links with the terminology of COG deconstruction. To avoid confusion, do not use both sets of terminology in the same analysis.

(2) <u>Commander's Intent</u>. If the CDR has expressed specific desired effects and success criteria, ensure they are captured in the CDR's intent and/or planning guidance.

(3) <u>Assessment</u>. Select and assess meaningful criteria as appropriate, though continue to maintain a broader perspective that cannot be conveyed by color-coded indicators.

f. <u>Tasks</u>. Theater tasks are those actions that the CCDR must execute in order to achieve the strategic effects which support overarching strategic objectives. These tasks are designed to be the broad theater strategic tasks from which subordinates derive their own operational and tactical tasks.

> **Task:** *An action or activity (derived from an analysis of the mission and concept of operations) assigned to an individual or organization to provide a capability.*

(1) Once the CDR and staff understand the objectives and effects that define the operation, they then match appropriate tasks to desired effects. Task determination begins during mission analysis, extends through COA development and selection, and provides the basis for the tasks eventually assigned to subordinate and supporting commands in the operation plan or order.

> **Objectives** prescribe friendly goals.
>
> **Effects** describe the state of system behavior in the operational environment.
>
> **Tasks** direct friendly action.

(2) The CDR emphasizes the development of effects-related tasks early in the planning process because of the obvious importance of these tasks to objective accomplishment. Each of these tasks aligns to one or more effects, reflects action on a specific system or node, is written in active voice, and can be assigned to an organization in the operation

plan or order. Support tasks such as those related to logistics and communications also are identified during mission analysis.

(3) Task Assessment. Mission success criteria describe the standards for determining mission accomplishment. The CDR may include these criteria in the planning guidance so that the staff and components better understand what constitutes mission success. When these criteria are related to the termination criteria, which typically apply to the end of a joint operation and disengagement by joint forces, this often signals the end of the use of the military instrument of national power. Mission success criteria can also apply to any joint operation, subordinate phase, and joint force component operation. These criteria help the CDR determine if and when to move to the next major operation or phase.

(a) The initial set of these criteria determined during mission analysis becomes the basis for assessment. Developing an assessment plan begins during plan initiation and is continuous.

(b) Assessment uses indicators, measures of performance (MOPs) and measures of effectiveness (MOEs) to shape a staff's collection effort and indicate progress toward achieving objectives. Indicators take the form of reports from subordinates, surveys and polls, and information requirements. Indicators help to answer the question "What is the current status of this MOE or MOP?" A single indicator can inform multiple MOPs and MOEs. Examples of indicators for the MOE "Decrease in insurgent activity" are – number of hostile actions per Province each week, number of munitions caches found per Province each week. If the mission is unambiguous and limited in time and scope, mission success criteria could be readily identifiable and linked directly to the mission statement. For example, if a JFC's mission is to evacuate U.S. personnel from the U.S. Embassy in Grayland, then the mission analysis could identify two primary success criteria: (1) all U.S. personnel are evacuated and (2) evacuation is completed before D-Day. However, more complex operations may require MOEs and MOPs for each task, effect, and phase of the operation. For example, if a JFC's specified tasks are to ensure friendly transit through the Strait of Gray, eject Redland forces from Grayland, and restore stability along the Grayland-Redland border, then mission analysis should indicate many potential success criteria — measured by MOEs and MOPs — some for each desired effect and task.

(c) Measuring the status of tasks, effects, and objectives becomes the basis for reports to senior commanders and civilian leaders on the progress of the operation. The CCDR can then advise the President and SecDef accordingly and adjust operations as required. Whether in a supported or supporting role, JFCs at all levels should develop their mission success criteria with a clear understanding of termination criteria established by the CJCS and SecDef.

(4) Assessment Process and Measures. The assessment process uses MOPs to evaluate task performance at all levels of war and MOEs to determine progress of operations toward achieving objectives.

(a) MOEs help answer questions like: "Are we doing the right things?" Are our actions producing the desired effects? Are alternative actions required? MOPs are closely associated with task accomplishment. MOPs help answer questions like: "Was the action taken?" Were the tasks completed to standard? How much effort was involved? Well-devised measures can help the commanders and staffs understand the causal relationship between specific tasks and desired effects. See Table 1 for assessment criterion example.

b) MOEs. Assess changes in system behavior, capability, or OE. They are the operational measures of changes in conditions, both positive and negative. They measure the attainment of an endstate, achievement of an objective, or creation of an effect; they do not measure task performance. These measures typically are more subjective than MOPs, and can be crafted as either qualitative or quantitative. MOEs can be based on quantitative measures to reflect a trend and show progress toward a measurable threshold. Examples of MOEs for the objective to "Provide a safe and secure environment" may include decrease in insurgent activity or increase in population trust of host-nation security forces.

MOE	MOP	INDICATOR
Answers the question: Are we doing the right things?	Answers the question: Are we doing things right?	Answers the question: What is the status of this MOE or MOP?
Measures purpose accomplishment	Measures task completion	Measures raw data inputs to inform MOEs and MOPs
Measures why in the mission statement	Measures what in the mission Statement	Information used to make measuring what or why possible
No hierarchical relationship to MOPs	No hierarchical relationship to MOEs	Subordinate to MOEs and MOPs
Often formally tracked in formal assessment plans	Often formally tracked in execution Matrixes	Often formally tracked in formal assessment plans
Typically challenging to choose the correct ones	Typically simple to choose the correct ones	Typically as challenging to select correctly as the supporting MOE or MOP

Assessment Terms
•A MOE is a criterion used to assess changes in system behavior, capability, or operational environment that is tied to measuring the attainment of an end state, achievement of an objective, or creation of an effect (JP 3-0).
•A MOP is a criterion used to assess friendly actions that is tied to Measuring task accomplishment (JP 3-0).
•In the context of assessment, an indicator is an item of information that provides insight into a measure of effectiveness or measure of performance.

Table 1. Assessment Criterion Example

(c) <u>MOPs. Measure Task Performance</u>. They generally focus on the friendly force and are generally quantitative, but also can apply qualitative attributes to task accomplishment. A MOP confirms or denies that a task has been properly performed. MOPs are commonly found and tracked at all levels in execution matrixes. They are measures that characterize physical or functional attributes relating to the [operation], system operation measured or estimated under specific conditions. MOPs are used in most aspects of combat assessment, since the latter typically seeks specific, quantitative data or a direct observation of an event to determine accomplishment of tactical tasks. But MOPs have relevance for noncombat operations as well (e.g., tons of relief supplies delivered or noncombatants evacuated). MOPs also can be used to measure operational and strategic tasks, but the type of measurement may not be as precise or as easy to observe. Evaluating task accomplishment using MOPs is relatively straightforward and often results in a yes or no answer; for example, Route XX cleared, $150,000 spent on dam construction, Al Jabber airfield secured.

> "Every attempt to make war easy and safe will result in humiliation and disaster."
> **General William T. Sherman**

UJTL Tasks (as Assessment Criteria)

*Note: As denoted in Table 2 the planning staff may find the Universal Joint Tasks in the Universal Joint Task List (UJTL CJCSM 3500.04) as a helpful starting point for crafting assessment criteria for an operation. For example, using the same Grayland Noncombatant Evacuation Operation (NEO) scenario mentioned earlier, the UJTL offers the Task descriptions and potential measures. Example follows below:

ST 8.4.3 Coordinate Evacuation and Repatriation of Noncombatants from Theater
Task Description: third-country resources to evacuate U.S. dependents, USG civilian employees, and private citizens (U.S. and third-country) from the theater and support the repatriation of appropriate personnel to the U.S. Such operations are conducted in support of the Department of State. Theater organizations at various echelons provide support (for example, medical, transportation, and security) to noncombatants (JP 3-0, 3-07, 3-07.5, 3-08v2, 3-10, 3-57, CJCSI 3110.14, CJCSM 3122.03).

Potential Measures offered by the UJTL

M1	Days	To organize and deploy fully operational JTF.
M2	Hours	To evacuate noncombatants (once CCDR directed to conduct evacuation).
M3	Hours	To evaluate situation and present recommendations to decision-maker(s).
M4	Percent	Of U.S. citizens and designated foreign nationals accounted for by name during evacuation.
M5	Percent	Of U.S. citizens and designated foreign nationals accounted for.
M6	Percent	Of U.S. citizens and designated foreign nationals evacuated.
M7	Percent	Of U.S. citizens desiring, evacuated.
M8	Percent	Of evacuees available and desiring evacuation, moved (IAW OPLAN).
M9	Yes/No	NEO plans include actions in the event of NBC attack.

Table 2. Universal Joint Task List (UJTL CJCSM 3500.04)

Where applicable, the theater tasks are linked to appropriate universal joint tasks in the UJTL. This association further develops the scope of these tasks while providing further linkage to Joint Capability Areas (JCA) association, capability gap identification, and CCDR resource/Planning Programming, Budgeting, and Execution (PPBE) planning.

> **THINKING CAP:** *Operations have always required assessment, this is nothing new. One should, however, be alert to the potential pitfalls encountered by a staff that becomes so enamored with color-coded MOE/MOP indicators that it fails to remember the nature of warfare. History is replete with these examples. The operational environment is constantly changing as enumerable human and environmental factors exert their influences. The enemy also has a vote, and is fighting to win! Friction has not disappearaed from the operational environment and friendly intelligence collection and analysis may be flawed. And, finally, as noted above, not everything is quantifiable – CDRs and their staffs will also have to rely on their experience and intuition.*

(5) The assessment process and related measures should be relevant, measurable, responsive, and resourced so there is no false impression of accomplishment. Quantitative measures can be helpful in this regard.

- Relevant. MOPs and MOEs should be relevant to the task, effect, operation, the OE, the endstate, and the CDR's decisions. This criterion helps avoid collecting and analyzing information that is of no value to a specific operation. It also helps ensure efficiency by eliminating redundant efforts.

- Measurable. Assessment measures should have qualitative or quantitative standards they can be measured against. To effectively measure change, a baseline measurement should be established prior to execution to facilitate accurate assessment throughout the operation.

*Note: Both MOPs and MOEs can be quantitative or qualitative in nature, but meaningful quantitative measures are preferred because they are less susceptible to subjective interpretation.

- Responsive. Assessment processes should detect situation changes quickly enough to enable effective response by the staff and timely decisions by the CDR. The JFC and staff should consider the time required for an action or actions to produce desired results within the OE and develop indicators that can respond accordingly. Many actions directed by the JFC require time to implement and may take even longer to produce a measurable result.

- Resourced. To be effective, assessment must be adequately resourced. Staffs should ensure resource requirements for data collection efforts and analysis are built into plans and monitored. Effective assessment can avoid duplication of tasks and unnecessary actions, which in turn can help preserve combat power.

6. <u>Key Step 6</u>: Determine Own and Enemy's Center(s) of Gravity (COG), Critical Factors and Decisive Points

> *"The adversary's COG… would be attacked carefully, with measured means, because its indiscriminate destruction, while useful in defeating military forces, might bring undesirable consequences in rebuilding that same society."*
>
> *Creating a New Center of Gravity: A New Model for Campaign Planning, Watson, Bryan C.*

a. <u>Center of Gravity</u>. Clausewitzian concepts such as friction, fog and culminating points, to name a few, abound in our military vernacular. But arguably, none has been discussed, debated nor written on more than the Clausewitzian concept of COG or main point. For the U.S. military, the origins of the COG concept are rooted in the Cold War. The COG concept matured in the American mindset largely during an era when the U.S. military was focused heavily (and almost exclusively) on producing doctrine that would win wars decisively against a conventional traditional military force and nation state, "especially in such places as the Fulda Gap."[27]

(1) The COG concept has served us for years as a giant lens for focusing strategic and operational efforts to achieve decisive results. However, for the current generation of military professionals, the conflicts in Iraq and Afghanistan have evoked a disquieting epiphany: battlefield victory is useless without an ensuing political victory.[28] The military

[27] *Rudolph M. Janiczek, A Concept At The Crossroads: Rethinking The Center Of Gravity, October 2007*

[28] *This enduring dictum also comes to us courtesy of Clausewitz: "The political object is the goal, war is the means of reaching it, and the means can never be considered in isolation from their purpose." On War, p. 87*

efforts in these countries [found] the U.S. military engaged in prolonged insurgencies and postwar reconstruction operations far removed from decisive battle. Furthermore, the strategic landscape suggests that the future for the U.S. military will be rife with other such "ambiguous and uncomfortable wars—and their aftermath."[29] This has evoked a corresponding renaissance in American doctrinal thinking and with it, not surprisingly, a number of proposals to redefine the COG.[30]

(2) As accepted definitions for COG are discussed in this text, planners must also strive to understand today's ambiguous environment and take the learned elements and acclimate and adjust ahead of an adaptable adversary.

(3) To understand the concept of COG we must begin with a discussion on the COG as a focal point for identifying critical factors; sources of strength as well as weaknesses and systemic vulnerabilities. Joint doctrine defines *center of gravity* as ***"the set of characteristics, capabilities, and sources of power from which a system derives its moral or physical strength, freedom of action, and will to act."***[31] This definition states in modern terms the classic description offered by Clausewitz, "the hub of all power and movement, on which everything depends," the point at which all our energies should be directed."

(4) COG can be categorized as either physical or moral, and they are not limited in scope to military forces. A *physical* COG, such as a capital city or a military force, is typically easier to identify, target, and assess. Physical COGs can often be influenced solely by military means. However, the U.S. experiences in Iraq and Afghanistan exemplify the notion that active conflict can outlast the neutralization of a perceived COG. Neither the demise of the Taliban nor the removal of Saddam Hussein's regime brought an end to violence in either theater. *As with the Lernaean Hydra from Greek mythology, for each head removed, two more grew back.* The U.S. found itself engaged with elements of the former regimes as well as a multitude of other groups with varying interests and motivations. At a minimum, the nature of the COG has changed in each case.[32] This change, or evolution of the COG, may be exploited by understanding the civil dimension or *moral* COG. The moral centers of gravity are dynamic and inherently related to human factors: civilian populations, a charismatic or key leader, powerful ruling elite or strong-willed populace. Influencing a moral COG is far more difficult and typically cannot be accomplished by military means alone. The intangible and complex nature of moral COG and their related vulnerabilities necessitates the collective, integrated efforts of the instruments of power. As an example, it is a common understanding that access to, and influence over, civilian populations is a source of strength for insurgent movements and arguably terrorist networks.[33] As noted in FM 3-24/ MCWP 3-33.5 "the ability to generate and sustain popular support, or at least acquiescence and tolerance, often has the greatest impact on the insurgency's long-term effectiveness. This ability is usually the insurgency's center of gravity." Engaging the civil COG and/or vulnerabilities utilizing a focused interagency approach during the planning process, the CDR may ultimately shape, mitigate a threat, or gain awareness over it, degrading the insurgency's long-term effectiveness.

[29] Max G. Manwaring, *The Inescapable Global Security Arena, Carlisle Barracks, PA: Strategic Studies Institute, U.S. Army War College, 2002, p. 2*

[30] Rudolph M. Janiczek, *A Concept At The Crossroads: Rethinking The Center Of Gravity, October 2007*

[31] *DOD Dictionary of Military and Associated Terms (DOD Dictionary), JP 5-0, Joint Planning*

[32] Sele, Richard K., *Engaging Civil Centers of Gravity and Vulnerabilities*

[33] *Ibid*

> "The source of strength, or civil COG, for those responsible for the genocide in Rwanda in 1994 was the civilian population. Planners of the genocide recruited Hutu Burundi refugees and militias from the lower economic classes. A combination of anti-Tutsi propaganda and physical threats fueled their massive participation in the slaughter. According to interviews with survivors of the massacres, most of the 50,000 recruited killers were peasants just like their victims. Any organization tasked to stop the killing would have had to influence the civil COG—the peasant population."
>
> Sele, Richard K., *Engaging Civil Centers of Gravity and Vulnerabilities.*

(5) At the strategic level, a COG might be an alliance, a political or military leader, a set of critical capabilities or functions, or national will. At the operational level a COG often is associated with the adversary's military capabilities — such as a powerful element of the armed forces — but could include other factors in the OE associated with the adversary's civil, political, economic, social, information, and infrastructure systems.

> During the 1990-91 Persian Gulf War the coalition itself was identified as a friendly strategic COG, and the CCDR took measures to protect it, to include deployment of theater missile defense systems.

(6) There is no certainty that a single COG will emerge at the strategic and operational levels. It is possible that no COG will emerge below the strategic level. At the tactical level, the COG concept has no utility; for us to speak of a tactical COG, the tactical level of war would have to exist independent of the operational and strategic level. It is commonly accepted that the tactical equivalent of the COG is the objective. Modern writings and understanding of the COG has evolved beyond the term's pre-industrial roots to include the possibility of multiple COGs existing at the strategic and operational levels.

(7) At all levels, COGs are interrelated. Strategic COGs have associated decisive points that may be vulnerable at the operational level, just as operational COGs may be vulnerable to tactical-level actions. Therefore, analysis of friendly and enemy COGs is a continuous and related process that begins during planning and continues throughout a major operation or campaign. Figure C below shows a number of characteristics that can be associated with a COG.

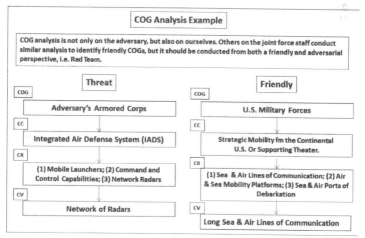

Figure C. Threat and Friendly Physical COG Analysis Example.[34]

[34] Joint Information Operations Planning Handbook

(8) The essence of "operational art" lies in being able to produce the right combination of effects in time, space, and purpose relative to a COG to neutralize, weaken, destroy (consistent with desired endstate/commanders' intent), or otherwise exploit it in a manner that best helps achieve military objectives and attain the military endstate. *In theory*, this is the most direct path to mission accomplishment. While doing this the CDR must also plan for protecting friendly potential COGs, such as agreements with neutral and friendly nations for transit of forces, information and networks, coalition relationships, and U.S. and international public opinion.

(9) COGs *are not vulnerabilities, however,* within every COG lies inherent vulnerabilities that when attacked can render those COGs weaker and even more susceptible to direct attack and eventual destruction. This process cannot be taken lightly, since a faulty conclusion resulting from a poor or hasty analysis can have very serious consequences, such as the inability to achieve strategic and operational objectives at an acceptable cost. Planners must continually analyze and refine COGs.

(10) An additional insight on the COG is provided by Antulio Echevarria in his Strategic Studies Institute paper, *Clausewitz's Center of Gravity; It's Not What We Thought.* Echevarria maintains that the COG needs to be redefined as a "focal point," not as a strength or weakness or a source of strength. A COG is more than a critical capability; it is the point where a certain centripetal force seems to exist, something that holds everything else together. As an example, he offers the following:

> " ...al-Qa'ida cells might operate globally, but they are united by their hatred of apostasy.[35] This hatred, not Osama bin Laden, is their CoG. They apparently perceive the United States and its Western values as the enemy CoG (though they do not use the term) in their war against "apostate" Muslim leaders. Decisively defeating al-Qa'ida will involve neutralizing its CoG, but this will require the use of diplomatic and informational initiatives more than military action.
>
> Commanders and their staffs need to identify where the connections— and the gaps—exist in the enemy's system as a whole before deciding whether a center of gravity exists. The CoG concept does not apply if enemy elements are not connected sufficiently. In other words, successful antiterrorist operations in Afghanistan may not cause al-Qa'ida cells in Europe or Singapore to collapse." [36]

b. The adversarial context pertinent to COG analysis takes place within the broader OE context. A system's perspective of the OE assists in understanding the adversary's COGs. In combat operations, this involves knowledge of how an adversary organizes, fights, and makes decisions, and of their physical and psychological strengths and weaknesses. Moreover, as in Figure D, the CCDR and staff must understand other OE systems and their interaction with the military system. This holistic understanding helps commanders and their staffs identify COGs, *critical factors*, and DP to formulate LOO[37] (LOO discussed in detail in Chapter 5-III *Concept Development*).

(1) Critical Factors. All COGs have inherent *"critical capabilities (CC)"* — those means that are considered crucial enablers for the adversary's COG to function and essential to the accomplishment of the adversary's assumed objective(s). These critical capabilities permit an adversary's COG to resist the military endstate. In turn, all critical capabilities have essential *"critical requirements (CR)"* — those essential conditions, resources, and means for a critical capability to be fully operational. *Critical vulnerabilities (CV)* are those

---

[35] *Al-Qa'ida (the Base)," International Policy Institute for Counter-Terrorism, on the World Wide Web at www.ict.org.il/inter_ter (accessed 3 April 2002)*

[36] *Echevarria, Antulio J. II, LtCol U.S. Army. Clausewitz's Center of Gravity; It's Not What We Thought, Naval War College Review, Winter 2003, Vol. LVI, No.1*

[37] *JP 3-0, Joint Operations*

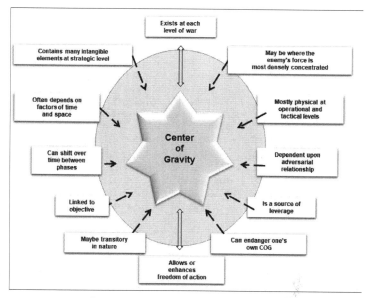

Figure D. Characteristics of Center Gravity.

> "All too often, the center of gravity is understood as being one of the
> enemy's vulnerabilities. On the contrary, a center of gravity is found
> invariably among critical strengths-never critical weaknesses or critical
> vulnerabilities."
>
> *Operational Warfare, Milan N. Vego, 2000*

aspects or components of the adversary's CR which are deficient or vulnerable to direct or indirect attack that will create decisive or significant effects disproportionate to the military resources applied. Collectively, these are referred to as *"critical factors."*

(a) Critical factors, including *CC and CV* become decisive points as they are identified and friendly forces can operate against them. They become apparent only after thoroughly analyzing the opponent's COG. Some of the critical factors may lie beyond the reach or ability of the military forces to affect them.

For example, the continuing sympathy of a broad ethnic group toward an insurgent group may be a critical capability for the insurgents. However, affecting that affinity lies beyond the capability of the military force in the JOA, and therefore it does not become a DP.

> "Another error is to confuse a decisive point with the center of gravity.
> Although closely related, decisive points are not necessarily sources
> of strength; they could be found among critical weaknesses. These
> weaknesses are only relevant if they are open to attack and facilitate
> an attack on the enemy center of gravity. Once the center of gravity is
> determined, decisive points are identified and targeted."
>
> *Operational Warfare, Milan N. Vego, NWC 2000 p.30;, Bruce L. Kidder,*
> *Center of Gravity: Dispelling the Myths (Carlisle Barracks, PA. US Army*
> *War College, 1996), p.12*

1 *Critical Capabilities (CC)* are the essential conditions, resources, and means required for a center of gravity to be fully operational. They are the attributes of a center of gravity that can affect friendly force operations within the context of a given scenario, situation, or mission. CC may vary between phases of an operation, function at various echelons, or change in character as an operation progresses. At the operational and tactical levels, CC represent a potential for action in that they possess an ability to inflict destruction, seize objectives or create effects, or prevent friendly forces from achieving mission success. For example, critical requirements can include the resources necessary to deploy a force into combat, the popular support a national leader needs to remain in power, or the perceived legitimacy required to maintain popular support. CC are also the primary means through which to isolate, dislocate, disintegrate, or destroy a COG.

2 *Critical Vulnerabilities (CV)* are aspects or components of those critical requirements that are deficient or vulnerable to direct or indirect attack in a manner that will create decisive or significant effects. CDRs may neutralize, weaken, or destroy an adversary's COG by attacking through CV. For example, an enemy armored Corps could represent an operational-level COG. Deficient mobile air defenses protecting the enemy corps could be a CV that friendly forces could exploit.

> Critical capabilities and vulnerabilities are interrelated. The loss of one critical capability may expose vulnerabilities in other critical capabilities; the loss of a critical capability may initiate a cascading effect that accelerates the eventual collapse of a COG. The analysis of a COG and its critical factors will reveal these systemic relationships and their inherent vulnerabilities.

(b) Critical factors are the keys to protecting or neutralizing COGs during a campaign or major operation. To achieve success, the force must possess sufficient combat power and operational reach to affect critical factors. Similarly, commanders must protect those friendly critical capabilities within the enemy's operational reach.

1 *Operational Approach. Direct versus Indirect. Operational approach* is the manner in which a CDR attacks the enemy COG. The essence of COG analysis lies in determining an approach to the COG that can neutralize, weaken, or destroy it.

a The *direct approach* is to apply combat power directly against the enemy's COG or principal strength. In most cases, however, the enemy COG is well-protected and not vulnerable to a direct attack.

b The *indirect approach* is to attack the enemy COG by applying combat power against a series of decisive points that avoid enemy strengths.

In theory, direct attacks against enemy COGs resulting in their neutralization or destruction is the most direct path to victory — if it can be done in a prudent manner (as defined by the military and political dynamics of the moment). Where direct attacks against enemy COGs mean attacking into an opponent's strength, CCDRs should seek an indirect approach until conditions are established that permit successful direct attacks. In this manner, the enemy's critical vulnerabilities can offer indirect pathways to gain leverage over its COGs. For example, if the operational COG is a large enemy force, the joint force may attack it indirectly by isolating it from its C2, severing its LOCs, and defeating or degrading its protection capabilities. In this way, CCDRs employ a synchronized and integrated combination of operations to weaken enemy COGs indirectly by attacking critical requirements, which are sufficiently vulnerable.

(2) COG *analysis* is thorough and detailed; faulty conclusions drawn from hasty or abbreviated analyses can adversely affect operations, waste critical resources, and expose forces to undue risk. A thorough, systemic understanding of the OE facilitates identifying and targeting enemy COGs. This understanding requires detailed knowledge of how adversaries and enemies organize, fight, and make decisions. Knowledge of their physical and psychological strengths and weaknesses is also needed. In addition, commanders should understand how military forces interact with the other systems (political, military, economic,

social, information, and infrastructure) within the OE. This understanding helps planners identify decisive points and COGs, and formulate lines of operation. Using Figure E as an example; as each center of gravity is identified, analysis is further refined by understanding the operational approach to each center of gravity and the decisive points associated with them.[38]

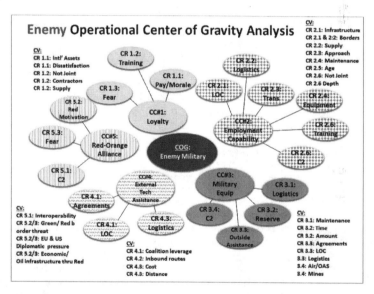

Figure E. Example of Enemy Operational Center of Gravity and Critical Factors.

(a) <u>Decisive Points (DPs)</u>. DPs emerge from an analysis of endstate, objectives, and COG. A *DP* is a place, key event, or enabling system that allows commanders to gain a marked advantage over an enemy and greatly influence the outcome of an operation. DPs are not COGs; they are keys to attacking or protecting their CR. When it is not feasible to attack a COG directly, commanders focus operations to weaken or neutralize the CR (CV) upon which it depends. These CV are DP, providing the indirect means to weaken or collapse the COG. Operational-level decisive points provide the greatest leverage on COGs.[39]

[38] *FM 3-0, Operations*

[39] *Ibid*

(b) Some DP are geographic, such as port facilities, transportation networks and nodes, and bases of operations. Other physical DP may include elements of an enemy force. Events, such as commitment of the enemy operational reserve or reopening a major oil refinery, may also be DP. A common characteristic of DP is their relative importance to the COG; the nature of a DP compels the enemy to commit significant effort to defend or marginalize it. The loss of a DP weakens a COG and may expose more vulnerabilities as it begins to collapse.

(c) DP assume a different character during stability or civil support operations. They may be less tangible and more closely associated with critical events and conditions. For example, they may include repairing a vital water treatment facility, establishing a training academy for national security forces, securing a major election site, or quantifiably reducing crime. In Hurricane Andrew relief operations in 1992, for example, reopening schools became a decisive point. None of these decisive points are physical, yet all are vital to establishing the conditions for a transition to a legitimate civil authority. In an operation predominated by stability tasks, this transition is typically a supporting condition of the desired endstate.

> "Critical vulnerabilities are those elements of one's military or nonmilitary sources of power open to enemy attack, control, leverage, or exploitation. For example; [In 1993, the United States allowed itself to be in a situation where its vital interests were not at stake but the very survival of the Somalian clan leader Mohamed Falah Aideed was. This dangerously asymmetrical situation allowed Aideed to indirectly attack the U.S. strategic center of gravity (will to fight) by exploiting the well-known U.S. critical vulnerability, an aversion to suffering high casualties. With no survival at stake, the Clinton administration was unwilling to take actions aimed at sustaining popular and political support, while Aideed's desire for independent power could be sustained indefinitely]."
> Joint Operational Warfare, Dr. Milan Vego, 2007, p. VII-16

(d) A situation typically presents more DP than the force can control, destroy, or neutralize with available resources. Through critical factors analysis, commanders identify the DP that offer the greatest leverage on COGs. They designate the most important DP as objectives and allocate enough resources to achieve the desired results on them. DP that enable commanders to seize, retain or exploit the initiative are crucial. Controlling these DP during operations helps commanders retain and exploit the initiative. If the enemy maintains control of a DP, it may exhaust friendly momentum, force early culmination, or facilitate a counterattack. DP shape the design of operations. They help commanders select objectives that are clearly decisive relative to the endstate; they ensure that vital resources are focused only on those objectives that are clearly defined, attainable, and directly contribute to establishing the conditions that comprise that endstate.

(e) DP are always oriented on the key vulnerabilities that can only be identified through the COG or another method of systems analysis. Generally, commanders attack adversary vulnerabilities at decisive points so that the results they achieve are disproportional to the military and other resources applied. Consequently, commanders and their staffs must analyze the OE and determine which systems' nodes or links or key events offer the best opportunity to affect the enemy's COGs or to gain or maintain the initiative. The CDR then designates them as decisive points, incorporates them in the LOO, and *allocates sufficient resources* to produce the desired effects against them.

> "However absorbed a commander may be in the elaboration of his own thoughts, it is sometimes necessary to take the enemy into consideration."
> Winston Churchill, The World Crisis, 1911-1918; 1923

c. No COG discussion is complete until we look at the whole OE and take a comprehensive look at all the systems in this environment relevant to the mission and operation at hand. A system is a functionally related group of elements forming a complex whole. A system's view to understanding the OE considers more than just an adversary's military capabilities, order of battle, and tactics. Instead, it strives to provide a perspective of interrelated systems PMESII, and others, that comprise the OE relevant to a specific operation. A system's perspective, as highlighted in Figure F from JP 5-0 provides the CCDR and staff with a common frame of reference for collaborative planning with other government agencies' (OGA) counterparts to determine and coordinate necessary actions that are beyond the CCDR's command authority.

> *"The first task... in planning for war is to identify the enemy's centers of gravity, and if possible, trace them back to a single one."*
> Carl von Clausewitz, On War, 1832

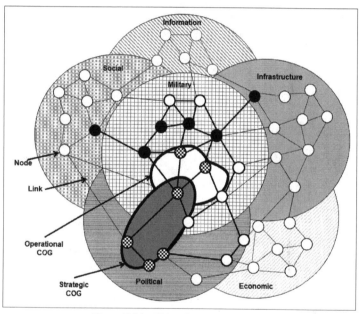

Figure F. Identifying Centers of Gravity.

Joint Planning Process

Understanding PMESII systems, their interaction with each other, and how system relationships will change over time will increase the JFC's knowledge of how actions within a system can affect other system components. Among other benefits, this perspective helps intelligence analysts identify potential sources from which to gain indications and warning, and facilitates understanding the continuous and complex interaction of friendly, adversary, and neutral systems. A systems understanding also supports operational design by enhancing elements such as centers of gravity, LOOs, and decisive points. This helps CDRs and their staffs visualize and design a broad approach to mission accomplishment early in the planning process, which makes detailed planning more efficient. *For example, Figure G (JP 5-0) depicts the critical capabilities, critical requirements, and critical vulnerabilities associated with two of the adversary's strategic and operational COGs.*

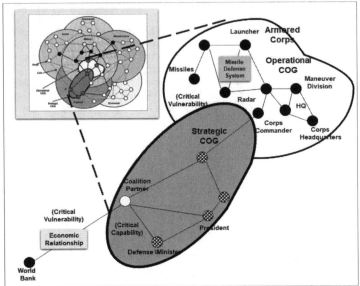

Figure G. Analyzing Critical Factors

"The traditional military-centric single center of gravity focus that worked so well in the cold war doesn't allow us to accurately analyze, describe, and visualize today's emerging networked, adaptable, asymmetric adversary. This adversary has no single identifiable 'source of all power.' Rather, because of globalization, the information revolution, and, in some cases, the non-state characteristic of our adversary, this form of adversary can only be described (and holistically attacked) as a system of systems."

Insights on Joint Operations: The Art and Science,
Gen (Ret) Gary Luck, September 2005.

d. <u>Center of Gravity Determination</u>.[40] While primarily a strategic and operational level concern, the identification of both the enemy and friendly COG is an essential element of any plan. If the staff gets this part wrong, the operation will at best be inefficient and, at worst, end in failure. The CDR and staff should be deeply involved in a dialogue with the higher joint force headquarters planning staff during this critical analysis. While tactical-level organizations may not be party to the formulation of a COG analysis, they most certainly will be participants in the execution of the resulting tactical objectives and tasks that are derived from the analysis. Therefore, even tactical commanders and their planning staffs should be familiar with the process and reasoning used for the COGs analysis in order to place their own operations in the proper context.

e. The purpose of this section is to provide the planner with a brief review of each of the information requirements displayed earlier in this chapter. This section is not intended to replace the extensive study of the nuances of COG analysis that all planners should strive to master; rather, it is intended to identify information requirements and to offer some considerations in the application of the collected data. The reader will note that the JPP

---

[40] *Sweeny, Dr. Patrick C., Professor with NWC and Citadel; Naval War College, NWC 4111H, Joint Operations Planning Process Workbook and NWP 5-01, Navy Planning.*

has the staff collecting information for both the enemy and friendly COGs. Neither can be identified nor considered in a vacuum—a common staff planning mistake. The struggle between opposing forces employing their unique means and ways to achieve their respective ends (objectives) is a dynamic that can only be appreciated if they are viewed collectively. While the explanations and examples provided below are for enemy COGs analysis, the process is the same for determining and analyzing friendly COGs. The only differences are in the planning actions taken once the analysis is completed. Planners develop courses of action that focus on defeating the enemy's COG while at the same time mitigating risks to their own COG.

f. The Center of Gravity Flow Chart, on the following page, illustrates the flow used to identify a COG and to determine the ways in which it can be attacked. Each step of the process, as it corresponds to the numbers in Figure H are described below. Later in this section an example, Desert Storm Enemy COG Analysis, is provided.

(1) Step 1: Identifying the Objective(s). Identifying the objective is a critical first step. Before one can determine a COG, the objective(s) must be identified. If this portion of the analysis is flawed, then the error infects the remainder of the process. The planner should first determine the ultimate (strategic or operational) objectives and then the supporting intermediate (operational or major tactical) objectives. The operational objectives should show a direct relationship to the strategic objectives. If this linkage between strategic and operational objectives cannot be established, the objectives are suspect. Objectives, and particularly strategic objectives, usually have requirements/tasks that fall primarily into the responsibility of instruments of power other than the military. These are still important to identify since the military may have a supporting role in their accomplishment.

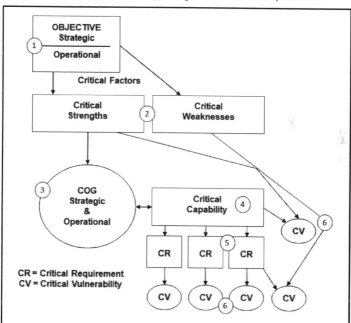

Figure H. Center of Gravity Flow Chart.

(2) Step 2: Identify Critical Factors. Critical factors are those attributes considered crucial for the accomplishment of the objective. These factors that in effect describe the environment (in relationship to the objective) must be identified and classified as either sufficient (critical strength) or insufficient (critical weakness). Critical factors are a cumulative term for critical strengths and critical weaknesses of a military or nonmilitary source of

power; they can be quantifiable (tangible) or unquantifiable (intangible). Critical factors are present at each level of war; they require constant attention because they are relative and subject to changes resulting from the actions of one's forces or of the enemy's actions. It is important while conducting the analysis for this step that planners maintain a sharp eye on the objectives identified in the first step—each level of war has critical factors that are unique to that level.

(a) The questions that should be asked when determining critical factors for the enemy are:

- "What are the attributes, both tangible and intangible, that the enemy has and must use in order to attain his strategic (operational) objective?" *These are critical strengths.* The second question is,

- "What are the attributes, both tangible and intangible, that the enemy has and must use in order to achieve his strategic (operational) objective, but which are weak and may impede the enemy while attempting to attain his objective?" *These are critical weaknesses.*

(b) The answers to these two questions will produce a range of critical strengths and critical weaknesses associated with specific levels of war. One should note that, like the close relationship expected to be found between strategic and operational objectives, there will undoubtedly be some critical strengths and critical weaknesses that have a similar close relationship between the corresponding critical factors. For example, a strategic critical weakness, such as a strategic leader having a tenuous communications link to their fielded forces, may also create an operational critical weakness for fielded forces unable to reliably communicate with their higher command.

(3) <u>Step 3: Identify the Centers of Gravity.</u> Joint doctrine defines a COG as "The source of power that provides moral or physical strength, freedom of action, or will to act."[41] While the definition is helpful for assisting in the identification of the operational COG, when considering the strategic COG, a planner should be alert to the fact that the definition is not focused upon only the military aspects of the analysis. In view of the discussion in the first step, when strategic objectives are being identified planners should consider the broader application of the definition, remembering that the role of instruments of power other than the military may prevail.

The COGs at each level of war should be found among the listed critical strengths identified within the critical factors of Step Two. While all of the identified strengths are critical, the planner must deduce which among those capabilities identified rise(s) above all others in importance in accomplishing the objective (that is, those tangible and intangible elements of combat power that would accomplish the assigned objectives). This critical strength is the COG. This does not diminish the importance of the other critical strengths; however, it forces the planner to examine closely the relationships of the various critical strengths to one another and the objective. This close examination of interrelationships could be improved by using a systems perspective of the OE. Such a study may well offer the planner an enhanced understanding of an adversary's COG and its interdependencies. See JP 5-0 for more information on the systems approach to COG refinement. This analysis of these relationships will prove important in the next step.

(4) <u>Step 4: Identify Critical Capabilities (CC).</u> Joint doctrine defines a critical capability as: *"a means that is considered a crucial enabler for a COG to function as such and is essential to the accomplishment of the specified or assumed objective(s)."[42]*

(a) If the COG is a physical force (often the case at the operational level), the CDR and staff may wish to begin their examination of CC by reviewing the integration, support, and protection elements of the enemy's combat power as they apply to the COG.

(b) Many of these elements are often found in the joint functions as described in the Universal Joint Task List (C2, intelligence, sustainment, protection, fires, and movement

[41] *DOD Dictionary of Military and Associated Terms (DOD Dictionary)*

[42] *Ibid*

and maneuver). Moreover, these capabilities often are located within the critical strengths and weaknesses identified in Step Two.

(c) The planner should be alert for two major considerations. First, although a capability is a critical strength, if it bears no relationship to the identified COG, it cannot be considered a CC. The second consideration is that although some capability may be perceived as a critical weakness, if it is an essential enabler for the enemy COG, then it is a critical capability, albeit weak in nature.

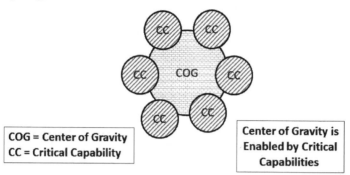

COG = Center of Gravity
CC = Critical Capability

Center of Gravity is Enabled by Critical Capabilities

(5) Step 5: Identify Critical Requirements. Once a COG's critical capabilities are identified, the next step is for the staff to identify those essential conditions, resources, and means for a critical capability to be fully operational.

These are the critical requirements that support each of the critical capabilities. This is essentially a detailed view of what comprises a critical capability.

> **Example:** a critical capability for an operational COG to accomplish its mission might be its ability to exert C2—its ability to receive direction as well as communicate directives to subordinates. The critical requirements might include tangible requirements such as: communication nodes, antennas, frequency bands, individual command posts, spare parts, bandwidth, specific satellites, and so forth. It may also include intangibles such as CDR's perceptions and morale.

Note: Planners should be cautious at this point. One is often presented with a wealth of potential targets or tasks as each critical capability is peeled back and the numerous supporting critical requirements are identified. There is often a temptation to stop at this point of the analysis and begin constructing target lists. Such an action could result in a waste of resources and may not be sufficient to achieve the desired effects. The planner should find the sixth step as a more effective way to achieve the defeat of a COG.

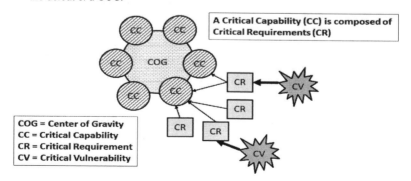

A Critical Capability (CC) is composed of Critical Requirements (CR)

COG = Center of Gravity
CC = Critical Capability
CR = Critical Requirement
CV = Critical Vulnerability

Joint Planning Process

(6) Step 6: Identify Critical Vulnerabilities (CV). Joint doctrine defines a critical vulnerability as "an aspect of a critical requirement which is deficient or vulnerable to direct or indirect attack that will create decisive or significant effects."[43]

The planner should contemplate those critical capabilities and their supporting critical requirements in this regard, keeping in mind that these weaknesses must bear a direct relationship to a COG and its supporting critical capabilities for it to be assessed as a critical vulnerability. Striking a weakness that bears no such relationship is simply a measure taken to harvest "low hanging fruit" that offers no decisive benefit. The planner should also take this opportunity to consider the previously assembled lists of critical strengths and critical weaknesses from Step Two to determine if there are any critical factors with a close relationship to the COG that were not captured in the previous CC/CR steps (steps four and five).

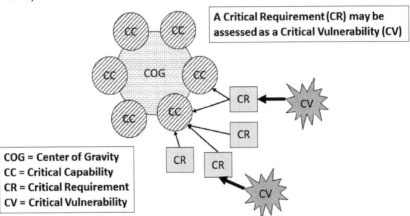

Note: While the planner first seeks critical weaknesses within the critical capabilities and supporting critical requirements as implied by the definition, there might be opportunities found in critical strengths that provide decisive or significant results disproportionate to the military resources applied. An example might be the integrated air defense (IAD) that is protecting an operational COG. While this critical capability might be assessed as strength, its neutralization and the subsequent opening of the COG to direct attack may be assessed by the CDR as more favorable in regard to the amount of resources and time expended to achieve the desired effects.

(7) Step 7: Identify Decisive Points. Joint doctrine defines decisive points as "a geographic place, specific key event, critical factor, or function that, when acted upon, allows CDRs to gain a marked advantage over an adversary or contribute materially to achieving success."[44] As with all previous steps, the value of a DP is directly related to its relationship to a COG and its objective:

(a) In the example shown in Figure I, from a friendly COG perspective, DPs 1 and 4, which provide access to the friendly COG, must be protected from attacks by the enemy COG. DP 2 and 3, which provide decisive access to the enemy COG, become friendly objectives or tasks. If there is no relationship, it is not a DP. *A DP is neutral in nature; that is, it is by definition as important to both the enemy and friendly CDRs.*

[43] JP 5-0, Joint Planning

[44] Ibid

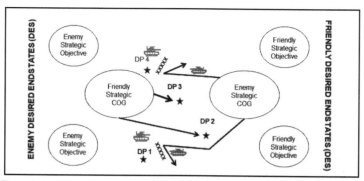

Figure I. Theoretical relationship of two opposing COGs and their Decisive Points.

(b) Using Figure J for example, an APOD/SPOD complex is a DP for a friendly CDR, enabling that CDR to project the COG through it on the way to the objective, then the enemy CDR will also assess the complex as a threat to the enemy COG and should attempt to deny the friendly force CDR control of the DP. In both cases, this DP, if within the capability of the force, will undoubtedly become an objective or task assigned to both enemy and friendly subordinate commands. Using this APOD/SPOD DP example, one might find the friendly JFC assigning the JFMCC the tactical task of "Seize Redland SPOD NLT D+2 in order to support the flow of JTF Blue Sword forces into Redland."

Note: The planner must remember that this is a dynamic process. Any changes in the information considered in the first two steps of this process require the staff to revalidate its conclusions and subsequent supporting operations. As objectives change, the sources of power required to achieve the desired endstate might also change. As new sources of strength appear in the OE, how do they interact?

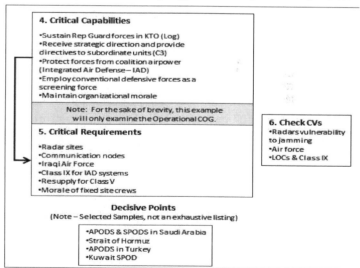

Figure J. Example APOD/SPOD DP

"How can one man say what he should do himself, if he is ignorant of what his adversary is about?"

Lt Gen Antoine-Henri,
Baron de Jomini, 1838

Enemy COG Determination (Example)

Table 3 below provides an example enemy COG analysis (note that the same must be done for the friendly COG to ensure measures are taken to protect one's own COG). This example is not intended to be exhaustive and serves only as an illustrative example, exploring only a single critical capability and its associated critical requirements, and offering simply a selection of DPs.

Enemy Center of Gravity Determination
(Desert Storm Enemy COG Analysis)

Identify

1a. Strategic Objectives (s)
• Retain Kuwait as 19th Province
• Enhance Saddam Hussein's hold on power
• Increase Iraq's political and military influence in the Arab world
• Increase Iraq's power and influence within OPEC

1b. Operational Objective(s)
• Defeat or neutralize a coalition attack to liberate Kuwait
• Prevent coalition forces from obtaining air superiority
• Prevent coalition forces from obtaining sea control in the northern part of the Persian Gulf

2a. Critical Strengths Integrated Air Defense (IAD)
• WMD
• Land-based ballistic missiles (Scuds)
• Republican Guards in the Kuwait Theater of Operations (KTO)
• Forces are in defensive positions
• Saddam and his strategic C2
• Combat experienced units and commanders
• Missile armed surface combatants
• Sea Mine inventories and delivery platforms

2b. Critical Weaknesses
• World Opinion: Arab world outrage
• Long and exposed Land
• LOCs from Iraq to KTO
• Combat skills and readiness of the Air Force
• Numerical and qualitative inferiority of naval forces
• Low morale and poor discipline of regular forces
• Class IX for weapon systems
• Inadequate forces to protect the Iraq-Iranian border

3a. Strategic COG
Saddam and his inner circle security apparatus

3b. Operational COG
Republican Guards in the Kuwait Theater of Operations (KTO)

Note: For the sake of brevity, this example will only examine the Operational COG

7. Key Step 7: Conduct Initial Force Structure Analysis (Apportioned and Assigned Forces)

a. <u>Conduct an Analysis of Available Forces and Assets</u>.

- Review forces that have been provided for planning and their locations (apportioned and assigned).
- Determine the status of reserve forces and the time they will be available.
- Referring back to your specified and implied tasks, now determine what broad force structure and capabilities are necessary to accomplish these tasks (e.g., is a Carrier Battle Group or a forcible entry capability required?).
- Identify shortfalls between the two.
  - Example:
    Task: Seize APOD.
    Observation: No forced entry capability (MEU, Airborne).

b. <u>Force Planning</u>. The purpose of force planning is to identify all forces needed to accomplish the supported CDR's CONOPS and plan the phase of the forces into the theater. It consists of determining the force requirements by operation phase, mission, mission priority, mission sequence, and OA. It includes major force phasing; integration planning; force list structure development; followed by force list development.

(1) Force planning is the responsibility of the CCDR supported by the Service component commanders, the Services, JS and JS JFC/JFPs (see Chapter 3, *GFM*).

(2) The primary objectives of force planning are to apply the right force to the mission while ensuring force visibility, force mobility, and adaptability.

(3) Force planning begins during CONOPS development. The supported CDR determines force requirements, develops a TPFDD and letter of instruction (LOI), and designs force modules to align and time-phase the forces in accordance with (IAW) the CONOPS.

(4) Major combat forces are selected for planning purposes from those apportioned for planning and included in the supported CDR's CONOPS by operation phase, mission, and mission priority. The supported CCMD Service components then collaboratively make tentative assessments of the combat support (CS) and combat service support (CSS) required IAW the CONOPS.

(5) During the planning stage, these forces are planning assumptions called "preferred forces."

(6) The CDR identifies and resolves or reports shortfalls.

(7) For some high priority plans, the JS JFC/JFPs may be tasked to contingency source the plan. Contingency sourcing is based on planning assumptions and contingency sourcing guidance. To the degree that the contingency sourcing planning assumptions and sourcing guidance mirrors the conditions at execution, contingency sourcing may validate the supported CCMD preferred forces and delineate associated risk. If the plan is directed to be executed, the force requirements are forwarded to the JS for validation and tasked to the JS JFC/JFPs to provide sourcing recommendations to the SECDEF. When the SECDEF accepts the sourcing recommendation, the FP is ordered to deploy the force to meet the requested force requirement. After the forces are sourced, the CCDR refines the force plan to ensure it supports the CONOPS, provides force visibility, and enables flexibility. However, the planner must realize that if the resources for the proposed plan are not available, the plan must change (see Chapter 6-III, *Plan Assessment*).

The CONOP for a plan generally covers how the combined forces will work together to accomplish the mission. When forwarding force requirements, the Force Providing Community, JS JFC/JFPs, Services and FPs, need to know the CONOP for each unit and what tasks and missions the specific requested force will be expected to accomplish. This information is necessary so the identified force can be trained, organized and equipped to maximize the chances of mission success. The overall operation CONOP and the CONOP for each unit are usually not the same.

    c. Availability of Forces for Joint Operations Planning.

      (1) Availability of forces for joint operation planning utilizes three terms as defined in the CPG, JSCP and GFMIG — *assigned, allocated, and apportioned* — to define the availability of forces and resources for planning and conducting joint operations.

      (2) For deliberate planning, contingency plans assigned through the JSCP process utilize force assignment and force apportionment (see Chapter 3, *GFM* for an in-depth discussion of assigned and allocated forces.)

      (a) Apportionment. Pursuant to the JSCP, "apportioned forces are types of combat and related support forces provided to CCDRs as a *starting point for planning purposes only.*" Forces apportioned for planning purposes may not be those allocated for execution.

      1 *Force Apportionment Guidance*. Apportioned forces are combat and related support capabilities provided to CCDR's for planning purposes only. Apportionment supports the overlapping requirements of the NDS and the NMS. "Available forces" are apportioned without consideration to readiness status; however, apportioned forces are what a CCDR can reasonably expect to be available, but not necessarily allocated for use when a contingency plan transitions to execution.

      2 *Force Apportionment*. Forces are grouped into three categories:

        • Forces committed (named operations). The SECDEF can dip into this category for high priority planning efforts.

        • Forces available for planning.

        • Homeland Defense.

      3 *Preferred Force Identification*. Preferred force identification spans the entire planning process from Strategic Guidance through Plan Assessment. Preferred forces are forces that are identified by the supported CCDR in order to continue employment, sustainment and transportation planning and assess risk. These forces are *planning assumptions* only, are not considered "sourced" units, and does not indicate that these forces will be contingency or execution sourced. CCDR Service and functional components are encouraged to work with the JS JFC/JFPs and their Service components to make the best possible assumptions with respect to generating preferred forces. The preferred forces identified for a plan by the CCDR should not be greater than the number of forces apportioned for planning unless the CJCS or his designee either grants permission or has provided amplifying planning guidance. To the degree the CCDR is able to make good planning assumptions when selecting preferred forces determines the feasibility of a plan, and may assist the JS JFC/JFPs in identifying forces should the plan be designated for contingency sourcing or transitions to execution. Preferred forces are based on planning assumptions; these units are still *not* considered "sourced."

    d. Force Sourcing and the GFM Process. Force sourcing covers a wide range of sourcing methodologies providing CCDRs with requested capabilities. The intent is to provide the CCDR with the most capable forces based on stated capability requirements, balanced against risks (operational, future challenges, force management, institutional) and global priorities. Force sourcing is broken into three broad categories: *execution and contingency sourced forces, and preferred force identification*, which was previously discussed under force apportionment guidance.

> *"It does not do to leave a live dragon out of your calculations, if you live near one."* – J.R.R. Tolkien

8. <u>Key Step 8</u>: Conduct Initial Risk Assessment

a. <u>Assessing Risk</u>. Risk assessment is initially conducted during mission analysis and is updated throughout the planning process. The concept of assessing and balancing risks has a long pedigree in defense planning. By its very nature, from the most basic tactical level to the realm of global grand strategy, military planning involves assessing risks. To the extent that resources are not sufficient to provide certainty of success, military plans and operations involve accepting varying degrees of risk and deciding whether one COA is, in view of what is at stake, worth the likely cost in blood and treasure, or is too risky to undertake, or whether another COA would entail more or less risk.[45]

War is inherently complex, dynamic, and fluid. It is characterized by uncertainty, ambiguity, and friction. Uncertainty results from unknowns or lack of information. Ambiguity is the blurring or fog that makes it difficult to distinguish fact from impression about a situation and the enemy. Friction results from change, operational hazards, fatigue, and fears brought on by danger. These characteristics cloud the operating environment; they create risks that affect a CDR's ability to fight and win. In uncertainty, ambiguity and friction, both danger and opportunity exist. Hence, a leader's ability to adapt and take risks are key traits.

b. <u>Risk</u>. Risk is the probability and consequence of an event causing harm to something valued, classified within one of four risk levels (*low, moderate, significant, or high*). CDRs and senior leaders should account for risk when evaluating the likelihood of mission success. Risk can be assessed through the cost imposed by, or the impact on, achievement of the objective. Risks may result from enemy action, incorrect assumptions, limited resources, lack of preparation, friendly force activities, environment and terrain, and public opinion, among others. The most serious risks are the ones that endanger mission success or pose significant threats to U.S. national interests. The most likely risks are commonly the security of forces, sustainability of equipment, or delayed timelines.

c. <u>Military Risk</u> – There are two categories of military risk: **Risk-to-Mission** and **Risk-to-Force.** Risk-to-Mission is the probability and consequence of current and contingency events causing harm to current or future military objectives, while Risk-to-Force is the probability and consequence of current and contingency events causing harm to the provision and sustainment of sufficient military resources. Both must be considered when calculating military risk. It involves balancing a CCMD's ability to attain steady state, current operations and contingency plan objectives against the Services' and JFP's ability to support CCMD missions (see Chapter 3, *GFM* and CJCSM 3105.01 *Joint Risk Analysis Methodology* for additional information).

(1) <u>Operational Risk (Risk-to-Mission)</u>: Operational risk is a function of the probability and consequence of failure to achieve mission objectives while protecting the force from unacceptable losses. It reflects the current force's ability to attain current military objectives called for by the current NMS, within acceptable human, material, and financial costs. This risk subset considers the ability to execute current, planned, and contingency operations in the force employment period. The TPFDD for each of these plans serves to identify and limit risk to the force. Plans without a verified TPFDD have more risk. CDRs consider the feasibility of these plans in conjunction with operational concerns, such as the potential for escalation, to assess risk to a threat adequately.[46]

(2) <u>Force Management Risk (Risk-to-Force)</u>: Force management risk is a function of the probability and consequence of not maintaining the appropriate force generation balance. It reflects a force provider's ability to generate ready forces within capacities to meet

[45] CRS, Quadrennial Defense Review 2010. The QDR has been superseded, however this observation is still relevant

[46] ATO 5-19 Risk Management, JP 5-0 Joint Planning, CJCJM 3105.01 Joint Risk Analysis Methodology

current campaign and contingency mission requirements. This risk subset considers the ability to execute plans today (execution) to contingency missions (e.g., potential conflict) over the force employment and force development periods.

d. Risk Management.

(1) Risk management (RM) is the process to identify, assess and control risks and make decisions that balance risk cost with mission benefits.[47]

(2) RM informs decisions to reduce or offset risk. Using this process increases operational effectiveness and the probability of mission accomplishment. It is a systematic way of identifying hazards, assessing them, and managing the associated risks. The process applies to all types of operations, tasks, and activities.

(3) All staff elements incorporate RM into their running estimates and provide recommendations for controls to mitigate risk within their areas of expertise. The CDR has overall responsibility for RM integration and is the risk acceptance authority. During planning, commanders, leaders, and individuals identify potential hazards and assess their likely impact. Steps 1 and 2 of RM—identifying and assessing hazards—provide a structure to enhance situational understanding and support developing sound courses of action and plans. Then, planners can state how forces will accomplish a mission within a predetermined level of risk. Making optimal use of planning time is essential for effective RM. The more thorough the planning, the more contingencies can be ready for implementation. During preparation, leaders balance the risks (such as readiness, political, economic, and environmental risks) against the costs of each COA. At the same time, planners develop actions that mitigate risk (controls), and leaders make risk decisions to eliminate unnecessary risks. CDRs and planners should continually assess the risk level and effectiveness of controls throughout execution. They should supervise the risk-related activities for which they are responsible and monitor other activities directly affecting risk during operations. Any time or reason risk levels appear to rise or new hazards occur, commanders and leaders should be prepared to order adjustments to activities, including the actions that mitigate risk. CDRs use continuous assessments to make adjustments. These cyclical processes support making adjustments where and when needed. Planners and staff capture lessons learned to benefit current and future operations.

e. Principles of Risk Management. RM outlines a disciplined approach to express a risk level in terms readily understood at all echelons. Except in time-constrained situations, planners complete the process in a deliberate manner—systematically applying all the steps and recording the results. Planners develop data and use charts, codes, and numbers to analyze probability and standardize the analysis of risk. They use this standardization to manage risk in a logical and controlled manner over time. However, the five-step process is compatible with intuitive and experience-based decision-making. In time-constrained conditions, commanders and planners use judgment to apply RM steps and principles. The principles of RM follow below:

- Integrate RM into all phases of missions and operations.
- Make risk decisions at the appropriate level.
- Accept no unnecessary risk.
- Apply RM cyclically and continuously.

(1) Integrate Risk Management into All Phases of Missions and Operations. CDRs must emphasize RM in planning processes; they must dedicate sufficient time and other resources to RM during planning to ensure forces manage risk effectively throughout all phases of missions and operations.

(2) Make Risk Decisions at The Appropriate Level. A risk decision is a CDR's determination to accept or not accept the risk(s) associated with an action he or she will take or will direct others to take. RM is only effective when the specific information about hazards and risks is passed to the appropriate level of command for a risk decision. Subordinates

[47] JP 5-00.2, JTF Planning Guidance and Procedures

must pass specific risk information up the chain of command. Conversely, the higher command must provide subordinates making risk decisions or implementing controls with the established risk tolerance—the level of risk the responsible CDR is willing to accept. RM application must be inclusive; those executing an operation and those directing it participate in an integrated process.

(3) <u>Accept No Unnecessary Risk</u>. An unnecessary risk is any risk that, if taken, will not contribute meaningfully to mission accomplishment or will needlessly endanger lives or resources. CDRs accept only a level of risk in which the potential benefit outweighs the potential loss. The process of weighing risks against opportunities and benefits helps to maximize unit capability, save lives, and preserve resources. The appropriate level of command makes risk acceptance decisions after applying RM and weighing potential gain against potential loss. CDRs need not be risk averse. Forces may undertake even high-risk endeavors when commanders determine that the sum of the benefits exceeds the sum of the costs. CDRs establish the basis for risk acceptance decisions through RM.

(4) <u>Apply Risk Management Cyclically and Continuously</u>. RM is a cyclical and continuous five-step process. CDRs, planners and staffs use this cyclical process (illustrated in Figure K) to identify and assess hazards; develop, choose, implement, and supervise controls; and evaluate outcomes as conditions change.

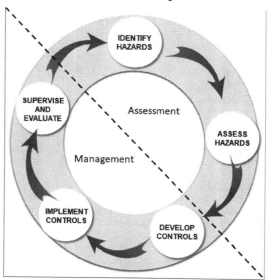

Figure K. Risk Management Process

f. <u>Application Levels of Risk Management</u>. The main factor that differentiates the RM approach is the amount of time available for planning. A deliberate approach is more analytical but takes more time; a real-time or time sensitive approach is more intuitive and tends to take less time. Regardless of the amount of time available, Joint forces manage risk throughout the operations process using the five steps of RM.

g. <u>Five Steps of Risk Management</u>. *Steps 1 and 2 of RM are assessment steps—risk assessment is the identification and assessment of hazards. Steps 3 through 5 of RM are management steps.* The goal of the process is managing risk. However, to manage risk, you must first identify and assess hazards. It is preferable to divide complex missions into subtasks and then identify and assess hazards and manage risk for each subtask.

- Step 1–Identify the hazards.
- Step 2–Assess the hazards.

- Step 3—Develop controls and make risk decisions.
- Step 4—Implement controls.
- Step 5—Supervise and evaluate.

(1) <u>Step 1—Identify the Hazards</u>. A hazard is a condition with the potential to cause injury, illness, or death of personnel; damage to or loss of equipment or property; or mission degradation. Hazards create the potential for harmful events that can cause degradation of capabilities or mission failure. Hazards lead to risk whenever people interact with equipment or their environment. Hazards exist in all types of environments and activities. When identifying hazards consider all aspects of mission, enemy, terrain, and weather, time, troops available and civilian (METT-TC) for current and future situations.

(2) <u>Step 2—Assess the Hazards</u>. To assess hazards, consider how identified hazards (conditions) could lead to harmful events and how those events would affect operations. They envision the potential for the events and their predictable effects. When hazards are assessed and risk levels are assigned, the resulting analysis is a measurement of risk—probability and severity driven chance of loss, caused by threat or other hazards. Risk levels reflect a combination of the probability of occurrence and the severity of the adverse impact. Probability and severity are independent measures of hazards. In other words, estimating probability has no direct relationship to estimating severity (See Table 4).

<u>1</u> <u>Probability and Consequences Defined</u>. In the context of RM, probability is the likelihood an event will occur; it is assessed as ***very likely, likely, very unlikely***. Severity is the expected consequences of an event in terms of injury, property damage, or other mission-impairing factors; it is assessed as ***extreme, major, modest, minor***. A risk level is a type of score that assesses the odds (probability) of something going wrong and the effect (severity) of the incident when it occurs.

<u>2</u> <u>Risk Levels</u>. Planners assess hazards (the conditions and the events that could result)—and assign associated risk levels—during mission analysis; COA development; COA analysis; and the orders production, dissemination, and transition steps of the JPP. CDRs and staff must consider aspects directly or indirectly related to the mission that could affect risk during operations. The result of this assessment is an initial estimate of a risk level for each identified hazard, expressed as— ***High, Significant, Moderate, Low***.

h. <u>Step 3—Develop Controls and Make Decisions</u>. In Step 3, develop and consider options for controls. During control development consider the mitigating effects of the proposed controls. Reassess the initial level of risk and determine a residual level of risk (risk after controls are implemented). Continue developing control options, considering their mitigating effects, and reassessing risk until they have determined the most effective controls. The CDR at the appropriate echelon determines the risk tolerance for the situation. The CDR makes risk decisions—to accept or not accept the risk based on the residual risk level. For example, the controls, when implemented, are expected to reduce the residual risk levels from moderate to low or from high to moderate. CDRs must always determine that the potential benefits of the action outweigh the potential cost.

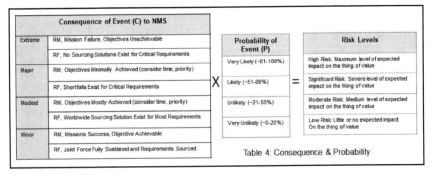

Table 4: Consequence & Probability

(1) Residual Level of Risk. After effective controls have been identified, planners return to the risk assessment matrix to determine the residual level of risk for each hazard and the overall residual risk for the operation. They should continue analyzing the hazards and proposing options to reduce or eliminate risks until they have identified the most effective controls. A given hazard's initial risk level may be lowered through the implementation of planned controls. For example, a hazard with an initial risk of high given effective controls emplaced may be lowered to a residual level of moderate. The appropriate level of command must approve the mission, making a final risk decision based on the residual level of risk. Planners should sort hazards and controls under consideration according to residual risk, placing the highest risk hazards first. This allows decision-makers at the appropriate level of command to identify the highest risk hazards easily. Decision-makers should keep in mind that the residual level of risk is valid (true) only if forces implement the controls.

(2) Make Risk Decisions. The purpose of RM is to provide a basis for commanders to make sound and informed risk decisions. To make those decisions, they must know the established risk tolerance and the potential gain. Ultimately, commanders are responsible for determining the risk tolerance within the command and for making risk decisions for operations, missions, or tasks. The appropriate level CDR must make risk decisions about specific hazards and controls, consistent with risk tolerance guidance. CDRs' decisions must balance risk against expected gains.

> "Sizing up opponents to determine victory, assessing dangers and distances is the proper course of action for military leaders."
> Sun Tzu, The Art of War, "Terrain"

i. Step 4—Implement Controls. Controls are normally implemented during the preparation activities of the operations process. CDRs establish how the controls will be implemented and who will manage them. They ensure selected controls are translated into briefings and curricula and then integrated with training. CDRs ensure subordinates fully understand and implement the controls. They ensure the implemented controls are maintained to standard.

j. Step 5—Supervise and Evaluate. Primarily, Step 5 involves ensuring that controls are implemented and performed to standard. CDRs, planners and staff apply this step to validate that selected controls support achieving the endstate. They identify weaknesses of controls and make changes or adjustments based on performance or changing situations, conditions, or events. However, supervision and evaluation are not limited to controls. Like other steps of RM, supervision and evaluation must occur throughout all phases of any operation or activity. CDRs supervise and evaluate all aspects of RM continuously.

(1) Supervise. Supervision is a primary means of regulating forces. Step 5 fully integrates supervision into RM. CDRs ensure supporting commanders responsible for implementing controls understand their responsibilities. They ensure subordinates understand how, when, and where to implement controls. CDRs supervise and monitor controls. They verify implementation and make sure controls remain in place.

(2) Evaluate. CDRs and planners conduct evaluation during all phases and activities of operations, including after AARs and other assessments at the end of an operation. Evaluation supports several goals, including but not limited to—

- Determining if risk levels changed during operations.
- Adapting to changes in the situation.
- Monitoring effectiveness of controls.
- Making corrections to control implementation.
- Improving the application of RM principles in current and future operations.

k. <u>Applying Risk Management</u>. Risk management requires a clear understanding of what constitutes unnecessary risk, when the benefits actually do outweigh costs, and guidance as to the command level to make those decisions. When a CDR decides to accept risk, the decision must be coordinated with the affected units; where and how the CDR is willing to accept risk are detailed in each COA.

(1) Risk management assists commanders in conserving lives and resources and avoiding or mitigating unnecessary risk, making an informed decision to execute a mission, identifying feasible and effective control measures where specific standards do not exist, and providing reasonable alternatives for mission accomplishment.

(2) While risk cannot be totally eliminated, it can be managed by a systematic approach that weighs the costs—time, personnel, and resources—against the benefits of mission accomplishment. CDRs have always risk-managed their actions: intuitively, by their past experiences, or otherwise. Risk-management won't prevent losses but, properly applied, it allows the CDR to take necessary and prudent risks without arbitrary restrictions and while maximizing combat capabilities.

(3) Risk management does not inhibit CDR's flexibility and initiative, remove risk altogether (or support a zero defects mindset), require a GO/NO-GO decision, sanction or justify violating the law, or remove the necessity for development of SOPs. Risk management should be applied to all levels of war, across the range of military operations, and all phases of an operation to include any branches and sequels of an operation. To alleviate or reduce risk, commanders may change the CONOPS or concept of fire support, execute a branch plan, or take other measures to reduce or bypass enemy capabilities.

(4) Safety is crucial to successful training and operations and the preservation of military power. High-tempo operations may increase the risk of injury and death due to mishaps. Command interest, discipline, risk mitigation measures, and training lessen those risks. The CDR reduces the chance of mishap by conducting risk assessments, assigning a safety officer and staff, implementing a safety program, and seeking advice from local personnel. Safety planning factors could include the geospatial and weather data, local road conditions and driving habits, uncharted or uncleared mine fields, and special equipment hazards.

(5) To assist in risk management, commanders and their staffs may develop or institute a risk management process tailored to their particular mission or operational area.

l. <u>Summary</u>. The CDR expresses guidance regarding risk in several ways. The CDR's risk estimate is based on the mission, his or her experience, higher headquarters' guidance and staff estimates. With these considerations, the CDR formulates initial staff guidance, followed by an intent statement during the mission analysis step. The CDR expresses an estimate of risk every time he or she provides guidance. Some risk factors permit quantitative analysis, while others will be wholly qualitative. Probability and statistics support risk analysis, but the CDR will have to address operational risk subjectively when supporting information is unavailable. See the Table 5 for an example Operational Risk Matrix.

Note: For more information refer to ATO 5-19 Risk Management, JP 5-0 Joint Planning and CJCJM 3105.01 Joint Risk Analysis Methodology.

"If you can recover from the loss, it's a risk. If not, it's a gamble"
Field Marshal Erwin Rommel

| Risk Identification | Mission Risk Assessment | | | Controls (mitigation) | Residual Mission Risk* |
	Probability	Severity	Risk		
Outside Interference with theater LOCs	Likely	Extreme	Significant	-Patrol -ISR -Escorts	Moderate
Blue infrastructure inadequate to support deployed forces	Unlikely	Minor	Low	-Site surveys -LOG ADVON -LOG afloat	Low
Blue forces conduct active asymmetric ops	Likely	Moderate	Moderate	-Force protection -IO/MISO -CMO	Low
Yellow/Orange conduct coordinated attacks with Red	Very Likely	Major	High	-Deter/Dissuade (DIME) -Force planning	Low
Air Strike and/or missile strike against Blue	Likely	Moderate	Moderate	-ISR -Air defense -Deter/Dissuade	Low

Table 5. Example: Operational Risk Matrix.

9. Key Step 9: Determine Commander's Critical Information Requirements (CCIR)

CCIRs are elements of information required by the CDR. They directly affect decision-making and are a key information management tool and help the CDR assess the operational environment and identify decision points throughout the conduct of operations. *CCIRs are established by the CDR* and should be developed and recommended by staffs as part of the planning process.

a. Characteristics of CCIRs result from the analysis of information requirements in the context of a mission, CDR's intent, and the concept of operation. CDRs designate CCIRs to let their staffs and subordinates know what information they deem necessary for decision-making. In all cases, the fewer the CCIRs, the better the staff can focus its efforts and allocate scarce resources. Staffs may recommend CCIRs; however, they keep the number of recommended CCIRs to a minimum. CCIRs *are not static*. CDRs add, delete, adjust, and update them throughout an operation based on the information they need for decision-making. To assist in managing CCIRs, CDRs should adopt a process to guide the staff. This process should include specific responsibilities for development, validation, dissemination, monitoring, reporting, and maintenance (i.e., modifying/deleting).[48]

> *"Tell them of us and say,*
> *For their tomorrow,*
> *We gave our today."*
> *– The Kohima Epitaph*

[48] JP 3-0, Joint Operations

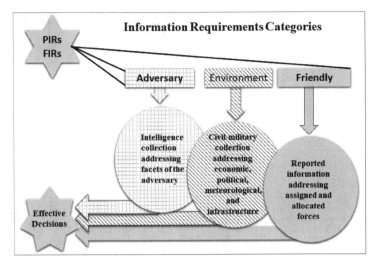

Figure L. Information Requirements Catergories.

b. <u>Commanders' Critical Information Requirement (CCIR)</u>.

(1) CCIRs comprise information requirements identified by the CDR as being critical to timely information management and the decision-making process that affect successful mission accomplishment. CCIRs result from an analysis of information requirements in the context of the mission and the CDR's intent. Described in the following sections are the two key subcomponents: Friendly Force Information Requirements (FFIR) and Priority Intelligence Requirement (PIRs) (See Figure L).

(a) *PIR*. Those intelligence requirements for which a CDR has an anticipated and stated priority in the task of planning and decision-making.[49]

(b) *FFIR*. Information that the CDR and staff need about the forces available for the operation.[50]

Commanders Critical Information Requirements

Commanders Critical Information Requirement.. An information requirement identified by the commander as being critical to facilitating timely decision-making. The two key elements are Friendly Force Information Requirements and Priority Intelligence Requirements.
JP 3-0, Joint Operations

•Two Categories:

•**Priority Intelligence Requirement (PIR)** – An intelligence requirement, stated as a priority for intelligence support, that the commander and staff need to understand the adversary or the environment.

•**Friendly Force Information Requirements (FFIR)** – Information the commander and staff need to understand the status of friendly force and supporting capabilities.

---

[49] *DOD Dictionary of Military and Associated Terms (DOD Dictionary)*
[50] *Ibid*

(c) The information needed to verify or refute a planning assumption is an example of a CCIR. As mentioned earlier, CCIRs are not static. They are situation-dependent, focused on predictable events or activities, time-sensitive, beg a decision by the CDR when answered, and always established by an order or plan.

(d) CCIR *previously* had three categories: (1) PIR, (2) Essential Elements of Friendly Information (EEFI), and (3) Friendly Force Information Requirement (FFIR). As discussed previously, PIR remains a category of CCIR, as does FFIR. The third category of CCIR, EEFI, has been eliminated as a category of CCIR. EEFI was defined as *"key questions likely to be asked by adversary officials and intelligence systems about specific friendly intentions, capabilities, and activities, so they can obtain answers critical to their operational effectiveness."* In reality, one can still expect to hear the term used in the operating forces because many commanders believe it is still necessary to provide guidance to their staff concerning operational security, therefore it is mentioned here.

(e) There is a proverbial saying that "a CCIR is so important that when answered, one must wake up the CDR." There are general criteria that apply to CCIR that planners must consider when proposing them to the CDR for approval: (1) answering a CCIR begs a decision by the CDR, and not necessarily by the staff, and (2) the information or intelligence necessary to answer or satisfy a CCIR must be critical to the success of the mission. Keep in mind that a CCIR cannot be a CCIR unless approved by the CDR. This criterion also pertains to PIRs, which is the subject of the next section.

Priority Intelligence Requirements:

• PIRs are the commander's statements of the force's critical intelligence needs.

  - The J-2 is responsible for assisting the commander in determining PIRs.

• Keep PIRs to a minimum.

---

c. Priority Intelligence Requirement.

(1) PIRs are normally identified during JIPOE and refined during mission analysis and COA development. JIPOE provides basic intelligence oriented on understanding the battlespace and adversary. The intelligence enables the J-2 to identify PIRs based on what intelligence critical to mission accomplishment is lacking.

(2) The J-2 will normally recommend PIRs to the CDR for approval. IO planners can have input into PIRs. Input can be submitted through several personnel, including J-2 personnel supporting the IO cell, the J-2 or Deputy J-2, the J-2 Joint Intelligence Operations Center (JIOC) CDR, J-2 Watch Officer in the Joint Operations Center (JOC), or Joint Intelligence Support Element (JISE) Watch Officer or Senior Analysis.

(3) PIRs are meant to *focus intelligence operations.* The J-2 will rarely, if ever, have all the intelligence collection assets and personnel necessary to satisfy all requirements. The preponderance of the level of effort for the J-2's organization and assets will direct their focus of effort on answering PIR. Therefore, it is important that the number of PIRs remain manageable. Otherwise, everything becomes a priority and PIRs no longer serve their intended purpose since there is no longer a focus for intelligence operations.

d. A Conceptual look at Intelligence Requirements. More often than not, intelligence requirements are worded in a manner that does not enable a collection asset or resource to answer it directly.

Intelligence requirement – any subject, general or specific, upon which there is a need for the collection of information, or the production of intelligence, and "a requirement for intelligence to fill a gap in the command's knowledge or understanding of the battlespace or threat forces."

(DOD Dictionary of Military and Associated Terms (DOD Dictionary)

---

e. If an intelligence requirement cannot be collected on directly, then it must be dissected into *information requirements.*

> **Information requirement** – "those items of information regarding the adversary and the environment that need to be collected and processed in order to meet the intelligence requirements of a commander."
>
> (DOD Dictionary of Military and Associated Terms (DOD Dictionary)

f. Information requirements are worded in a manner that enables a collection manager to task an asset to collect on it directly. It is basically an intelligence requirement broken down into "*bite-size chunks*," and when an asset collects on and answers these "bite-size chunks," an answer to the intelligence requirement can be determined. If an intelligence requirement is worded in a manner that enables the collection manager to task a collection asset to collect directly on it, then it is not required to be broken down into information requirements. Otherwise, one or more information requirements will normally be drafted to support answering an intelligence requirement.

(1) There are various methods that can be used to break down an intelligence requirement into information requirements. Either the submitter of a Request for Information (RFI) can attempt to break it down, or the J-2 analysts can do so upon receipt of a validated intelligence requirement. Either way can be effective, but if the first method is attempted, one can expect J-2 personnel to modify the information requirements to facilitate collection or processing of the intelligence necessary to answer the intelligence requirement.

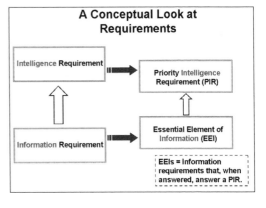

(2) Since a PIR is an intelligence requirement, it too may have to be dissected in the same manner as a standard intelligence requirement. The information requirements that must be answered to answer a PIR are called *Essential Elements of Information*, or EEI. They are simply information requirements that receive a special title due to their importance, i.e., because they are tied to a PIR. According to the Joint Pub 2-0, *Joint and National Intelligence Support to Military Operations*, "those

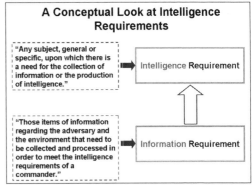

information requirements that are most critical or that would answer a PIR are known as essential elements of information (EEIs)."

(3) A useful unclassified reference documents for drafting intelligence and information requirements are the *Intelligence Requirements Handbooks* published by the Marine Corps Intelligence Activity (MCIA). They are also available electronically on the MCIA portal. MCIA has produced a *Generic Intelligence Requirements Handbook* (GIRH) specifically for IO.

(4) These examples below contain a select sample of EEIs; however, there could be any number of EEIs necessary to support answering a PIR.

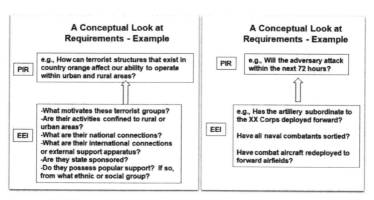

g. <u>Requests for Information (RFI)</u>. Many RFI's are between supporting and supported headquarters. It is important to remember that RFI's are broader than just intelligence based, however, we will discuss intelligence RFI's since they seem prevalent.

(1) If we cannot answer intelligence or information requirements with existing re-sources, then we must generate an RFI. However, keep in mind that the need for friendly information must be satisfied at the LOWEST possible level – that is you! If you cannot find the information that satisfies your intelligence or information requirement, then begin generating an RFI.

(a) If you determine that you must submit an RFI, there are certain responsibilities associated with doing so. *First*, you must conduct the initial research. While this sounds trivial, it is often overlooked. The answer to your question may lie in a readily available document or on the Internet - whether it is classified or unclassified. While it may be easier to simply "pass the buck" and let someone else do the work for you, it will always delay the answer and slow down the entire process.

> ### Request for Information
> Any specific time-sensitive ad hoc requirement for intelligence information or products to support an ongoing crisis or operation not necessarily related to standing requirements or scheduled intelligence production. A request for information can be initiated to respond to operational requirements and will be validated in accordance with the theater command's procedures.
>
> *Two excellent sources that can immediately be visited to begin initial research are the INTELINK homepage on both SIPRNET and JWICS. Each page contains a Google™ search capability just as the NIPRNET does if you choose to use open sources. Depending on your search string, you can significantly increase the timeframe by which you can find an answer to your RFI.*

(b) *Second*, if you are unsuccessful in your initial search, then you must clearly articulate your requirement. Be as specific as possible, but avoid using language associ-ated with a specific occupation that may appear foreign to the person who must review and validate your RFI. Additionally, *avoid asking for a particular intelligence collection asset to satisfy your requirement.* It is the Collection Manager's job to determine the most effective manner to satisfy your requirement.

(c) *Third*, you must justify your request. Clearly articulate why your request is so important compared to the hundreds of others. Stating, "because the boss wants it" is not good enough. You must clearly and accurately justify your request for it to be accepted and prioritized. If you can tie your requirement to a PIR, it may be prioritized higher than other requirements which may not be.

(d) *Fourth*, you must truthfully determine and document the "Latest Time Information is of Value," or LTIOV. We have already mentioned this is the time beyond which the information will no longer be useful to you. When identifying this time, ensure it is accurate. Obviously, you will want your answer immediately, but if you really don't need it for a week, then state that. This helps the collection manager to prioritize requests and free collection assets to perform other priority missions until assets or resources are available to support your requirement.

RFI Responsibilities:
- Conduct Initial Research.
- Clearly State the Requirement.
- Justify the Request.
- Provide an Accurate LTIOV.

(e) Prior to submitting an RFI via the means established by the J-2, you should ask four questions to validate your efforts. The questions represent validation criteria that should be met prior to submitting the RFI. When this criterion is met, then your RFI can be submitted.

<u>1</u> Requestor's Questions:
- Is the answer to the question relevant to executing or planning my mission?
- Have I thoroughly searched existing, accessible repositories for this information?
- Have I provided concise, yet sufficient detail to enable the RFI Manager to understand the intelligence requirement that I need answered?
- Is my stated LTIOV realistic?

<u>2</u> The RFI Manager will receive your RFI and ask questions to determine if it will be validated or not.

<u>3</u> RFI Manager's Questions:
- Does the information justify the dedication of scarce intelligence resources?
- Does it duplicate an existing requirement?
- Has the requirement been previously satisfied?
- Does the RFI pertain to only one intelligence requirement?
- Can I retrieve the information required by the LTIOV?

(2) If the RFI Manager decides it cannot be validated, you should expect a phone call or email explaining why. If your RFI is validated, then it will be prioritized against other validated "intelligence" or "information" requirements. The RFI will then evolve into either a production requirement or a collection requirement. A *production requirement* is submitted when new, finished intelligence derived from original research is required to satisfy all or a portion of the RFI. A *collection requirement* is submitted when insufficient information exists to answer an RFI. A production requirement is submitted to the "Analysis & Production Cell," and a collection requirement is submitted to the "Collection Management Cell" for action. The RFI Manager will normally confer with each cell (or section) to receive their estimate of supportability prior to submitting it either to one cell or the other as a whole or a portion thereof to each of them. If it takes the efforts of both cells to satisfy the requirement, the RFI Manager will work with personnel from each cell to determine which portion of the requirement each will satisfy. Each cell will in turn take their portion of the RFI for action.

(3) The RFI Manager could be co-located with the Collection Management Cell, or they could occupy a separate and distinct RFI Management Cell. To learn more about the RFI process and the organization that supports it, planners should direct their questions to either the J-2, Deputy J-2, or directly to the RFI Manager or Collection Manager. Each can offer insight into how the process is managed and how the organization was established to support it.

h. Requesting National Intelligence Support. If organic or attached intelligence capabilities within a JTF cannot satisfy an RFI, it will likely be forwarded to the higher headquarters, i.e., to the CCMD JIOC, requesting that they satisfy the RFI with their organic or attached intelligence capabilities. RFI's are normally submitted to higher via *Community On-line Intelligence System for End Users and Managers* (COLISEUM). Likewise, if the CCDR JIOC cannot satisfy the RFI, it is then submitted via COLISEUM to the Defense JIOC (DJIOC), requesting that it be satisfied with national intelligence capabilities.

> COLISEUM is a production management tool used throughout the Defense Intelligence Production Community. The tool allows all users to register and track requests for information and production requirements, perform research for existing requirements or answers, and manage/ account for production resources. COLISEUM is a web-based application available throughout the Intelligence Community.

i. The CCMD JIOC is the primary focal point for providing intelligence support to the CCMD. The CCMDs JIOCs are the primary intelligence organizations providing support to joint forces at the operational and tactical levels. The JIOC fuses the in-theater capabilities of all Director of National Intelligence (DNI), Service, combatant support agency, and combat command ISR assets into a central location for intelligence planning, collection management, tasking, analysis and support. The JIOC concept seamlessly combines all intelligence functions, disciplines, and operations in a single organization, ensures the availability of all sources of information from both CCMD ISR assets and national intelligence resources, and fully synchronizes and integrates intelligence with operation planning and execution. Although a particular JIOC cannot be expected to completely satisfy every RFI, it can coordinate support from other intelligence organizations, both lower, higher, and laterally.

j. As the lead DOD intelligence organization for coordinating intelligence support to meet CCMD requirements, the DJIOC coordinates and prioritizes military intelligence requirements across the CCMDs, CSA's, RC, and Service intelligence centers. The DJIOC formulates recommended solutions to de-conflict requirements for national intelligence with USSTRATCOM's Joint Functional Component Command for Intelligence, Surveillance and Reconnaissance (JFCC-ISR) and DNI representatives to ensure an integrated response to CCMD needs. It is the channel through which joint force's crisis-related and time-sensitive intelligence requirements are tasked to the appropriate national agency or command, when they cannot be satisfied using assigned or attached assets.

k. National Augmentation Support. One of the most effective means that a CCDR or JTF J-2 can facilitate having crisis-related RFI's satisfied in a relatively rapid manner is to request and exploit the capabilities of a National Intelligence Support Team, or NIST. At the request of a CCDR, the DJIOC may deploy a NIST to support a CDR, JTF (CJTF), during a crisis or operation. The NIST is a nationally sourced team composed of intelligence and communications experts from DIA, CIA, National Geospatial-Intelligence Agency (NGA), NSA, or other IC agencies as required. The NIST mission is to provide a tailored, national level, all-source intelligence team to deployed commanders during crisis or operations. NIST supports intelligence operations at the JTF HQ and is traditionally collocated with the J-2. In direct support of the JTF, the NIST will perform functions as designated by the J-2. The NIST is designated to provide a full range of intelligence support to a CJTF, from a single agency element with limited ultra-high frequency voice connectivity to a fully equipped team with the Joint Deployable Intelligence Support System (JDISS) and Joint Worldwide Intelligence Communications System (JWICS) video-teleconferencing capabilities. NIST provides coordination with national intelligence agencies, analytical expertise, Indications and Warning (I&W), special assessments, targeting support, and access to national data bases, and facilitates RFI management.

l. Decision Support.[51] CCIR support the CDR's future decision requirements and are often related to Measures of Effectiveness and Measures of Performance. PIR are often expressed in terms of the elements of PMESII while FFIR are often expressed in terms of DIME. All are developed to support specific decisions the CDR must make.

10. <u>Key Step 10</u>: Develop Mission Statement

a. <u>Mission Statement</u>. One product of the mission analysis process is the mission statement. Your initial mission analysis as a staff will result in a *tentative mission state-ment.* This tentative mission statement is a *recommendation* for the CDR based on mission analysis. It will serve to identify the broad options open to the CDR and to orient the staff. This recommendation is presented to the CDR for approval normally during the mission analysis brief. It must be a clear, concise statement of the **essential tasks** to be accomplished by the command and the **purpose** of those tasks. Although several tasks may have been identified during the mission analysis, the proposed mission includes only those that are essential to the overall success of the mission. The tasks that are routine or inherent responsibilities of a CDR are not included in the proposed mission. The proposed mission becomes the focus of the CDR's staff's estimates. It should be continually reviewed during the planning process to ensure planning is not straying from this critical focus (or that the mission requires adjustment). It is contained in Paragraph 1 of the CDR's Estimate and Paragraph 2 of the basic OPLAN or OPORD.

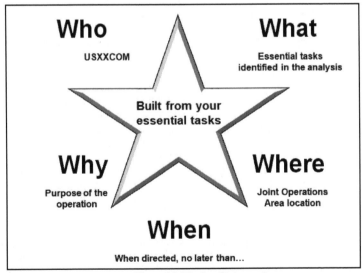

Figure M. Mission Statement

b. The mission statement should be a short sentence or paragraph that describes the organization's essential task (or tasks) and purpose — a clear (*brevity and clarity*) state-ment of the action to be taken and the reason for doing so. As denoted in Figure M the mission statement contains the elements of who, what, when, where, and why, but seldom specifies how. These five elements of the mission statement answer the questions:

[51] *JP 2-01, Joint and National Support to Military Operations*

- Who will execute the tasks (unit/organization)?
- What is the essential task(s) (mission task)?
- When will the operation begin (time/event i.e., O/O, when directed)?
- Where will the operation occur (AO, objective)?
- Why will we conduct the operation (for what purpose)?

c. Clarity of the joint force mission statement and its understanding by subordinates, before and during the joint operation, is vital to success. *The mission statement along with the commanders' intent, provide the primary focus for subordinates during planning, preparations, execution and assessment.*[52]

d. No mission statement should be written and not revisited thereafter; it's important to revisit it during the entire plan development process to ensure that it meets the needs of the CDR and the national leadership. A sample CCDR's mission statement could look like this:

> *"When directed, CDRUSXXCOM deters regional aggressors; if deterrence fails, CDRUSXXCOM defends the country of X and defeats external aggressors and conducts stability and support operations in order to protect U.S. interests and the Government of X."*

e. The who, where, when of the mission statement is straightforward. The *what* and *why*, however, are more challenging to write clearly and can be confusing to subordinates.

The **what** is a task and is expressed in terms of action verbs (for example, deter, defeat, deny, conduct, provide, contain, isolate, etc.). These tasks are measurable and can be grouped by actions by friendly forces and effects on adversary forces/capabilities. The what in the mission statement is the essential task(s) to be accomplished. It may be expressed in terms of either actions by a friendly force or effects on an adversary force. CDRs should utilize doctrinal-approved tasks. These tasks have specific meaning, are measurable, and often describe results or effects of the tasks relationship to the adversary and friendly forces.

The **why** puts the task(s) into context by describing the reason for conducting the task(s). It provides the mission purpose to the mission statement-why are we doing this task(s)? The purpose normally describes using a descriptive phrase and is often more important than the task because it provides clarity to the task(s) and assists with subordinate initiatives.

Example: Task Lists

Terrain:
- Seize
- Secure
- Clear
- Occupy
- Retain
- Recon

Enemy:
- Disrupt
- Defeat
- Destroy
- Block
- Contain
- Fix
- Canalize
- Delay
- Interdict
- Isolate
- Penetrate
- Suppress
- Neutralize
- Feint
- Demonstration
- Ambush
- Bypass

Friendly:
- Screen
- Guard
- Cover
- Withdraw
- Attack by Fire
- Support by Fire
- Follow & Assume
- Follow & Support
- Breach
- Disengage
- Exfiltrate
- Infiltrate

Purpose (in order to...)
- Allow
- Cause
- Create
- Deceive
- Deny
- Divert
- Enable
- Influence
- Open
- Prevent
- Protect
- Restore
- Support
- Surprise

Joint Planning Process

[52] *JP 3-0, Joint Operations*

11. <u>Key Step 11</u>: Develop and Conduct Mission Analysis Brief

Upon conclusion of the Mission Analysis and JIPOE, the staff will present a Mission Analysis Brief to the CDR. The purpose of the Mission Analysis Brief is to provide the CDR with the results of the preliminary staff analysis, offer a forum to surface issues that have been identified, and an opportunity for the CDR to give their guidance to the staff and to approve or disapprove of the staff's analysis. However, modifications to this brief may be necessary based on the CDR's availability of relevant information. There is no set format for the Mission Analysis Brief, however Figure N contains two *tested examples* of Mission Analysis Briefings that work, just tailor them to your needs.

a. The mission analysis briefing should not be a unit readiness briefing. Staff officers must know the status of subordinate and supporting units and brief relevant information as it applies to the situation.

b. The mission analysis briefing is given to both the CDR and the staff. This is often the only time the entire staff is present, and the only opportunity to ensure that all staff members are starting from a common reference point. Mission analysis is critical to ensure thorough understanding of the task and subsequent planning.

c. The briefing focuses on relevant conclusions reached as a result of the mission analysis. This helps the CDR and staffs develop a shared vision of the requirements for the OPLAN and execution.

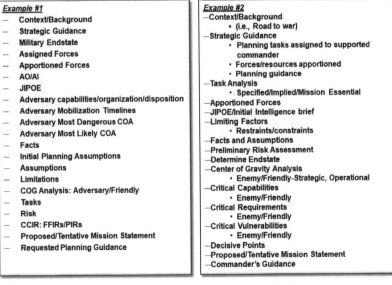

Figure N. Sample: Mission Analysis Briefs.

12. <u>Key Step 12</u>: Prepare Initial Staff Estimates

As discussed earlier, the development of an effective CDR's estimate must be supported by mission analysis, planning guidance, and *staff estimates*.

a. Battle rhythm is a deliberate daily cycle of command, staff, and unit activities intended to synchronize current and future operations.[53] The battle rhythm facilitates integration and collaboration. The Chief of Staff (COS) normally manages the headquarters battle rhythm. This battle rhythm serves several important functions, to include: establishing a routine for staff interaction coordination, facilitating interaction between CDR and staff, synchronizing activities of the staff in time and purpose and facilitating planning by the staff and decision-making by the CDR.

b. Early staff estimates are frequently given as oral briefings to the rest of the staff. They are continually ongoing and updated based on changes in the situation. In the beginning, they tend to emphasize information collection more than analysis. CJCSM 3122.01 contains sample formats for staff estimates.

c. The role of the staff is to support the CDR in achieving situational understanding, making decisions, disseminating directives, and following directives through execution. The staff's effort during planning focuses on developing effective plans and orders and helping the CDR make related decisions. The staff does this by integrating situation-specific information with sound doctrine and technical competence. The staff's planning activities initially focus on mission analysis, which develops information to help the CDR, staff, and subordinate commanders understand the situation and mission. Later, during COA development and comparison, the staff provides recommendations to support the CDR's selection of a COA. Once the CDR approves a COA, the staff coordinates all necessary details and prepares the plan or order.

d. Throughout planning, staff officers prepare recommendations within their functional areas, such as system, weapons, and munitions capabilities, limitations and employment; risk identification and mitigation; resource allocation and synchronization of supporting assets; and multinational and interagency considerations. Staff sections prepare and continuously update staff estimates that address these and other areas continuously throughout the JPP. The staff maintains these estimates throughout the operation, not just during pre-execution planning.

e. Not every situation will require or permit a lengthy and formal staff estimate process. During a crisis situation the CDR may review the assigned mission, receive oral staff briefings, develop and select a COA informally, and direct that plan development commence. However, deliberately planning will demand a more formal and thorough process. Staff estimates should be shared collaboratively with subordinate and supporting commanders to help them prepare their supporting estimates, plans, and orders. This will improve parallel planning and collaboration efforts of subordinate and supporting elements and help reduce the planning times for the entire process.[54]

13. <u>Key Step 13</u>: Approval of Mission Statement, Develop Commander's Intent and Publish Initial Planning Guidance

a. <u>Restated Mission Statement</u>. Immediately after the mission analysis briefing, the CDR approves a restated mission. This can be the staff's recommended mission statement, a modified version of the staff's recommendation, or one that the CDR has developed personally. Once approved, the restated mission becomes the unit mission.

b. <u>Commander's Intent</u>. The intent statement is the CDR's personal vision of how the campaign will unfold. Generally, the CDR will write his own intent statement. Frequently the staff will provide substantial input(s). The CDR's intent is a clear and concise expression of the purpose of the operation and the military endstate. It provides focus to the staff and helps subordinate and supporting commanders take actions to achieve the military

[53] JP 3-33, Joint Task force Headquarters

[54] JP 5-0, Joint Planning

endstate without further orders, even when operations do not unfold as planned. It also includes where the CDR will accept risk during the operation. At the theater strategic level, CDR's intent must necessarily be much broader – it must provide an overall vision for the campaign that helps staff and subordinate commanders understand the intent for integrating all elements of national power and achieving unified action. The CCDR must *envision* and *articulate* how military power and joint operations will dominate the adversary and support or reinforce the interagency and our allies in accomplishing strategic success. Through their intent, the CDR identifies the major unifying efforts during the campaign, the points and events where operations must dominate the enemy and control conditions in the OE, and where other elements of national power will play a central role. The intent must allow for decentralized execution.

(1) It ***provides the link between the mission and the concept of operations*** by stating the method that, along with the mission, are the basis for subordinates to exercise initiative when unanticipated opportunities arise or when the original concept of operations no longer applies. If the CDR wishes to explain a broader purpose beyond that of the mission statement, they may do so. The mission and the CDR's intent must be understood two echelons down. The intent statement at any level must support the intent of the next higher CDR.

(2) The initial intent statement normally contains the purpose and military endstate as the initial impetus for the planning process. It could be stated verbally when time is short. The CDR refines the intent statement as planning progresses. The CDR's approved intent is written in the "*Execution*" paragraph as part of the operation plan or order.

(3) A well-devised intent statement enables subordinates to decide how to act when facing unforeseen opportunities and threats, and in situations where the concept of operations no longer applies. This statement deals primarily with the military conditions that lead to mission accomplishment, so the CDR may highlight selected objectives and effects. The statement also can discuss other instruments of national power as they relate to the mission and the potential impact of military operations on these instruments. The CDR's intent may include the CDR's assessment of the adversary CDR's intent and an assessment of where and how much risk is acceptable during the operation.

(4) Remember, the CDR's intent is not a summary of the CONOPs. It should not tell specifically how the operation is being conducted, but should be crafted to allow subordinate commanders sufficient flexibility and freedom to act in accomplishing their assigned mission(s) even in the "*fog of war.*"

> "It is the one overriding expression of will by which everything in the order and every action by every commander and soldier in the army must be dominated, it should, therefore, be worded by the commander himself."
>
> Field Marshal Viscount William Joseph Slim, Defeat into Victory, Batting Japan in Burma and India, 1942-1945, Cooper Square Press, p. 211.

While there is no specified joint format for CDR's intent, a generally accepted construct includes the purpose, method, and endstate:

- <u>Purpose</u>: The purpose is the *reason* for the military action with respect to the mission of the next higher echelon. It explains why the military action is being conducted. This helps the force pursue the mission without further orders, even when actions do not unfold as planned. Thus, if an unanticipated situation arises, participating commanders understand the purpose of the forthcoming action well enough to act decisively and within the bounds of the higher CDR's intent.

- Method: The "*how*," in doctrinally concise terminology, explains the offensive form of maneuver, the alternative defense, or other action to be used by the force as a whole. Details as to specific subordinate missions are not discussed.

- Endstate: The endstate describes what the CDR wants to see in military terms after the *completion of the mission* by the friendly forces.

Commander's Intent (Purpose)

Maintain Green's sovereignty and territorial integrity.

Commander's Intent (Method)

USXXCOM forces will secure LOCs to ensure a rapid build-up of forces in the JOA. The utilization of HN support will be maximized. Forces will deploy into theater under the auspices of participation in C/J (Coalition/ Joint) exercises demonstrating C/J force capabilities. IO will be optimized to communicate capability and coalition resolve against Red aggression. During these exercises, C/J forces will be positioned throughout the JOA with the capability to rapidly project full spectrum combat power against Red forces violating Green sovereignty.

Commander's Intent (Endstate)

Red forces withdrawn from forward staging bases and postured at their peacetime locations. These forces will be incapable of conducting rapid force build-up (>7 days) threatening Green.

Sample Commander's Intent Statement

c. Initial Planning Guidance. After approving the mission statement and issuing their intent, commanders provide the staff (and subordinates in a collaborative environment) with enough additional guidance (including preliminary decisions) to focus the staff and subordinate planning activities during COA development. As a minimum, the initial planning guidance should include the **mission statement; assumptions; operational limitations; a discussion of the national strategic endstate; termination criteria; military endstate military objectives; and the CDR's initial thoughts on desired and undesired effects.** The planning guidance should also address the role of agencies and multinational partners in the pending operation and any related special considerations as required.[55]

(1) The CDR approves the derived mission and gives the staff (and normally subordinate commanders) initial *planning guidance*. This guidance is essential for timely and effective COA development and analysis. The guidance should precede the staff's preparation for conducting their respective staff estimates. The CDR's responsibility is to *implant a desired vision* of the forthcoming combat action into the minds of the staff. Enough guidance (preliminary decisions) must be provided to allow the subordinates to plan the action necessary to accomplish the mission consistent with his and the SecDef's intent. The CDR's guidance must focus on the essential tasks and associated objectives that support the accomplishment of the assigned national objectives. It emphasizes in broad terms when, where, and how the CDR intends to employ combat power to accomplish the mission within the higher CDR's intent.

(2) The CDR may provide the planning guidance to the entire staff and/or subordinate commanders or meet each staff officer or subordinate unit CDR individually as the situation and information dictates. The guidance can be given in a written form or orally. No format for the planning guidance is prescribed. However, the guidance should be sufficiently detailed to provide a clear direction and to avoid unnecessary efforts by the staff or subordinate commanders.

[55] JP 5-0, Joint Planning

(3) The content of planning guidance varies from CDR to CDR and is dependent on the situation and time available. Planning guidance may include:

- Situation.
- The derived mission, including essential task(s) and associated objectives.
- Purpose of the forthcoming military action.
- Information available (or unavailable) at the time.
- Forces available for planning.
- Limiting factors (constraints and restraints) – including time constraints for planning.
- Pertinent assumptions.
- Tentative COAs under consideration: friendly strengths to be emphasized or enemy weaknesses the COAs should attack, or specific planning tasks.
- Preliminary guidance for use (or non-use) of nuclear weapons.
- Coordinating instructions.
- Acceptable level of risk to own and friendly forces.
- Information Operations guidance.

(4) Planning guidance can be very explicit and detailed, or it can be very broad, allowing the staff and/or subordinate CDR's wide latitude in developing subsequent COAs. However, no matter its scope, the content of planning guidance must be arranged in a logical sequence to reduce the chances of misunderstanding and to enhance clarity. Moreover, one must recognize that all the elements of planning guidance are *tentative only*. The CDR may issue successive planning guidance during the decision-making process. Yet, the focus of the staff should remain upon the framework provided in the initial planning guidance. The CDR should provide subsequent planning guidance during the rest of the plan development process.

(5) Initial planning guidance includes *Conflict Termination Criteria* and *Mission Success Criteria*. These criteria become the basis for assessment and include measures of performance (MOP) and measures of effectiveness (MOE).

14. Summary

The CCDR and staff conduct mission analysis to better understand the situation and problem, and identify what must be accomplished, when and where it must be done, and most importantly why—the purpose of the operation. In this step, the supported CCDR's analysis of his/her tasking from Strategic Direction and Guidance results in a mission statement. The mission statement is a short sentence or paragraph that describes the organization's essential task(s), purpose, and action containing the elements of who, what, when, where, and why. The mission statement provides the task or set of tasks, together with purpose, that clearly indicates the action to be taken and the reason for doing so. Key considerations for a supported CCDR during mission analysis are the national strategic end state, and appropriately, theater strategic end state. Based on this understanding, CDRs issue their initial CDR's intent and planning guidance to guide the staff in COA Development.

III(a). Concept Development (Function II)

> *"True genius resides in the capacity for evaluation of uncertain, hazardous, and conflicting information."*
>
> Winston Churchill

1. Function II — Concept Development: Overview

a. During the concept development step, CCDRs develop, analyze, and compare viable COAs and develop staff estimates that are coordinated with the Military Departments when applicable. Analysis includes wargaming, operational modeling, and initial feasibility assessments.

b. As you work through the concept development function you will be visualizing and thinking through the entire operation or campaign from end to start, start to end. It's important to emphasize here, as discussed in Chapter 1-III *Sequencing Actions*, operations and campaigns are sequenced into phases which are a way to view and conduct a complex joint operation in manageable parts. A phase can be characterized by the focus that is placed upon it and you will determine requirements in terms of *forces, resources, time, space and purpose*. The main purpose of phasing is to integrate and synchronize related activities, thereby enhancing flexibility and unity of effort during execution. Reaching the endstate often requires arranging a major operation or campaign into several phases which will assist CCDRs and their staffs in visualizing, discerning and defining requirements in terms of forces, resources, time, space, and purpose. Phases are designed to be conducted sequentially, but activities from a phase may continue into subsequent phases.

c. The staff writes (or graphically portrays) the CONOPS in sufficient detail so that subordinate and supporting CDRs understand their mission, tasks, and other requirements and can develop their supporting plans accordingly. During CONOPS development, the CDR and staff determines the best arrangement of simultaneous and sequential actions and activities to accomplish the assigned mission consistent with the approved COA. This arrangement of actions dictates the sequencing of forces into the OA, *providing the link between the CONOPS and force planning*. The link between the CONOPS and force planning is preserved and perpetuated through the TPFDD structure. This structure must ensure unit integrity, force mobility, and force visibility as well as the ability to rapidly transition to branches or sequels as operational conditions dictate. Planners ensure that the CONOPS, force plan, deployment plans, and supporting plans provide the flexibility to adapt to changing conditions, and are consistent with the CCDR's intent.

d. If the scope, complexity, and duration of the military action you contemplate warrants a campaign, then the staff outlines the series of military operations and associated objectives and develops the CONOPS for the preliminary part of the campaign in sufficient detail to impart a clear understanding of the CDR's concept of how the assigned mission will be accomplished.

e. During CONOPS development, the CCDR must assimilate many variables under conditions of uncertainty to determine the essential military conditions, sequence of actions, and application of capabilities and associated forces to create effects and achieve objectives. CCDRs and their staffs must be continually aware of the higher-level objectives and associated effects that influence planning at every juncture. If operational objectives are not linked to strategic objectives, the inherent linkage or "nesting" is broken and eventually tactical considerations can begin to drive the overall strategy at cross-purposes.

2. Concept Development

a. <u>Concept Development</u>. ***The purpose of concept development is to develop and refine COAs based upon the outputs from the strategic guidance function.*** Your operational design conducted earlier brings the objective or end into focus and bounds the available means and proposes and evaluates multiple ways to achieve these ends. At the strategic level, developed and analyzed COAs inform and provide multiple alternatives, including escalation controls to manage crises and conflicts and inform the decision-making of the CDR and national leaders and shape subsequent detailed plan development. A continuous and iterative dialogue, at all levels throughout the chain of command, is maintained during concept development to ensure the COAs being considered conform to evolving strategic direction, meets DoD and national-level leader needs, and reflects the changing strategic environment. During planning, the CDR develops several COAs with each containing an initial CONOPS. Each CONOPS should identify major capabilities, authorities required, task organization, major operational tasks to be accomplished by components, a concept for employment and sustainment and assessment of risk. Each COA may contain embedded multiple alternatives to accomplish designated objectives as conditions change (e.g., OE, problem, strategic direction). In time-sensitive situations, a WARNORD may not be issued, and a PLANORD, ALERTORD, or EXORD might be the first directive the supported CDR receives with which to initiate planning. Using the guidance included in the directive and the CCDR's mission statement, planners solicit input from supporting and subordinate commands to develop COAs based upon the outputs of the strategic guidance planning function. The figure below depicts the concept development planning function.

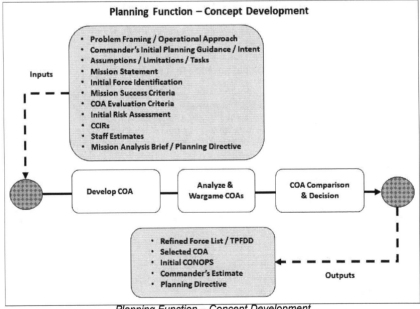

Planning Function – Concept Development

(1) During concept development, if an IPR is required, the CDR outlines the COA(s) and makes a recommendation to higher authority for approval and further development.

(a) The CDR recommends a COA (Chapter 5-III(b), *COA Development*) that is most appropriate for the situation.

(b) Concept development should consider a range of COAs that integrate robust options to provide greater flexibility and to expedite transition during a crisis. CCDRs should be prepared to continue to develop multiple COAs to provide national-level leadership options should the crisis develop.

(c) For CCP, CCDRs should address resource requirements, expected changes in the environment, and how each COA supports achieving national objectives.

(d) The CCDR also requests SecDef's guidance on interorganizational planning and coordination and makes appropriate recommendations based on the interorganizational requirements identified during mission analysis and COA development.

(2) One of the main products from the concept development function is **approval for continued development of one or more COAs**. Detailed planning begins upon COA approval in the concept development function.

b. Develop COAs. COAs are potential methods of accomplishing an assigned mission. Their development allows a planning team to further advance one or more solutions for subsequent analysis and comparison. COAs are developed with enough detail to enable planning by subordinate commands. The CDR may limit the number and details of the COAs developed, especially if the staff is operating under time constraints. All developed COAs should be developed with the following content and characteristics:

(1) Content. Each COA contains ends, ways, and means for accomplishing the mission and align with the CDR's intent. Each COA should contain an initial CONOPS that may identify a description of the OE, objectives, key tasks and purpose, forces/capabilities required, integrated timeline, task organization, operational concept, supporting concepts, initial force availability and readiness, and risks. Each COA may also contain embedded alternatives that describe alternatives to accomplish designated objectives as conditions change (e.g., OE, threat, strategic direction).

(a) Force Requirements Refinement. In addition to initial forces identified during the strategic guidance planning function, the planning team continues to refine the force requirements and documents them in the associated force list for each COA. Planning team coordination with CCMD Service components, Services, JFPs and the JS JFC may provide subject matter expertise to facilitate the proper characterization of force and capability requirements. Planners must communicate precisely the requisite information to those documenting the requirements into the force list to ensure the capabilities required are captured accurately (Chapter 3, *GFM*).

1 Force List Refinement. The initial force list is refined during concept development. Force requirements should be documented in a format that enables feasibility analysis or assessment and execution sourcing, to include force allocation decisions by the Secretary, should the plan transition to execution. If development of a TPFDD is necessary either as a directed planning requirement or to support crisis deployment, a force list may be entered into the JOPES IT system.

2 TPFDD Guidance. When documenting force requirements as a TPFDD in JOPES IT, the supported CCMD should publish guidance specifying the procedures for entering the plan's force requirements. Typically, this is done via a JOPES Newsgroup. Supplementary Instructions (SI) or Letter of Instruction (LOI) to the planning team, supporting and subordinate commands, and Services. TPFDD development guidance should include:

a Plan identification (PID). Informs planners of the PID to be used for TPFDD development within JOPES IT.

b Force modules. Organizes force requirements in JOPES IT into a logical structure which facilitates planning and analysis. For contingency planning, force modules are developed to correspond to major combat forces identified in the force apportionment tables.

c Priority of movement and movement windows. Movement windows are defined in JOPES by an earliest arrival date (EAD) and latest arrival date (LAD). Movement windows correspond to the priority of movement IAW developing the plan's CONOPS.

Joint Planning Process

d Standard points of embarkation/debarkation. Prescribes specific points of embarkation and points of debarkation (POE/PODs) to be used when documenting force requirements within JOPES IT.

e TPFDD development points of coordination. Identifies planners at supported, supporting, and subordinate commands responsible for documenting plan force requirements.

Note: Further guidance for development of an LOI can be found in CJCSM 3122.O2 Series, *Joint Operation Planning and Execution System (JOPES) Volume III Time-Phased Force and Deployment Data Development and Deployment Execution.*[1]

3 Consideration of non-DoD capabilities. Separate from GFM, planners should identify non-DoD capability requirements to include interagency, contractors, NGOs, and coalition partners during force planning. Key identified non-DoD capabilities may be addressed during COA development and may generate subsequent requirement for supporting DoD capabilities. For example, coalition or interagency capabilities may require liaison teams or operational contract support (OCS) may require a contracting staff.

4 Identify Mobilization Requirements. If any RC forces are identified as potential sourcing solutions, planners should consider the authorities and decision timelines for mobilization and reflect them in the developed COAs. The timeline for mobilization is dependent upon the scale of force mobilization and the level of decision authority required. The supported CCDR works with their Service components and aligned Service HQs (or USSOCOM for RC SOF) to develop realistic estimates for mobilization timelines of RC forces. Supported commands should identify such requirements and present mobilization requests to the appropriate level of mobilization authority (JP 4-05, *Joint Mobilization Planning* discuss mobilization planning in further detail.)

(b) Integrated Planning of Force Requirements. The supported CCDR for planning will lead integrated planning for their problem sets with all supporting CCMDs. This includes ensuring identified force requirements are within the constraint of the Force Apportionment Tables for all plans associated with the problem set. The supported CCDR should attempt to resolve any force requirement conflicts with supporting CCDRs. The JS J-5 will facilitate unresolved force requirement issues as necessary. Planners should also consider that the interdependence of plan force requirements is not exclusive to problem sets and may impact other plans. The CPG and JSCP discuss problem sets and integrated planning in further detail.

(c) Other Resource Requirements. Planners should consider other resource requirements for COAs being developed. These requirements inform the planning staff and shape refinement of staff estimates to identify the capacity and feasibility of satisfying resource requirements. Strategic guidance provides separate guidance for resource planning for both campaign and contingency planning. Planners should consider guidance contained in JP 4-0, *Joint Logistics* when developing logistics resource requirements.

(d) Planning Assumptions. While introduced during the strategic guidance function additional assumptions may need to be included as the detail of the plan increases. The validity of the assumptions should be reviewed and evaluated throughout planning, tied to a CCIR, and discussed via civilian-military dialogue and during IPRs. Planners may consider developing additional COAs or branch plans to address the potential invalidation of one or more assumptions.

(e) Preferred Force Identification. Preferred force identification is a specific type of planning assumption that can be done throughout the planning process and provides the basis for resource informed plan development. The supported CCDR works with their Service components and their aligned Service HQs (or USSOCOM for SOF), JS JFC and JFPs to identify preferred forces. ***Preferred forces are specific units identified as planning***

---

[1] *CJCSM 3122.02 Series, "Joint Operation Planning and Execution System (JOPES) Volume III Time-Phased Force and Deployment Data Development and Deployment Execution" and CJCSG 3122, "Time-Phased Force and Deployment Data (TPFDD) Primer," will be rescinded upon publication of CJCSM 3130.04 Series, Deployment Policies and Procedures.*

assumptions only to aid in development and resource estimation and are not considered "sourced" units. Different units may be identified during contingency or execution sourcing. The preferred forces identified for a plan by the CCDR should not be greater than the forces apportioned for planning unless granted permission to do so by the Chairman or designated representative. The Preferred Force Generator (PFG) is one tool that can be used to identify available candidate forces for planners to choose from. CJCSM 3130.06 Series, *Global Force Management Allocation Policies and Procedures* discusses preferred forces in more detail, as does Chapter 3, *Global Force Management.*

(2) <u>Characteristics</u>. All COAs selected for analysis should be valid. A valid COA is one that is adequate, feasible, acceptable, distinguishable, and complete. During COA development, staff estimates are refined to address the validity of each COA from their functional perspective. The evaluation of each of these characteristics can be subjective and a COA may be considered that requires mitigation of some criteria. JP 5-0, *Joint Planning* and Chapter 5-III(b), *COA Development*, discusses COA validity characteristics in further detail.

(a) <u>Adequate</u>. The COA should be able to accomplish the mission within the provided planning guidance. At the strategic level, the military objectives of a COA should support the accomplishment of national strategic objectives.

(b) <u>Feasible</u>. The COA should be able to accomplish the mission within the established time, space, and resource limitations. A COA should be resource informed in order to be developed into an executable plan.

(c) <u>Acceptable</u>. The COA should balance cost and risk. While acceptability is a judgement of the CDR, at the strategic level, political acceptability is shaped through civilian-military dialogue and confirmed during IPRs.

(d) <u>Distinguishable.</u> The COA should be sufficiently different from the other developed COAs. A selection of COAs should offer alternatives for the CDR and senior leaders.

(e) <u>Complete</u>. The COA should contain (1) objectives and tasks to be performed, (2) major forces required, (3) concepts for deployment, employment, and sustainment, (4) estimated times for achieving objectives, and (5) criteria for military objective and mission success. Incomplete development of a COA precludes accurately determining its other validity characteristics.

c. <u>COA Analysis and Wargaming (Chapter 5-IV)</u>. The purpose of COA analysis and wargaming is to evaluate developed COAs against common criteria, assess them against an adversary and within an operational environment. The adversary and operational environment can be hypothetical or assumed for contingency planning or based upon current JIPOE for planning under crisis conditions. This analysis enables a better understanding of each COA in preparation for potential further plan development. COA analysis is effectively the operational activity of assessment applied during concept development.

(1) <u>Analysis</u>. COA analysis seeks to confirm that a COA meets all the requisite validity characteristics, and is tailored to the time available. It allows the staff and subordinate CDRs to gain a common understanding of friendly and adversary COAs. This allows them to determine the advantages and disadvantages of each friendly COA and forms the basis for the comparison and approval decision.

(a) <u>Staff analysis</u>. Staff sections, informed by their updated staff estimates, should analyze each COA from their functional area perspective.

(b) <u>Subordinate analysis.</u> Subordinate commands analyze each COA from the perspective of their COA tasks. Subordinate components can provide subject matter expertise of Service capabilities and subordinate functional components can provide expertise of an operating domain.

(c) <u>Supporting analysis</u>. Supporting commands, Services, and DoD Agencies provide their regional or functional perspective to COA analysis.

(2) <u>Method of Analysis</u>. The method of analysis should be developed so as to produce a meaningful and objective measurement of a COA's strengths, weaknesses, and associ-

ated risks for subsequent comparison with other COAs. Analysis may include simulation or other means of modeling the problem. The method of analysis may be limited with respect to available time, personnel, and resources and scaled accordingly. Wargaming is a common method of analyzing one or more COAs.

(3) Wargaming. Time permitting, the CDR may choose to wargame one or more proposed COAs as a tool to assess and refine them and identify problems, risks, capability requirements, and gaps. The wargaming process provides the CDR and staff with a projected mission flow of a given COA with respect to assumed variables. Several COAs may be developed for follow-on wargaming and comparison. The degree to which a COA achieves the essential tasks within the limitations (including resource levels) set by higher headquarters (HHQs) allows the CDR to determine which COA is optimum with respect to adequacy, feasibility, acceptability, distinguishability, and completeness based on the available time, space, and resources. The CDR will select the developed COAs he wants wargamed and provide wargaming guidance and evaluation criteria (evaluation criteria) based on circumstances relative to the situation. The CDR's evaluation criteria address specific issues that the CDR requires the staff to determine for each validated COA during the conduct of the wargame. The staff should plan to counter all adversary COAs identified during the JIPOE process. The effectiveness of each COA is wargamed against the selected adversary COAs and the OE using the CDR's evaluation criteria. There are multiple wargaming options available. Depending on the time and resources available, a wargame may consist of a simple table top exercise or a complex simulation involving multiple commands and Services leveraging wargaming IT systems. The results of COA wargaming include:

(a) The refinement of a COA.

(b) A refined COA resource or authority requirements.

(c) The identification of risk(s).

(d) An identified need to develop additional COA(s) as potential branches or sequels.

(e) In the event of a major change in a functional concept of support, the planning team may be required to re-wargame the affected COAs to ensure they remain valid and to consider any potential unintended consequences.

d. COA Comparison and Decision.(Chapter 5-V and 5-VI)

(1) Comparison. The results of the wargaming analysis are used to conduct COA comparisons with the goal to identify and recommend the COA with the highest probability of mission success. The staff evaluates each COA against the established COA evaluation criteria and briefs the CDR on the results of the wargaming analysis, the results of the comparison of each COA, and the COA they recommend for plan development. The CDR considers the advantages and disadvantages of each of the COAs and compares and weighs them against each other.

(2) Decision. With the results of COA comparison and the advice of the planning team, the CDR can select or modify a COA for subsequent development or direct the planning team to develop additional COAs with revised guidance.

(a) After comparing COAs, the staff recommends a COA to the CDR. The CDR can select an evaluated COA, modify a COA with changes identified during analysis, combine COAs to create a new COA, or reject all COAs and start over with COA development or mission analysis. Alternately, if none of the analyzed COAs are satisfactory to the CDR, the planning team may be directed to return to develop additional COAs with updated guidance. Once the CDR has made a decision, the selected COA is reviewed against the mission statement to ensure it includes all required essential tasks. As required, the selected COA will be recommended to HHQs for approval.

(b) Updated Commander's Estimate. The CDR's estimate is updated based upon the results of planning and an improved understanding of the operating environment informed by staff estimates. The updated CDR's estimate provides a concise summary of how the CDR intends to accomplish the mission and provides the necessary focus for plan development. It reflects the analysis of the problem and usually provides options and recommendations to the directing authority. CDRs may present more than one valid COA for approval as appropriate. While the CDR will tailor the content of the estimate based upon the situation, it should address:

<u>1</u> An overview of the CDR's mission analysis and problem framing to include assessed strategic and operational centers of gravity, the CDR's intent and desired endstate.

<u>2</u> The CDRs' recommended COA and supporting rationale.

<u>3</u> Brief descriptions and assessments of alternate COAs and the rationale for not selecting them.

<u>4</u> The estimated level and duration of the operation.

<u>5</u> The nature, purpose, time-phasing, and inter-relationship of operations.

<u>6</u> Proposed command relationships for forces to be employed.

<u>7</u> A gross estimate of force, support, and deployment feasibility.

<u>8</u> Potential interagency and/or multinational involvement.

<u>9</u> Branches, sequels, or other options, including warning and response times, that involve scenarios likely to confront the command.

e. <u>Concept Development Outcomes</u>. The concept development planning function produces the following outcomes distinctive to each level of command.

(1) <u>National Level</u>. Produces options for consideration by the President, Secretary, or designated representative within the supported CCDR's operational approach, in support of national strategic objectives. This may be accomplished through informal dialog or an IPR based upon the supported CCDR's estimate. This discussion allows the CCDR to convey COAs with a higher-fidelity discussion of ends, ways, means, and risk than done previously during the strategic guidance planning function. It also allows for the identification of other instruments of national power that require synchronization with the plan development. The President, Secretary, or designated representative, selects one or more COAs for plan development and provide further guidance to shape continued planning. If none of the presented COAs are acceptable, the supported CCDR may be directed to return to concept development with updated guidance to develop additional COAs. Under crisis conditions national level discussion may be significantly abbreviated or potentially virtual dialog to more rapidly develop a plan ready for execution.

(2) <u>Chairman of the Joint Chiefs of Staff.</u> For CCDR-level planning, the Chairman, in consultation with the supported and supporting CCDRs and other members of the JCS, reviews and analyzes the supported CCDR's estimate and provides recommendations and advice to the President, Secretary, or designated representative for their COA decision. As required, the Chairman will release a planning directive formally conveying the selected COA.

(3) <u>Supported Command</u>. For planning directed at the national-level, the CCDR's estimate and a COA decision may be made by the President, Secretary, or designated representative. Below the national-level, COAs are selected by the authority that directed a planning effort. Once a COA is selected, the supported CDR produces a planning directive to subordinate CDRs. The planning directive conveys the selected COA, associated initial CONOPS, and a further refined force list which provide a progressively more detailed framework for plan development by the supported command, subordinate commands, and supporting commands.

(4) <u>Supporting Commands, Services and CSAs</u>. The selected COA provides increasing plan fidelity of ways and means that allows supporting commands, Services, and DoD Agencies to conduct their own planning aligned with the supported CDR. While not directive of supporting commands, the supported CCDR's operational approach and outputs from the national level planning dialog inform supporting plans. Supporting CCDR, Service, or DoD Agency planning continues in parallel based upon planning direction from the supported CCDR.

(5) Subordinate Commands. This planning function produces detailed direction from HHQ that shapes subordinate command planning. The subordinate CDR and planning staff conduct their own concept development informed by the selected COA and CCDR's guidance or planning directive. The initial CONOPS and further refined force list and/or TPFDD provide a starting point for subordinate command plan development.

3. COA Development Preparation and Considerations

- Time Available.
- Political Considerations.
- Flexible Deterrent Options.
- Lines of Operation.

a. Time Available. The CDR and the nature of the mission will dictate the number of COAs to be considered. Staff sections continually inform COA development by an ongoing staff estimate process to ensure suitability, feasibility, acceptability, and compliance with Joint Doctrine (deviations from Joint Doctrine should be conscious decisions and not the result of a lack of knowledge of doctrinal procedures). Additionally, staffs ensure completeness (answers Who, What, When, Where, Why, How).

b. Political Considerations. Planning for the use of military forces includes a discussion of the political implications of their transportation, staging, and employment. The CCDR's political advisor is a valuable asset in advising the CCDR and staff on issues crucial to the planning process, such as overflight and transit rights for deploying forces, basing, and support agreements. Multinational and coalition force concerns and sensitivities must also be considered.

(1) Political objectives drive the operation at every level from strategic to tactical. There are many degrees to which political objectives influence operations: ROE restrictions and basing access and overflight rights are examples. Two important factors about political primacy stand out. *First*, all military personnel should understand the political objectives and the potential impact of inappropriate actions. Having an understanding of the political objective helps avoid actions which may have adverse political effects. It is not uncommon today for junior leaders to make decisions which have significant political implications. *Secondly*, CDRs should remain aware of changes not only in the operational situation, but also to changes in political objectives that may warrant a change in military operations. These changes may not always be obvious.

(2) The integration of U.S. political and military objectives and the subsequent translation of these objectives into action have always been essential to success at all levels of operation. The global environment today with threats and challenges from adversaries in every operating domain requires an even greater integration and cooperation between political and military objectives.

(3) Attaining our national objectives requires the efficient and effective use of the diplomatic, informational, military, and economic (DIME) instruments of national power and systems taxonomy of the multi-dimensional Political, Military, Economic, Social, Information and Infrastructure (PMESII). This situational understanding supported by and coordinated with that of our allies and various intergovernmental, nongovernmental, and regional security organizations is critical to success.

(4) Military operations must be strategically integrated and operational and tactically coordinated with the activities of other agencies of the USG, IGOs, NGOs, regional organizations, the operations of foreign forces, and activities of various HN agencies. Sometimes the CDR draws on the capabilities of other organizations; sometimes the CDR provides capabilities to other organizations; and sometimes the CDR merely deconflicts their activities with those of others. These same organizations may be involved in pre-hostilities operations, activities during combat, and in the transition to post-hostilities activities. Roles and relationships among agencies and organizations, CCMDs, U.S. state and local

governments, and overseas with the U.S. Chief of Mission (COM), and country team in a U.S. embassy, must be clearly understood. Interagency coordination forges the vital link between the military and the diplomatic, informational, and economic instruments of national power. Successful interagency, IGO, and NGO coordination helps enable the USG to build international support, conserve resources, and conduct coherent operations that efficiently achieve shared goals.

c. Flexible Deterrent Options (FDOs). Flexible deterrent options are preplanned, deterrence-oriented actions carefully tailored to send the right signal and influence an adversary's actions. They can be established to dissuade actions before a crisis arises or to deter further aggression during a crisis. FDOs are developed for each instrument of national power — diplomatic, informational, military, economic, and others (financial, intelligence and law enforcement-DIMEFIL) — but they are most effective when used in unison or as a combination with other instruments of national power.

(1) FDOs facilitate early strategic decision-making, rapid de-escalation and crisis resolution by laying out a wide range of interrelated response paths. Examples of FDOs for each instrument of national power are listed in the figures on the following pages. *Key goals of FDOs are:*

- Deter aggression through communicating the strength of U.S. commitments to treaty obligations and peaceful development.
- Confront the adversary with unacceptable costs for its possible aggression.
- Isolate the adversary from regional neighbors and attempt to split the adversary coalition.
- Rapidly improve the military balance of power in the OA.

(2) FDOs Implementation. The use of FDOs must be consistent with U.S. national security strategy (i.e., the instruments of national power are normally used in combination with one another), therefore, continuous coordination with interagency partners is imperative. All operation plans have FDOs, and CCDRs are tasked by the JSCP to plan requests for appropriate options using all instruments of national power.[2]

(3) Military FDOs. Military FDOs underscore the importance of early response to a crisis. Deployment timelines, combined with the requirement for a rapid, early response, generally requires military FDO force packages to be light; however, military FDOs are not intended to place U.S. forces in jeopardy if deterrence fails (risk analysis should be an inherent step in determining which FDOs to use, and how and when to use them). Military FDOs are carefully tailored to avoid the classic "too much, too soon" or "too little, too late" responses. They rapidly improve the military balance of power in the OA, especially in terms of early warning, intelligence gathering, logistic infrastructure, air and maritime forces, information operations, and force protection assets, without precipitating armed response from the adversary. Military FDOs are most effective when used in concert with the other instruments of power. They can be initiated before or after, and with or without unambiguous warning.[3]

[2] *JP 3-0, Joint Operations*
[3] *JP 5-0 Joint Planning*

Example Flexible Deterrent Options (FDOs)

Example of Requested Diplomatic Flexible Deterrent Options

- Alert and introduce special teams (e.g., public diplomacy).
- Reduce international diplomatic ties.
- Increase cultural group pressure.
- Promote democratic elections.
- Initiate noncombatant evacuation procedures.
- Identify the steps to peaceful resolution.
- Restrict activities of diplomatic missions.
- Prepare to withdraw or withdraw U.S. embassy personnel.
- Take actions to gain support of allies and friends.
- Restrict travel of U.S. citizens.
- Gain support through the United Nations.
- Demonstrate international resolve.

Example of Requested Informational Flexible Deterrent Options

- Promote U.S. policy objectives through public statements.
- Ensure consistency of strategic communications themes and messages.
- Encourage Congressional support.
- Gain U.S. and international public confidence and popular support.
- Maintain open dialogue with the news media.
- Keep selected issues as lead stories.
- Increase protection of friendly critical information structure.
- Impose sanctions on communications systems technology transfer.
- Implement psychological operations.

Example of Requested Military Flexible Deterrent Options

➢ Increase readiness posture of in-place forces.
➢ Upgrade alert status.
➢ Increase intelligence, surveillance, and reconnaissance.
➢ Initiate or increase show-of-force actions.
➢ Increase training and exercise activities.
➢ Maintain and open dialogue with the news media.
➢ Take steps to increase U.S. public support.
➢ Increase defense support to public diplomacy.
➢ Increase information operations.
➢ Deploy forces into or near the potential operational area.
➢ Increase active and passive protection measures.
➢ Ensure consistency of strategic communications messages.

Example of Requested Economic Flexible Deterrent Options

➢ Freeze or seize real property in the U.S. where possible.
➢ Freeze monetary assets in the U.S. where possible.
➢ Freeze international financial institutions to restrict or terminate financial transactions.
➢ Encourage U.S. and international financial institutions to restrict or terminate financial transactions.
➢ Encourage U.S and international corporations to restrict transactions.
➢ Embargo goods and services.
➢ Enact trade sanctions.
➢ Enact restrictions on technology transfer.
➢ Cancel or restrict U.S.-funded programs.
➢ Reduce security assistance programs.

Joint Planning Process

4. Lines of Operations (LOO)

So far, we've discussed the process of operational design in the following steps:

- *Endstate* (in terms of desired strategic political-military outcomes).
- *Objectives* that describe the conditions necessary to meet the endstate.
- Desired *effects* that support the defined objectives.
- Friendly and enemy *center(s) of gravity (COG)* using a systems approach.
- *Decisive points* (DPs) that allow the joint force to affect the enemy's COG and look for DPs necessary to protect friendly COGs.

Now let's look at identifying *lines of operation* that describe how decisive points are to be achieved and linked together in such a way as to overwhelm or disrupt the enemy's COG.

A *line of operations* is a line that defines the directional orientation of a force in time and space in relation to the enemy and links the force with its base of operations and objectives. LOO can be thought of as the analytical bridge between the outcomes of the mission analysis process and the development of COAs. It is important to conduct LOO analysis prior to COA development to ensure COAs achieve military objectives. As CDRs visualize the design of the operation, they may use several LOO to help visualize the intended progress of the joint force toward achieving operational and strategic objectives. LOOs connect a series of DPs that lead to control of a geographic or force-oriented objective. Operations designed using LOO generally consist of a series of actions executed according to a well-defined sequence. Major combat operations are typically designed using LOO with these lines tying offensive and defensive tasks to the geographic and positional references in the OA. CDRs synchronize activities along complementary lines of operations to achieve the endstate.

a. In operational design, LOO describe how DPs are linked to operational objectives. Joint doctrine defines *LOO* as "lines that define the orientation of the force in time, space, and purpose in relation to an adversary or objective."[4] They connect the force with its base of operations and its objectives.

- CDRs establish the military conditions and endstate for each operation, developing LOO that focus efforts to create the conditions that produce the endstate.
- Subordinate CDRs adjust the level of effort and missions along each LOO. LOO are formulated during COA development and refined through continual assessment.[5]

b. <u>LOO must be Derived from Decisive Points</u>. The kinds of DPs related to a LOO define the description of the LOO. This is why DPs must be determined first before defining LOO. LOO are the least understood portion of operational design and therefore tend to be misapplied. The importance of well-defined and understood LOO is basic to linking DPs, COG, objectives, and endstate. Properly defined, LOO provide clarity and distinction and provide the rationale for everything that the joint force does. Therefore, poorly defined LOO weaken the plan and lead to confusion. LOO should be broadly defined to encompass a more flexible way of thinking.

c. Normally, joint operations require CDRs to synchronize activities along multiple and complementary LOO working through a series of military strategic and operational objectives to attain the military endstate. There are many possible ways to graphically depict LOO, which can assist planners to visualize/conceptualize the joint operation from beginning to end and prepare the OPLAN or OPORD accordingly.

[4] *DOD Dictionary of Military and Associated Terms (DOD Dictionary)*
[5] *FM 3-0, Operations*

d. From the perspective of unified action, there are many diplomatic, economic, and informational activities that can affect the sequencing and conduct of military operations along both physical and logical LOO. Planners should consider depicting relevant actions or events of the other instruments of national power on their LOO diagrams.

(1) A LOO connects a series of DPs over time that lead to control of a geographic objective or defeat of an enemy force as illustrated in Figures A and B. *CDRs use LOO to connect the force with its base of operations and objectives when positional reference to the enemy is a factor.*

Operations designed using LOO generally consist of a series of cyclic, short-term events executed according to a well-defined, finite timeline. Major combat operations are typically designed using LOO. These tie offensive and defensive operations to the geographic and positional references of the AO. CDRs synchronize activities along complementary LOO to attain the endstate. LOO may be either *interior* or *exterior*.[6]

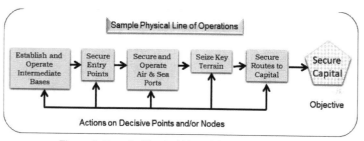

Figure A. Sample Physical Line of Operation (JP 5-0).

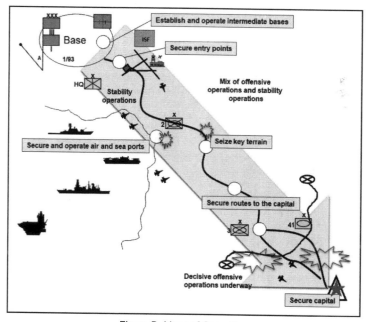

Figure B. Lines of Operation.

[6] *FM 3-0, Operations*

(a) <u>Interior and Exterior Lines</u>. The concept of interior and exterior lines applies to both maneuver and logistics. If a force is interposed between two or more adversary forces, it is said to be operating on interior lines. Thus the force is able to move against any of the opposing forces, or switch its resources over a shorter distance than its adversary. Such a concept depends on the terrain and the state of mobility of both sides. In Figure C below the defending force (Force B) has a shorter distance to move in order to reinforce its force elements in contact. The attacking force (Force A) has a greater distance to travel to switch resources across its three operations (A1, A2, A3).[7]

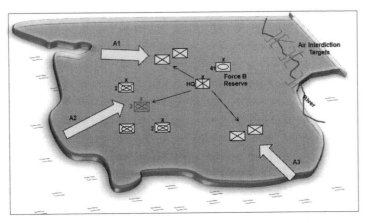

Figure C. Interior and Exterior Lines.[8]

(b) A force operates on *interior lines* when its operations diverge from a central point and when it is therefore closer to separate adversary forces than the latter are to one another. Interior lines benefit a weaker force by allowing it to shift the main effort laterally more rapidly than the adversary, and provide increased security to logistical support operations. Interior lines usually represent central position, where a friendly force can reinforce or concentrate its elements faster than the enemy force can reposition. With interior lines, friendly forces are closer to separate enemy forces than the enemy forces are to one another. Interior lines allow an isolated force to mass combat power against a specific portion of an enemy force by shifting capabilities more rapidly than the enemy can react.

(c) A force operates on *exterior lines* when its operations converge on the adversary. Successful operations on exterior lines require a stronger or more mobile force, but offer the opportunity to encircle and annihilate a weaker or less mobile opponent. Assuring strategic mobility enhances exterior LOO by providing the JFC greater freedom of maneuver.[9]

<u>1</u> The relevance of interior and exterior physical lines depends on the relationship of time and distance between the opposing forces. Although an adversary force may have interior lines with respect to the friendly force, this advantage disappears if the friendly force is more agile and operates at a higher operational tempo. Conversely, if a smaller force maneuvers to a position between larger but less agile adversary forces, the friendly force may be able to defeat them in detail before they can react effectively.

[7] *Joint Doctrine Publication 01 (JDP 01), United Kingdom*

[8]*FM 3-0, Operations*

[9] *JP 5-0, Joint Planning*

<u>2</u> A joint operation may have *single* or *multiple* physical LOO.

<u>a</u> A *single LOO* has the advantage of concentrating forces and simplifying planning.

<u>b</u> *Multiple LOO*, on the other hand, increases flexibility and creates opportunities for success. Multiple LOO also makes it difficult for an adversary to determine the objectives of the campaign or major operation, forcing the adversary to disperse resources to defend against multiple threats. The decision to operate on multiple lines will depend to a great extent on the availability of resources.

(2) LOO reflect the more traditional linkage of DPs, objectives and endstate. However, using LOO alone does not project the operational design beyond the defeat of the enemy force. **Combining LOO with Lines of Effort (LOE) allows CDRs to project operational design beyond the current phase of the operation to set the conditions for an enduring peace.** It allows them to consider the less tangible aspects of the OE in which the other instruments of national power are predominant. CDRs can visualize post-hostility operations from a far more conceptual perspective. The resulting operational design reflects the thorough integration of full spectrum operations across the spectrum of conflict.[10]

> "Having determined in order, endstate, objectives and effects, center(s) of gravity, decisive points, and lines of operation, planners then can link lines of operation to decisive points and examine the how and where certain decisive points support multiple lines of operation."
>
> Dr. Keith D. Dickson, Operational Design: A Methodology for Planners, Professor of Military Studies, Joint Forces Staff College

5. Lines of Effort (LOE)

A LOE links multiple tasks and missions using the logic of purpose—cause and effect— to focus efforts toward establishing operational and strategic conditions. LOE are essential to operational design when positional references to an enemy or adversary have little relevance. In operations involving many nonmilitary factors, lines of effort may be the only way to link tasks, effects, conditions, and the desired endstate as illustrated in Figures D and E on the following pages. LOE are often essential to helping CDRs visualize how military capabilities can support the other instruments of national power. They are a particularly valuable tool when used to achieve unity of effort in operations involving multinational forces and civilian organizations, where unity of command is elusive, if not impractical.

a. **CDRs use LOE to link the logic of purpose – conditions and effects – with a series of conceptual DPs or objectives to the conditions that define the endstate.** Operations designed using LOE are typically *focused on conditions rather than physical objectives*. LOE combine the complementary, long-term effects of stability tasks with the cyclic, short-term events characteristic of combat operations. *LOE also help CDRs visualize how military means can support nonmilitary instruments of national power.*

b. CDRs use LOE to describe how they envision their operations creating the more intangible endstate conditions. These LOE show how individual actions relate to each other and to achieving the endstate. Ideally, LOE combine the complementary long-term effects of stability or civil support tasks with the cyclic, short-term events typical of offensive or defensive tasks. Using LOE, CDRs develop tasks and missions, allocate resources, and assess the effectiveness of the operation. The CDR may specify which LOE represent the decisive operation and which are shaping operations. CDRs synchronize activities along multiple LOE to achieve the conditions that compose the desired endstate. [11]

[10] FM 3-0, Operations

[11] FM 3-0, Operations

Example Lines of Effort

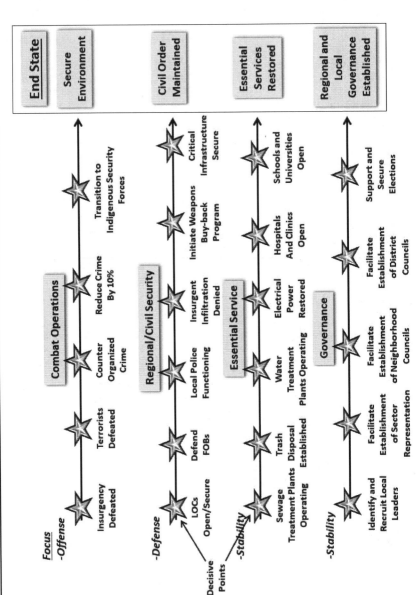

Figure D. Example Lines of Effort (Stability) (FM 3-0).

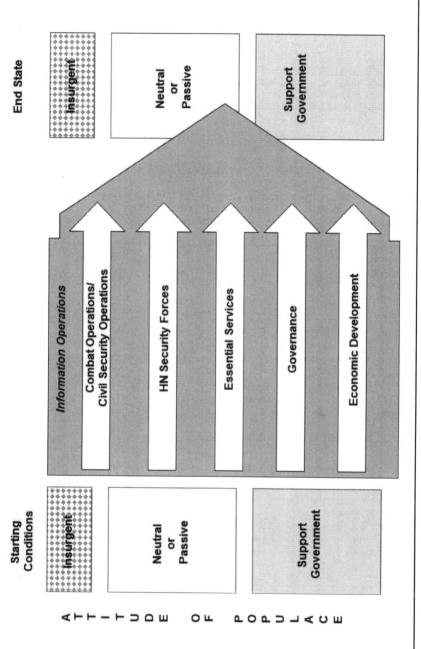

Figure E. Example Lines of Effort (Counterinsurgency) (FM 3-24).

c. CDRs at all levels may use LOE to develop missions and tasks and to allocate resources. CDRs may designate one LOE as the decisive operation and others as shaping operations. CDRs synchronize and sequence related actions along multiple LOE. Seeing these relationships helps CDRs assess progress toward achieving the endstate as forces perform tasks and accomplish missions.

d . CDRs typically visualize stability and civil support operations along LOE as previously seen in Figures D and E. For stability operations, CDRs may consider linking primary stability tasks to the corresponding Department of State post-conflict technical sectors. These stability tasks link military actions with the broader interagency effort across the levels of war. A full array of LOE might include offensive and defensive lines, as well as a line for information operations. Information operations typically produce effects across multiple LOE.

e. As operations progress, CDRs may modify the LOE after assessing conditions and collaborating with multinational military and civilian partners. LOE typically focus on integrating the effects of military operations with those of other instruments of national power to support the broader effort. Each operation, however, is different. CDRs develop and modify LOE to keep operations focused on achieving the endstate, even as the situation changes.

f. LOE are directly related to one another and success in one LOE reinforces successes in the others. They connect objectives that, when accomplished, support achieving the endstate. They connect objectives that, when accomplished, support achieving the endstate. Operations designed using LOE typically employ an extended, event-driven timeline with short-, mid-, and long-term goals.

(1) CDRs determine which LOE apply to their AO and how the LOE connect with and support one another.

(2) CDRs at all levels should select the LOE that relate best to achieving the desired endstate in accordance with the CDR's intent. The following list of possible LOE is not all inclusive. However, it gives us a place to start:

- Conduct information operations.
- Conduct combat operations/civil security operations.
- Train and employ HN security forces.
- Establish or restore essential services.
- Support development of better governance.
- Support economic development.

(3) These lines can be customized, renamed, changed altogether, or simply not used. CDRs may combine two or more LOE or split one LOE into several. For example, in Figure F on the following page Information Operations (IO) is integrated into all LOE; however, CDRs may designate a separate LOE for IO if necessary to better describe their intent. Likewise, some CDRs may designate separate LOE for combat operations and civil security operations.[12]

[12] FM 3-24/MCWP 3-33.5, Counterinsurgency

Example of Goals and Objectives along Lines of Effort

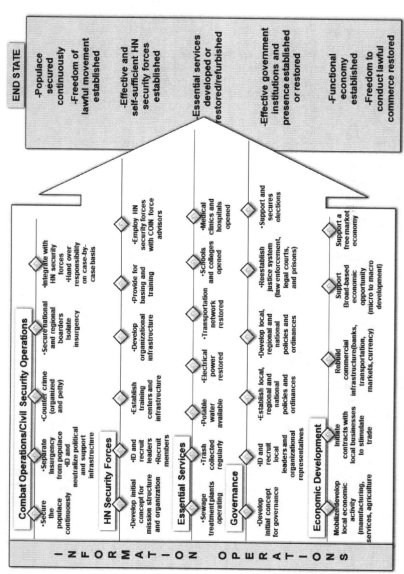

Figure F. Example of Goals and Objectives along Lines of Effort (FM 3-24).

Joint Planning Process

6. Combining LOO and LOE

CDRs may use both LOO and LOE to connect objectives to a central, unifying purpose. Continuous assessment gives CDRs the information required to revise and adjust LOO and LOE. LOO portray the more traditional links between objectives, DPs, and COG. However, LOO do not project the operational design beyond defeating enemy forces and seizing terrain. Combining LOO and LOE allows CDRs to include nonmilitary activities in their operational design. This combination helps CDRs incorporate stability tasks that set the endstate conditions into the operation. It allows CDRs to consider the less tangible aspects of the OE where the other instruments of national power dominate. CDRs can then visualize concurrent and post-conflict stability activities. Making these connections relates the **tasks and purposes** of the elements of full spectrum operations with joint effects identified in the campaign plan. The resulting operational design effectively combines full-spectrum operations throughout the campaign or major operation.

> *For example; CDRs may conduct offensive and defensive operations along a LOO to form a shield behind which LOE for simultaneous stability operations can maintain a secure environment for the populace. Accomplishing the objectives of combat operations/civil security operations sets the conditions needed to achieve essential services and economic development objectives. When the populace perceives that the environment is safe enough to leave families at home, workers will seek employment or conduct public economic activity. Popular participation in civil and economic life facilitates further provision of essential services and development of greater economic activity. Over time such activities establish an environment that attracts outside capital for further development. Neglecting objectives along one LOO or LOE risks creating vulnerable conditions along another that the enemy can exploit. Achieving the desired endstate requires linked successes along all lines.*

a. CDRs may describe an operation along LOO, LOE, or a combination of both. Irregular warfare, for example, typically requires a deliberate approach using LOO complemented with LOE; the combination of them may change based on the conditions within the operational area. An operational approach using both LOO and LOE reflects the characteristics and advantages of each. With this approach, CDRs synchronize and sequence actions, deliberately creating complementary and reinforcing effects. The lines then converge on the well-defined, commonly understood endstate outlined in the commander's intent.[13]

b. With this approach, it is vital for the CDR to synchronize actions across the LOO, creating complementary and reinforcing effects. This ensures that the LOO converge on a well-defined, commonly understood endstate that is composed of the set of conditions initially outlined in the CDR's intent (see Figure G on following page for example of an Operational Design Schematic and Figure H for a sample Operational Approach Schematic).[14]

---

[13] *FM 3-0, Operations*

[14] *Ibid*

Operational Design Schematic

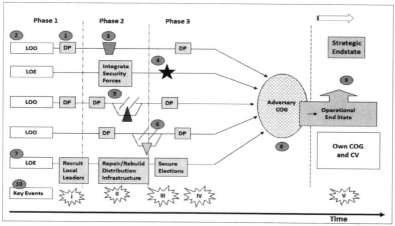

Figure G. Operational Design Schematic.

1. DPs are sequenced in time and space along LOO. This sequencing can be assisted by Phases. DPs are the key to unlocking the COG, and without proper identification, the COG cannot be defeated nor neutralized.

2. LOO can be environmental or functional or a mixture of both. They should not be decided until the DPs have been derived and the critical path identified.

3. Operational Pauses may be introduced where necessary. Momentum must be maintained elsewhere.

4. Culmination Point is reached when an operation or battle can just be maintained, but not developed to any great advantage.

5. Branches are deliberate plans which can be introduced to LOO whenever necessary, and are continuously refined as the campaign develops.

6. Sequels are contingency plans introduced when phases are not completed as planned.

7. LOE typically visualize stability and civil support operations, although CDRs select the lines of effort that relate best to achieving the desired endstate.

8. The adversary COG at the operational level is that which most resists the endstate. Without the neutralization or destruction of the adversary's COG, the endstate cannot be reached. Activity, necessary to finally achieve the endstate conditions, may take place after its destruction or neutralization, but this will not be decisive or critical. It may be useful to show own COG as it is the thing that needs protecting most, and is therefore that which the adversary is likely to direct their efforts against.

9. The endstate provides the focus for campaign planning and all activities should be judged against their relevance to its achievement. The operational endstate will usually be given by the Military Strategic Authority and may be a list of objectives or a statement. It needs analysis in order to identify measurable conditions which together indicate that the endstate has been achieved.

10. It may be useful to include a line showing key events. These might be the deadline for compliance with a UN resolution, the date an adversary 2nd Echelon force might be ready for combat, the estimated time for the completion of mobilization, or the holding of the first free and fair elections.

Sample Operational Approach

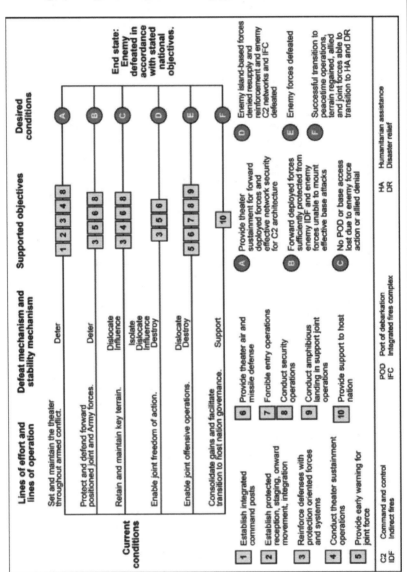

Figure H. Sample Operational Approach (FM 5-0)

III(b). Course of Action (COA) Development

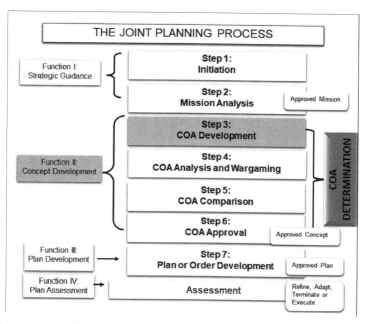

THE JOINT PLANNING PROCESS

Function I: Strategic Guidance
- Step 1: Initiation
- Step 2: Mission Analysis — Approved Mission

Function II: Concept Development
- Step 3: COA Development
- Step 4: COA Analysis and Wargaming
- Step 5: COA Comparison
- Step 6: COA Approval — Approved Concept

COA DETERMINATION

Function III: Plan Development → Step 7: Plan or Order Development — Approved Plan

Function IV: Plan Assessment → Assessment — Refine, Adapt, Terminate or Execute

1. Step 3 to JPP — COA Development

a. <u>COA Development</u>. A COA is any force employment option open to a CDR that, if adopted, would accomplish the desired strategic and military ends (mission). The staff supports the CDR through in-depth analysis and presentation of a range of options for future military and non-military actions. One-way staffs help CDRs refine their visualization is to develop *alternative* COA to execute the CDR's envisioned operational approach and achieve the objectives. For each COA, the CDR must envision the employment of own/friendly forces and assets as a whole, taking into account externally imposed limitations, the AO, and the conclusions previously drawn during the mission analysis and the CDR's guidance. Prior to actually developing the alternative COA, we must also consider the time available, any political considerations, planned or on-going FDOs and LOO (see Chapter 5-III(a), *Concept Development*).

b. Defining the COA. Each COA is a broad statement of a possible way to accomplish the mission. A COA consists of the following information:

- WHO (type of forces) will execute the tasks/take the action?
- WHAT type of action or tasks are contemplated?
- WHEN will the tasks begin?
- WHERE will the tasks occur?
- WHY (for what purpose) the action is required (relate to endstate)?
- HOW will the available forces be employed?

The staff converts the approved COA into a CONOPS. COA determination consists of four primary activities: *COA development, analysis and wargaming, comparison, and approval.*

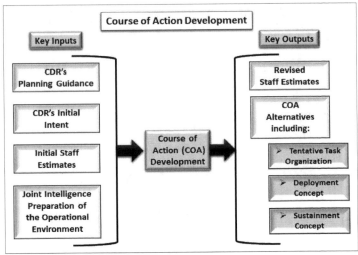

Course of Action Development.

2. Tentative Courses of Action

The output of COA development are *tentative* COAs in which the CDR describes for each COA, in broad but clear terms, what is to be done, the size of forces deemed necessary, and time in which force needs to be brought to bear. Tentative COAs allow for initial conceptualization and broad descriptions of potential approaches to the conduct of operations that will accomplish the desired endstate. The CCDR gives the staff their preliminary thoughts on possible and acceptable military actions early in the planning process to provide focus to their efforts, allowing them to concentrate on developing COAs that are the most appropriate.

a. A *tentative* COA should be simple, brief, yet complete. The COA sketch contains the general arrangement of forces, the anticipated movement or maneuver of those forces, a brief description of the concept of operation, and *major tasks* to components as displayed in Figure A on the following page. The COA's should have descriptive titles. Distinguishing factors of the COA may suggest titles that are descriptive in nature. A COA should answer the following questions:

- How much force is required to accomplish the mission?
- Generally, in what order should coalition forces be deployed?
- Where and how should coalition air, naval, ground and special operation forces be employed in theater?
- What major tasks must be performed and in what sequence?
- How is the coalition to be sustained for the duration of the campaign?
- What are the command relationships?

b. To develop *tentative* COAs, the staff must focus on key information necessary to make decisions and assimilate the data in mission analysis. Usually, the staff develops no more than three COAs to focus their efforts and concentrate valuable resources on the

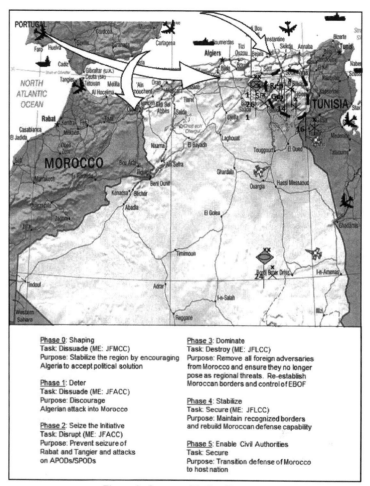

Phase 0: Shaping
Task: Dissuade (ME: JFMCC)
Purpose: Stabilize the region by encouraging Algeria to accept political solution

Phase 1: Deter
Task: Dissuade (ME: JFACC)
Purpose: Discourage Algerian attack into Morocco

Phase 2: Seize the Initiative
Task: Disrupt (ME: JFACC)
Purpose: Prevent seizure of Rabat and Tangier and attacks on APODs/SPODs

Phase 3: Dominate
Task: Destroy (ME: JFLCC)
Purpose: Remove all foreign adversaries from Morocco and ensure they no longer pose as regional threats. Re-establish Moroccan borders and control of EBOF

Phase 4: Stabilize
Task: Secure (ME: JFLCC)
Purpose: Maintain recognized borders and rebuild Moroccan defense capability

Phase 5: Enable Civil Authorities
Task: Secure
Purpose: Transition defense of Morocco to host nation

Figure A. Course of Action Example.

most likely scenarios. All COAs selected for analysis must be valid. A valid COA is one that is *adequate, feasible, acceptable, distinguishable, and complete.*

(1) Adequate - Can accomplish the mission within the CDR's guidance.

(2) Feasible - Can accomplish the mission within the established time, space, and resource limitations.

(3) Acceptable - Must balance cost and risk with the advantage gained.

(4) Distinguishable - Must be sufficiently different from the other courses of action.

(5) Complete - Must incorporate:

- Objectives (including desired effects) and tasks to be performed.
- Major forces required.
- Concepts for deployment, employment, and sustainment.
- Time estimates for achieving objectives.
- Military endstate and mission success criteria.

c. The staff should reject potential *tentative* COAs that do not meet all five criteria. A good COA accomplishes the mission within the CDR's guidance and positions the joint force for future operations and provides flexibility to meet unforeseen events during execution. It also gives components the maximum latitude for initiative. Planners can develop different COAs for using joint force capabilities (operational fires and maneuver, joint force organization, etc.) by varying the elements of operational design (such as phasing, LOOs, and so forth).

d. <u>Risk</u>. During COA development, the CDR and staff continue risk assessment, focusing on identifying and assessing hazards to mission accomplishment. The staff also continues to revise intelligence products. Generally, at the theater level, each COA will constitute a theater strategic or operational concept and should outline the following:

(1) Major strategic and operational tasks to be accomplished in the order in which they are to be accomplished.

(2) Capabilities required.

(3) Task organization and related communications systems support concept.

(4) Sustainment concept.

(5) Deployment concept.

(6) Estimate of time required to reach mission success criteria or termination criteria.

(7) Concept for maintaining a theater reserve.

e. <u>Tentative Courses of Action TTP's</u>. Following is listed a logical flow of Tactics, Techniques and Procedures (TTP's) that will help focus the staff while conceptualizing Tentative COA's:

(1) <u>Review</u>. Review information contained in the mission analysis and CDRs' guidance. The staff should review once again the mission statement it developed and the CDR approved during mission analysis. All staff members should understand the mission and the tasks that must be accomplished to achieve mission success. Following this review or upon the receipt of new information or tasking(s) from higher headquarters, if the mission statement appears inadequate or outdated, then the staff should recommend appropriate changes. The CDR has also given the staff their planning guidance which is directly linked to the CDR's operational design and how the CDR visualizes the operation unfolding.

(2) <u>Determine the COA Development Technique</u>. A critical first decision in COA development is whether to conduct simultaneous or sequential development of the COAs. Each approach has distinct advantages and disadvantages. The advantage of simultaneous development of COAs is potential time savings. Separate groups are simultaneously working on different COAs. The disadvantage of this approach is that the synergy of the JPG may be disrupted by breaking up the team. The approach is manpower intensive and requires component and directorate representation in each COA group, and there is an increased likelihood that the COAs will not be distinctive. While there is potential time to be saved, experience has demonstrated that it is not an automatic result. The simultaneous COA development approach can work, but its inherent disadvantages must be addressed and some risk accepted up front. The recommended approach if time and resources allows is the sequential method.

(3) Planning cells with land, maritime, air, space and special operations planners as well as Joint Interagency Coordination Group (JIACG) reps (and others as necessary) should initially develop ways to accomplish the *essential tasks*.

(a) Regardless of the eventual COA, the staff should plan to accomplish the higher CDR's intent by understanding its essential task(s) and purpose and the intended contribution to the higher CDR's mission success. The staff must ensure that all the COAs developed will fulfill the command mission and the purpose of the operation by conducting a review of all essential tasks developed during mission analysis.

---

[1] per JP 5-0, *Joint Planning*

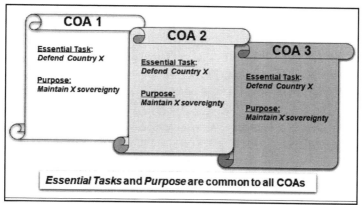

COA 1

Essential Task:
Defend Country X

Purpose:
Maintain X sovereignty

COA 2

Essential Task:
Defend Country X

Purpose:
Maintain X sovereignty

COA 3

Essential Task:
Defend Country X

Purpose:
Maintain X sovereignty

Essential Tasks and *Purpose* are common to all COAs

(b) They should then consider ways to accomplish the other tasks. A technique is for these planners to think two levels down (e.g., how could the MARFOR's component commands, MEF, or appropriate subordinate, accomplish the assigned tasks?).

(4) Once the staff has begun to visualize a *tentative* COA, it should see how it can best synchronize (arrange in terms of time, space, and purpose) the actions of all the elements of the force. The staff should estimate the anticipated duration of the operation. One method of synchronizing actions is the use of phasing as discussed earlier. Phasing assists the CDR and staff to visualize and think through the entire operation or campaign and to define requirements in terms of forces, resources, time, space, and purpose. Planners should then *integrate and synchronize* these ideas (which will essentially be Service perspectives) by using the joint architecture of maneuver, firepower, protection, support, and command and control (see the taxonomy used in the Universal Joint Task List). See the questions below.

(a) Land Operations. What are ways land forces can integrate/synchronize maneuver, firepower, protection, support, and command and control with other forces to accomplish their assigned tasks? Compare friendly forces against enemy forces to see if there are sufficient land forces to accomplish the tasks.

(b) Air Operations. What are ways air forces can integrate/synchronize maneuver, firepower, protection, support, and command and control with other forces to accomplish their assigned tasks? Compare friendly forces against enemy forces to see if there are sufficient air forces to accomplish the tasks.

(c) Maritime. What are ways maritime forces can integrate/synchronize maneuver, firepower, protection, support, and command and control with other forces to accomplish their assigned tasks? Compare friendly forces against enemy forces to see if there are sufficient maritime forces to accomplish the tasks.

(d) Special Operations. What are ways special operations forces can integrate/synchronize maneuver, firepower, protection, support, and command and control with other forces to accomplish their assigned tasks? Compare friendly forces against enemy forces to see if there are sufficient special operations forces to accomplish the tasks.

(e) Space Operations. What are the major ways that space operations can support maneuver, firepower, protection, support and establishment of command and control?

(f) Information Operations (IO). What are the ways joint forces can integrate the core capabilities of electronic warfare, computer network operations, psychological operations, military deception, and operations security, in concert with specified supporting and related capabilities, to influence, disrupt, corrupt or usurp adversarial human and automated decision-making while protecting our own.

(5) The *tentative* COAs should focus on where COGs and DPs (or vulnerabilities, e.g., "keys to achieving desired effect on centers of gravity") may occur. The CDR and the

staff review and refine their COG analysis begun during mission analysis based on updated intelligence, JIPOE products and initial staff estimates. The refined enemy and friendly COGs and critical vulnerabilities are used in the development of the initial COAs. The COG analysis helps the CDR orient on the enemy and compare friendly strengths and weakness to those of the enemy. The staff takes the CDR's operational design, reviews it, and focuses on the friendly and enemy COGs and critical vulnerabilities.

(6) By looking at friendly COG's and vulnerabilities, the staff understands the capabilities of their own force and those critical vulnerabilities that will require protection. Protection resource limitations will probably mean that the staff cannot plan to protect each asset individually, but rather look at developing overlapping protection techniques. The strength of one asset or capability may provide protection from the weakness of another.

(7) Identify the Sequencing (simultaneous/sequential/or combination) of the operation for each COA.

(8) Identify Main and Supporting Efforts by phase, the purposes of these efforts, and key supporting/supported relationships within phases.

(9) Identify Component Level Mission/Tasks (who, what and where) that will accomplish the stated purposes of main and supporting efforts. Think of component tasks from the perspective of movement and maneuver, firepower, protection, support and C2. Display them with graphic control measures as much as possible. The LOO/LOE that you completed earlier will help identify these tasks.

(10) Recognize Military Deception (MILDEC) Planning. Results of deception operations may influence any COA. The CCDR has a resident MILDEC planner on staff charged with developing the CDR's Deception Plan. Access to this plan will be as required and as directed by the CDR. Recognize that by design MILDEC planning is just behind operational planning (see JP 3-13.4, *Military Deception*).

(11) Task-Organization. The staff should develop a detailed task organization (two levels down) to execute the COA. The CDR and staff determine appropriate command relationships to include operational mission assignments and support relationships.

(12) Logistics. No COA is complete without a plan to sustain it properly. The logistic concept is more than just gathering information on various logistic functions. It entails the organization of capabilities and resources into an overall theater campaign or operation sustainment concept. It concentrates forces and material resources strategically so that the right force is available at the designated times and places with the essential equipment to conduct decisive operations. It assists thinking through a cohesive sustainment for joint, single Service and supporting forces relationships, in conjunction with multinational, interagency, non-governmental, or international organizations.

"Logistics is the Ball and Chain of Warfare"
General Heinz Wilhelm Guderian

---

(13) Develop Initial COA Sketches and Statements. Answer the questions:

- WHO (type of forces) will execute the tasks?
- WHAT is the task?
- WHERE will the tasks occur? (Start adding graphic control measures, e.g., areas of operation, amphibious objective areas).
- WHEN will the tasks begin?
- HOW (but do not usurp the components' prerogatives) should the CCDR provide "operational direction," so the components can accomplish "tactical actions?"
- WHY (for what purpose) will each force conduct its part of the operation?

(14) Test the Validity of each *Tentative* COA. (See facing page for further discussion).

(14) Test the Validity of each Tentative COA

(a) Tests for adequacy/suitability
- Does it accomplish the mission?
- Does it meet the CCDR's intent?
- Does it accomplish all the essential tasks?
- Does it meet the conditions for the endstate?
- Does it take into consideration the enemy and friendly COG?

(b) Preliminary test for feasibility
- Does the CCDR have the force structure and lift assets (means) to carry it out? The COA is feasible if it can be carried out with the forces, support, and technology available, within the constraints of the physical environment and against expected enemy opposition.
- Although this process occurs during COA analysis and the test at this time is preliminary, it may be possible to declare a COA infeasible (for example, resources are obviously insufficient). However, it may be possible to fill shortfalls by requesting support from the CCDR or other means.

(c) Preliminary test for acceptability
- Does it contain unacceptable risks? (Is it worth the possible cost?) A COA is considered acceptable if the estimated results justify the risks. The basis of this test consists of an estimation of friendly losses in forces, time, position, and opportunity.
- Does it take into account the limitations placed on the CCDR (must do, cannot do, other physical limitations)?
- Acceptability is considered from the perspective of the CCDR by reviewing the strategic objectives.
- COAs are reconciled with external constraints, particularly ROE.
- Requires visualization of execution of the COA against each enemy capability. Although this process occurs during COA analysis and the test at this time is preliminary, it may be possible to declare a COA unacceptable if it violates the CCDR's definition of acceptable risk.

(d) Test for variety
Is it fundamentally different from other COAs? They can be different when considering:
- The focus or direction of main effort.
- The scheme of maneuver (land, air, maritime, and special operation).
- Sequential vs. simultaneous maneuvers.
- The primary mechanism for mission accomplishment.
- Task organization.
- The use of reserves.

(e) Test for completeness
Does it answer all of the questions WHO, WHAT, WHERE, WHEN, HOW and WHY?

(15) <u>Determine Command Relationships and Organizational Options</u>.

(a) <u>Joint Force Organization and Command Relationships</u>. Organizations and relationships are based on the campaign design, complexity of the campaign, and degree of control required. Establishing *command relationships* includes determining the types of subordinate commands and the degree of authority to be delegated to each. Clear definition of command relationships further clarifies the intent of the CCDR and contributes to decentralized execution and unity of effort. The CCDR has the authority to determine the types of subordinate commands from several doctrinal options, including Service components, functional components, and subordinate joint commands. The options for delegating authority emanate from COCOM and range from command to support relationships. Regardless of the command or support relationships selected, it is the CCDR's (or JFC's) responsibility to ensure that these relationships are understood and clear to all subordinate, adjacent and supporting HQs. The following are considerations for establishing Joint Force Organizations:

<u>1</u> CCDRs (JFCs) will normally designate JFACCs and organize special operations forces into a functional component.

<u>2</u> Joint Forces will normally be organized with a combination of Service and functional components with operational responsibilities.

<u>3</u> Functional component staffs should be joint with Service representation in approximate proportion to the mix of subordinate forces. These staffs will be required to be organized and trained prior to employment in order to be efficient and effective, which will require advanced planning.

<u>4</u> CCDRs may establish supporting/supported relationships between components to facilitate operations.

<u>5</u> CCDRs define the authority and responsibilities of functional component CDRs based on the strategic concept of operations and may alter their authority and responsibility during the course of an operation.

<u>6</u> CCDRs must balance the need for centralized direction with decentralized execution.

<u>7</u> Major changes in the joint force organization are normally conducted at phase changes.

(b) <u>Operational Objectives and Subordinate Tasks</u>.

<u>1</u> The theater and supporting *operational objectives* assigned to subordinates are critical elements of the theater-strategic design of the campaign. They establish the conditions necessary to reach the desired endstate and achieve the national strategic objectives. The CCDR carefully defines the objectives to ensure clarity of theater and operational intent, and identify specific tasks required to achieve those objectives. Tasks are shaped by the CONOPS—intended sequencing and integration of air, land, sea, special operations, and space forces. Tasks are prioritized in order of criticality while considering the enemy's objectives and the need to gain advantage.

<u>2</u> One of the fundamental purposes of a campaign plan is to synchronize employment of all available military (land, sea, air, and special operations, as well as space, information and protection) forces and capabilities. This overwhelming application of military capabilities can be achieved by assigning the appropriate tasks to components for each phase, though supporting CDRs will also contribute with their own capabilities. These tasks can be derived from an understanding of how component and supporting forces interrelate, not only among themselves, but also with respect to the enemy.

(16) <u>Refining the theater design/operational area and initial battlespace architecture</u> (e.g., control measures). The Theater Design is normally a legally/politically binding document which will initiate planning and negotiations throughout the CCMD, interagency and internationally. It will provide flexibility/options and/or limitations to the CCDR. The theater design must be *precise*. Specifics are required to negotiate basing and overflight with DOS being the lead agency here. Theater design is also *resource sensitive*. Limited infrastruc-

ture resources must be optimized, (i.e., APOE/DS/SPOE/DS), and when utilizing a host nation's resources, negotiations for sharing those resources is common.

(a) *Operational area (OA)*. An OA is an overarching term encompassing more descriptive terms for geographic areas in which military operations are conducted. OAs include, but are not limited to, such descriptors as AOR, theater of war, theater of operations, joint operational area (JOA), amphibious objective area (AOA), joint special operations area (JSOA), and AOS. Except for AOR, which is normally assigned in the UCP, the GCC and other JFCs designate smaller OAs on a temporary basis. OAs have physical dimensions comprised of some combination of air, land, and maritime domains. The GCC and JFCs define these areas with geographical boundaries, which facilitate the coordination, integration, and deconfliction of joint operations among joint force components and supporting commands. The size of these OAs and the types of forces employed within them depend on the scope and nature of the crisis and the projected duration of operations.

(b) CCMD-Level Areas. GCCs conduct operations in their assigned AORs across the range of military operations. When warranted, the President, SecDef, or GCCs may designate a theater of war and/or theater of operations for each operation. GCCs can elect to control operations directly in these operational areas, or may establish subordinate joint forces for that purpose, allowing themselves to remain focused on the broader AOR.[2]

1 *Area of Responsibility*. An AOR is an area established by the President and SecDef on an enduring basis that defines geographic responsibilities for a GCC. A GCC has authority to plan for operations within the AOR and conduct those operations approved by the President or SecDef.

2 *Theater of War*. A theater of war is a geographical area comprised of some combination of air, land, and maritime domains established for the conduct of major operations and campaigns involving combat. A theater of war is established primarily when there is a formal declaration of war or it is necessary to encompass more than one theater of operations (or a JOA and a separate theater of operations) within a single boundary for the purposes of C2, logistics, protection, or mutual support. A theater of war does not normally encompass a GCC's entire AOR, but may cross the boundaries of two or more AORs.

3 *Theater of Operations*. A theater of operations is a geographical area comprised of some combination of air, land, and maritime domains established for the conduct of joint operations. A theater of operations is established primarily when the scope of the operation in time, space, purpose, and/or employed forces exceeds what can normally be accommodated by a JOA. One or more theaters of operations may be designated. Different theaters of operations will normally be geographically separate and focused on different missions. A theater of operations typically is smaller than a theater of war, but is large enough to allow for operations in depth and over extended periods of time. Theaters of operations are normally associated with major operations and campaigns.

4 *Combat Zones and Communications Zones (COMMZ)*. Geographic CCDRs also may establish combat zones and COMMZs. The combat zone is an area required by forces to conduct combat operations. It normally extends forward from the land force rear boundary. The COMMZ contains those theater organizations, LOCs, and other agencies required to support and sustain combat forces. The COMMZ usually includes the rear portions of the theaters of operations and theater of war (if designated) and reaches back to the CONUS base or perhaps to a supporting CCDR's AOR. The COMMZ includes airports and seaports that support the flow of forces and logistics into the OA. It usually is contiguous to the combat zone but may be separate — connected only by thin LOCs — in very fluid, dynamic situations.[3]

(17) Prepare the COA Concept of Operations Statement (or Tasks), Sketch, and Task Organization.

(a) COA concept of operations statements (or tasks) answer WHO, WHAT, WHERE, WHEN, HOW, and WHY.

[2] *JP 3-0, Joint Operations*

[3] *Ibid*

(b) Finalize COA sketches.

(c) Finalize the task organization.

(18) <u>Conduct COA Development Brief to CCDR</u>. Figure B on the facing page is a suggested sequence.

(19) <u>CCDR Provides Guidance on COAs</u>.

(a) Review and approve COAs for further analysis.

(b) Direct revisions to COAs, combinations of COAs, or development of additional COA(s).

(c) Directs priority for which enemy COA will be used during wargaming of friendly COA(s).

(20) <u>Continue the Staff Estimate Process</u>. The staff must continue to conduct their staff estimates of supportability for each COA.

(21) <u>Conduct Vertical and Horizontal Parallel Planning</u>.

(a) Discuss the planning status of staff counterparts with both CCDR's and JFC components' staffs.

(b) Coordinate planning with staff counterparts from other functional areas.

(c) Permit adjustments in planning as additional details are learned from higher and adjacent echelons, and permit lower echelons to begin planning efforts and generate questions (e.g., Requests for Information/Intelligence).[4]

f. There are several planning techniques available to you during COA development. See Figure C for a step-by-step approach utilizing the *backwards planning technique* (reverse planning).

3. Planning Directive Published

Joint Planning Process

The Planning Directive identifies planning responsibilities for developing CCMD plans. It provides guidance and requirements to the HQ staff and subordinate commands concerning coordinated planning actions for plan development. The CCDR normally communicates initial planning guidance to the staff, subordinate CDRs, and supporting CDRs by publishing a planning directive to ensure that everyone understands the CDR's intent and to achieve unity of effort. Generally, the J-5 coordinates staff action for contingency (deliberate) planning. The J-5 staff receives the CCDR's initial guidance and combines it with the information gained from the initial staff assessments. The CCDR, through the J-5, may convene a preliminary planning conference for members of the JPEC who will be involved with the plan. This is the opportunity for representatives to meet face-to-face. At the conference, the CCDR and selected members of the staff brief the attendees on important aspects of the plan and may solicit their initial reactions. Many potential conflicts can be avoided by this early exchange of information.[5]

4. Staff Estimates

Continuous staff estimates are the foundation for the CCDR's selection of a COA. Up until this point our COAs have been developed from initial impression based on limited knowledge. Staff estimates determine whether the mission can be accomplished and also determine which COA can best be supported. The staff divisions analyze and refine each COA to determine its supportability. These estimates can form the cornerstone for staff annexes to orders and plans.

[4] *CJCSM 3500.05A, JTFHQ Master Training Guide*

[5] *CJCSM 3122.01A, Joint Operation Planning and Execution System Vol I: Planning, Policies, and Procedures, Enclosure T, Appendix A, contains sample formats for the Planning Directive*

COA Development Brief to CCDR (Recommended Briefing Sequence)

J3
- Context/Background - (i.e., road to war)
- Initiation- review guidance for initiation in general
- Strategic guidance - planning tasks assigned to supported CDR, forces/resources apportioned, planning guidance, updates, defense agreements, USG Security Cooperation Plan, Theater Security Plan, JSCP
- Forces Apportioned/Assigned

J2
- Joint Intelligence Preparation of the Operational Environment (JIPOE)
- ECOA - Most dangerous, most likely; strengths and weaknesses

J3
- Update Facts and Assumptions
- Mission Statement
- Commander's Intent (purpose, method, endstate)
- Endstate: political/military
  - Termination criteria
- Center of Gravity Analysis results: Critical Factors Strategic/Operational
- JOA/Theater of Ops/COMMZ sketch
- Phase 0 Shaping Activities recommended (for current theater security plan)
- FDO's with desired effect (DIME)
- COA Sketch and Statement by phase (3 COA's minimum)
  - Task organization
  - Component tasking
  - Timeline (I+5=W Day/MEU (SOC) in Gulf, etc.)
  - Recommended C2 by phase
  - Operational Design (LOO)
  - COA Risks
- COA summarized distinctions (COA 1, 2, and 3)
- COA Priority for Wargaming

Commander's Guidance

Figure B. Recommended Briefing Sequence.

Joint Planning Process

Backwards Planning Technique (Reverse Planning)[6]

There are several planning techniques available to you during COA development. The step-by-step approach below utilizes the backwards planning technique (reverse planning):

Step 1
Determine how much force will be needed in the theater at the end of the campaign, what those forces will be doing, and how those forces will be postured geographically. Use troop to task analysis. Draw a sketch to help you visualize the forces and their location.

Step 2
Looking at your sketch and working backwards, determine the best way to get the forces you just postured in Step 1 from their ultimate locations at the end of the campaign to a base in friendly territory. This will help you formulate your desired basing plan.

Step 3
Using your mission statement as a guide, determine the tasks the force must accomplish enroute to their ultimate positions at the end of the campaign. Draw a sketch of the maneuver plan. Make sure your force does everything the SecDef has directed the CCDR to do (refer to specified tasks from the mission analysis steps).

Step 4
Determine the basing required to posture the force in friendly territory, and the tasks the force must accomplish to get to these bases. Sketch this as part of a deployment plan.

Step 5
Determine if the force you just considered is enough to accomplish all the tasks the SecDef has given you. Adjust the force strength to fit the tasks. You should now be able to answer the first question.

Step 6
Given the tasks to be performed, determine in what order you want the force to be deployed into theater. Consider force categories such as combat, C4ISR, protection, sustainment, theater enablers, and theater opening. You can now answer the second question.

Step 7
You now have all the information necessary to answer the rest of the questions regarding force employment, major tasks and their sequencing, sustainment and command relationships.

Figure C. Backwards Planning Technique

[6]*Army War College, Campaign Planning Primer*

a. Staff estimates must be comprehensive, continuous, and must visualize the future, but at the same time they must optimize the limited time available and not become overly time-consuming. *Comprehensive* estimates consider both the quantifiable and the intangible aspects of military operations. They translate friendly and enemy strengths, weapons systems, training, morale, and leadership into combat capabilities. The estimate process requires a clear understanding of weather and terrain effects and, more important, the ability to visualize the battle or crisis situations requiring military forces. Estimates must provide a timely, accurate evaluation of the unit, the enemy, and the unit's area of operations at a given time (Figure D).

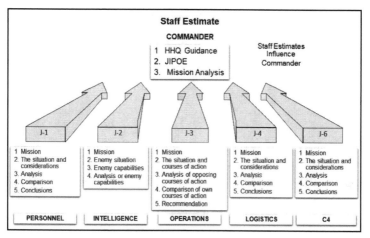

Figure D. Example: Staff Estimates.

b. Estimates must be as thorough as time and circumstances permit. The CDR and staff must constantly collect, process, and evaluate information. They update their estimates:

- When the CDR and staff recognize new facts.
- When they replace assumptions with facts or find their assumptions invalid.
- When they receive changes to the mission or when changes are indicated.

c. Estimates for the current operation can often provide a basis for estimates for future missions as well as changes to current operations. Technological advances and near-real-time information estimates ensure that estimates can be continuously updated. Estimates must *visualize the future* and support the CDR's battlefield visualization. They are the link between current operations and future plans. The CDR's vision directs the endstate and each subordinate unit CDR must also possess the ability to envision the organization's end-state. Estimates contribute to this vision. Failure to make staff estimates can lead to errors and omissions when developing, analyzing, and comparing COA's.

d. Not every situation will allow or require an extensive and lengthy planning effort. It is conceivable that a CDR could review the assigned task, receive oral briefings, make a quick decision, and direct writing of the plan to commence. This would complete the process and might be suitable if the task were simple and straightforward.

e. Most CDRs, however, are more likely to demand a thorough, well-coordinated plan that requires a complete staff estimate process. Written staff estimates are carefully prepared, coordinated, and fully documented.

f. Again, the purpose of the staff estimates is to determine whether the mission can be accomplished and to determine which COA can best be supported. This, together with the supporting discussion, gives the CDR the best possible information to select a COA. Each staff division:

- Reviews the mission and situation from its own staff functional perspective.
- Examines the factors and assumptions for which it is the responsible staff.
- Analyzes each COA from its staff functional perspective.
- Concludes whether the mission can be supported and which COA can be best supported from its particular staff functional perspective.

g. Because of the unique talents of each joint staff division, involvement of all is vital. Each staff estimate takes on a different focus that identifies certain assumptions, detailed aspects of the COAs, and potential deficiencies that are simply not known at any other level, but nevertheless must be considered. Such a detailed study of the COAs involves the corresponding staffs of subordinate and supporting commands.

h. The form and the number of COAs under consideration may change during this step. These changes result in refined COAs.

i. The product of this step is the sum total of the individual efforts of the staff divisions. Complete, fully documented staff estimates are extremely useful to the J-5 staff, which extracts information from them for the CDR's estimate. The estimates are also valuable to planners in subordinate and supporting commands as they prepare supporting plans. Although documenting the staff estimates can be delayed until after the preparation of the CDR's estimate, they should be sent to subordinate and supporting CDRs in time to help them prepare annexes for their supporting plans.

j. The principal elements of the staff estimates normally include *mission, situation and considerations, analysis of opposing COAs, comparison of friendly COAs, and conclusions.* The coordinating staff and each staff principle develop facts, assessments, and information that relate to their functional field. Types of estimates generally include, but are not limited to, operations, personnel, intelligence, logistics, civil-military operations, special staff, etc. The details in each basic category vary with the staff performing the analysis. The principal staff divisions have a similar perspective — they focus on friendly COAs and their support-ability. However, the Intelligence Directorate (J-2) estimates on intelligence (provided at the beginning of the process) concentrate on the adversary: adversary situation, including strengths and weaknesses, adversary capabilities and an analysis of those capabilities, and conclusions drawn from that analysis. The analysis of adversary capabilities includes an analysis of the various COAs available to the adversary according to its capabilities, which include attacking, withdrawing, defending, delaying, etc. The J-2's conclusion will indicate the adversary's most likely COA and identify adversary COGs.[7]

k. In many cases the steps in the concept development phase are not separate and distinct, as the evolution of the refined COA illustrates. During planning guidance and early in the staff estimates, the initial COAs may have been developed from initial impressions and based on limited staff support. But as concept development progresses, COAs are refined and evolve to include many of the following considerations:
- What military operations are considered?
- Where will they be performed?
- Who will conduct the operation?
- When is the operation planned to occur?
- How will the operation be conducted?

l. *An iterative process of modifying, adding to, and deleting from the original tentative list is used to develop these refined COAs.* The staff continually evaluates the situation as the planning process continues. Early staff estimates are frequently given as oral brief-ings to the rest of the staff. In the beginning, they tend to emphasize information collection more than analysis. It is only in the later stages of the process that the staff estimates are expected to indicate which COAs can be best supported.

---

[7]*CJCSM 3122.01, Joint Operation Planning and Execution System Vol. I: Planning, Policies, and Procedures, Enclosure S contains sample formats for staff estimates*

IV. Course of Action (COA) Analysis & Wargaming

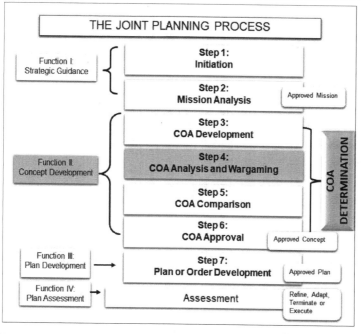

THE JOINT PLANNING PROCESS

Function I:
Strategic Guidance

Step 1:
Initiation

Step 2:
Mission Analysis

Approved Mission

Function II:
Concept Development

Step 3:
COA Development

Step 4:
COA Analysis and Wargaming

Step 5:
COA Comparison

Step 6:
COA Approval

Approved Concept

COA DETERMINATION

Function III:
Plan Development

Step 7:
Plan or Order Development

Approved Plan

Function IV:
Plan Assessment

Assessment

Refine, Adapt,
Terminate or
Execute

1. Step 4 to JPP — COA Analysis and Wargaming

a. COA analysis and wargaming allows the CDR, staff and subordinate CDRs and their staffs to gain a common understanding of friendly and threat COAs. This common understanding allows them to determine the advantages and disadvantages of each COA and forms the basis for the CDR's comparison (Step 5) and approval (Step 6). *COA Wargaming involves a detailed assessment of each COA as it pertains to the enemy and the operational environment*. Each friendly COA is wargamed against selected threat COAs. The CDR will select the COAs to be wargamed and provide wargaming guidance along with *evaluation criteria*.

b. While time-consuming, this procedure reveals strengths and weaknesses of each friendly COA, anticipates battlefield events, synchronizes warfighting functions, determines task organization for combat, identifies decision points, informs potential branches and sequels, and identifies cross-Service, interagency or component support requirements. Wargaming should also answer these questions:

- Does the COA achieve the **purpose** of the mission?
- Is the COA supportable?
- What if?

> **Wargaming Process**
>
> 1. Assemble the necessary tools and information.
> 2. Establish specific analysis rules to follow.
> 3. Wargame the COAs.
> 4. Record vital wargaming activities.
> 5. Identify advantages/disadvantages.
> 6. Refine each COA based on wargaming results.

2. Analysis of Opposing COAs

The heart of the CDR's estimate process is the *analysis of opposing COAs*. Analysis is nothing more than wargaming—either manual or computer assisted. In the previous steps of the staff estimate, *enemy* COAs (ECOAs) and COAs were examined relative to their basic concepts. ECOAs were developed based on enemy capabilities, objectives, and our estimate of the enemy's intent, and COAs developed based on friendly mission and capabilities. In this step we conduct an analysis of the probable effect *each ECOA has on the chances of success of each friendly COA*. The aim is to develop a sound basis for determining the *feasibility and acceptability* of the COAs. Analysis also provides the planning staff with a greatly improved understanding of their COAs and the relationship between them. The COA analysis identifies which COA best accomplishes the mission while best positioning the force for future operations. It helps the CDR and staff to:

- Determine how to maximize combat power against the enemy while protecting the friendly forces and minimizing collateral damage.
- Have as near an identical visualization of the operation as possible.
- Anticipate events in the operational environment and potential reaction options.
- Determine conditions and resources required for success.
- Determine when and where to apply the force's capabilities.
- Focus intelligence collection requirements.
- Determine the most flexible COA.

3. The Conduct of Analysis and Wargaming

a. The wargame is a disciplined process, with rules and steps that attempt to visualize the flow of the operation. The process considers friendly dispositions, strengths, and weaknesses; enemy assets, probable COAs and characteristics of the physical environment. It relies heavily on joint doctrinal foundation, tactical judgment and operational experience while focusing the staff's attention on each phase of the operation in a logical sequence. It is an iterative process of action, reaction, and counteraction that stimulates ideas and provides insights that might not otherwise be discovered while highlighting critical tasks, and providing familiarity with operational possibilities otherwise difficult to achieve. Wargaming is a critical portion of the planning process and should be allocated more time than any other step. Each retained COA should, at a minimum, be wargamed against both the *most likely* (ML) and *most dangerous* (MD) ECOAs.

b. During the wargame, the staff takes a COA statement and begins to add more detail to the concept, while determining the strengths or weaknesses of each COA. Wargaming tests a COA and can provide insights that can be used to improve upon a developed COA. The CDR and staff (and subordinate CDRs and staffs if the wargame is conducted collaboratively) may change an existing COA or develop a new COA after identifying unforeseen critical events, tasks, requirements, or problems.

4. Examining and Testing the COAs

For the wargame to be effective, the CDR should indicate what aspects of the COA the CDR desires to be examined and tested. Wargaming guidance may include a list of friendly COAs to be wargamed against specific threat COAs (e.g., COA 1 against the enemy's most likely (ML), most dangerous (MD)), the timeline for the phase or stage of the operations, a list of critical events and level of detail (i.e., two levels down).

a. Analysis of the proposed COAs should reveal a number of criteria including:

- Potential DPs.
- Task organization adjustment.
- Identification of plan branches and sequels.
- Identification of high-value targets.
- Risk assessment.
- COA advantages and disadvantages.
- Recommended CCIR.

b. COA Analysis Considerations. Before we move to making key decisions for wargaming, we need to collate and review a few important items. A few are listed below:

- Evaluation criteria.
- Friendly and enemy forces.
- RFI's.
- Assumptions.
- Known critical events.

c. In the context of COA Analysis, we will look at evaluation criteria and critical events.

(1) Evaluation Criteria. Determining the evaluation criteria to be used is a critical requirement that begins with wargaming (Figure A is an example of possible CDRs evaluation criteria.) The CDR and staff choose the evaluation criteria during wargaming that will be used to select the COA that will become the CONOPS. CDRs establish evaluation criteria based on judgment, personal experience, METT-T and those criteria the staff uses to measure the effectiveness and efficiency of one COA relative to other COAs following the wargame. These evaluation criteria help focus the wargaming effort and provide the framework for data collection by the staff. **They are those aspects of the situation (or externally imposed criteria) that the CDR deems critical to the accomplishment of the mission.** Potential influencing criteria include elements of the CDR's Guidance and/or CDR's Intent, selected principles of war, external constraints, and even anticipated future operations for involved forces or against the same objective. Evaluation criteria change from mission to mission. Though these criteria will be applied in the next step when the COAs are compared, it will be helpful during this wargaming step for all participants to be familiar with the criteria so that any insights into a given COA which influence a criterium are recorded for later comparison. The criteria may include anything the CDR desires. If not received directly from the CDR, they are often derived from the CDR's intent statement.

The criteria should look at both what will create success and what will cause failure. They may be used to determine the criteria of success for comparing the COAs in Step 5.

(2) Developing Evaluation Criteria. Evaluation criteria do not stand alone. Each must have a clearly-defined definition. Defining the criteria in precise terms reduces subjectivity and ensures the interpretation of each remains constant. The following list provides a good starting point for developing a COA comparison criteria list.

Figure A. Possible Commander's Evaluation Criteria.

Some possible sources for determining criteria are:

- The CDR's Guidance and CDR's Intent.
- Mission accomplishment at an acceptable cost.
- The principles of war — Joint operations/SSTR (MOOSEMUSS).
- Doctrinal fundamentals for the type of operation(s) being conducted.
- The level of residual risk in the COA.
- Implicit significant criteria relating to the operation (e.g., need for speed, security).
- Each staff member may identify criteria relating to that staff function.
- Elements of Operational Art/Design (Table 1).
- Other criteria to consider: political constraints, risk, financial costs, flexibility, simplicity, surprise, speed, mass, sustainability, C2, infrastructure survivability, operational design elements, etc.

Operational Design Elements		
Termination	Military Endstate	Objectives
Effects	Center of Gravity	Decisive Points
Lines of Operations & Effort	Direct vs Indirect Approach	Operational Reach
Forces and Functions	Anticipation	Culmination
Arraigning Operations		

Table 1. Operational Design Elements

> **Example: Evaluation Criteria for Humanitarian Assistance Operation:**
>
> -Protects the force — Delivers resources — Complements HN
>
> -Facilitates enterprise transition — Targets COG's — Enduring Partnerships
>
> -Meet critical partner nation stabilization/development needs that only U.S./DOD can/will provide

(3) <u>List Known Critical Events</u>. These are essential tasks, or a series of critical tasks, conducted over a period of time that require detailed analysis (e.g., the series of component tasks to be performed on D-Day). This may be expanded to review component tasks over a phase(s) of an operation (e.g., lodgment phase) or over a period of time (C-Day through D-Day). The planning staff may wish at this point to also identify decision points (those decisions in time and space that the CDR must make to ensure timely execution and synchronization of resources). These decision points are most likely linked to a critical event (e.g., commitment of the JTF Reserve force).

5. Wargaming Decisions

There are two key decisions to make before COA analysis begins. The *first* decision is to decide *what type of wargame will be used*. This decision should be based on CDR's guidance, time and resources available, staff expertise, and availability of simulation models. Note that at this point in the planning process, there may be no phases developed for the COA; Pre-hostilities, Hostilities, and Post-hostilities may be your only considerations at this point. Phasing comes later when the planner begins to flesh out the selected COA into a strategic concept. Some considerations are:

- Information Review: Mission Analysis, CDR's intent, planning guidance, CCDR's orders.
- Gather tools, materials, personnel and data:
  - Friendly COA to be analyzed.
  - ECOA against which you will evaluate the friendly COAs.
  - Representations of the operational area such as maps, overlays, etc.
  - Representations of friendly and enemy force dispositions and capabilities.
  - Subject matter experts (INTEL, SJA, POLAD, Log, IW, C4, PAO, etc.).
  - Red cell.
  - Scribe/recorder.
- Keep discussions elevated to the theater level.
- Balance between stifling creativity and making progress.

The *second* decision is to prioritize the ECOAs the wargame is to be analyzed against. In time-constrained situations, it may not be possible to wargame against all COAs.

6. Wargaming

Wargaming provides a means for the CDR and participants to analyze a tentative COA and obtain insights that otherwise might not have occurred. An objective, comprehensive analysis of tentative COAs is difficult even without time constraints. Based upon time available, the CDR should wargame each tentative COA against the most probable and the most dangerous adversary COAs (or most difficult objectives in non-combat operations) identified through the JIPOE process.

Some considerations are:

- Refine wargaming methodology.
- Pre-conditions or start points and endstate for each phase.
- Advantages/disadvantages of the COA.
- Unresolved issues.
- COA modifications or refinements.
- Estimated duration of critical events/phases.
- Major tasks for components.
- Identify critical events and DPs.
- Identify branches and sequels.
- Identify risks.
- Recommended EEIs and supporting collections plan priorities.
- Highlight ROE requirements.

a. Wargaming is a conscious attempt to visualize the flow of the operation, given joint force strengths and dispositions, adversary capabilities and possible COAs, the OA, and other aspects of the operational environment. Each critical event within a proposed COA should be wargamed based upon time available using the action, reaction, counteraction method of friendly and/or opposition force interaction. Here, the friendly force will make two moves because this activity is intended to validate and refine the friendly forces COA, not the adversaries. The basic wargaming method (modified to fit the specific mission and environment) can apply to noncombat as well as combat operations.

b. Wargaming stimulates thought about the operation so the staff will obtain ideas and insights that otherwise might not have occurred. This process highlights tasks that appear to be particularly important to the operation and provides a degree of familiarity with operational-level possibilities that might otherwise be difficult to achieve.

c. The wargaming process can be as simple as a detailed narrative effort that describes the action, probable reaction, counteraction, assets, and time used. A more comprehensive version is the "sketch-note" technique, which adds operational sketches and notes to the narrative process in order to gain a clearer picture. The most sophisticated form of wargaming is modern, computer-aided modeling and simulation. The figure on the facing page provides a list of key inputs and outputs for wargaming.

d. <u>Where to Begin</u>. See Figure B, "Wargaming Steps" on facing page.

(a) <u>Manual Wargaming</u>:

1 <u>Deliberate Timeline Analysis</u> — Consider actions day by day or in other discrete blocks of time. This is the most thorough method when time permits for detailed analysis.

2 <u>Operational Phasing</u> — Used as a framework for COA analysis. Identify significant actions and requirements by functional area and/or JTF component.

3 <u>Critical Events/Sequence of Essential Tasks</u> — The sequence of essential tasks, also known as the critical events method, highlights the initial shaping actions necessary to establish a sustainment capability and to engage enemy units in the deep battle area. At the same time, it enables the planners to adapt if the enemy executes a reaction that necessitates the reordering of the essential tasks. This technique also allows wargamers to concurrently analyze the essential tasks required to execute the CONOPS.

a *Focus on specific critical events that encompass the essence of the COA.* If time is particularly limited, focus only on the principal defeat mechanism. It is important to identify a measure of effectiveness (MOE) that attempts to quantify the achievement of that defeat mechanism. This MOE should enable a consistent comparison of each COA against

Wargaming Steps

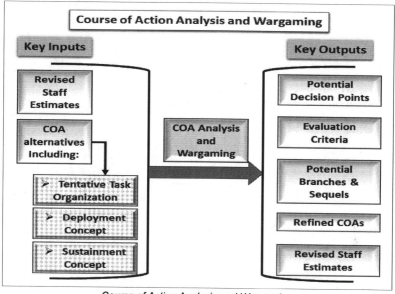

Course of Action Analysis and Wargaming.

Step 1— Prepare for the Wargame
- Gather tools.
- List and review friendly forces.
- List and review enemy forces.
- List known critical events.
- Determine participants.
- Determine enemy COA to oppose.
- Select wargame method (manual, computer).
- Select method to record and display wargaming results (narrative, sketch and notes, wargame worksheets, synchronization matrix).

Step 2 — Conduct the Wargame and Assess the Results
- Purpose of the Wargame (identify gaps, visualization, etc.).
- Basic methodology (action, reaction, counteraction).
- Type of method to be used: Based on CDR's guidance.

Step 3— Output of Wargame
- Results of the Wargame Brief: potential decision points, evaluation criteria (those criteria that the CDR deems critical to mission accomplishment), potential branches and sequels.
- Revised staff estimates.
- Refined COAs.
- Feedback through the COA Decision Brief.

Figure B. Wargaming Steps

each ECOA for each specific critical event. If necessary, different MOE should be developed for assessing different types of critical events (e.g., destruction, blockade, air control, neutralization, ensure defense). As with the focus on operational phasing, the critical events discussion identifies significant actions and requirements by functional area and/or by JTF component.

(b) <u>Computer Assisted</u>. There are many forms of computer-assisted wargames. Most wargames require a significant spin-up time to load scenarios and then to train users. However, the potential to utilize the computer model for multiple scenarios or blended scenarios makes it valuable.

e. <u>Preparing for the Wargame.</u>

(1) <u>Room Organization</u>. Contemplate how to organize the players in a logical manner. Figure C below is an *example* of a manual wargame room organization. Note that along one wall are blown-up slides for CCIRs, Facts and Assumptions, Friendly Forces, Enemy COA, etc.. On the screens in the front of the wargame table are the working Synchronization Matrix and Decision-Support Template. There is no limit to the information you can display around the room for quick reference. However, it's important to limit those items displayed to essential items that facilitate the wargame, not distract from it.

Room Organization

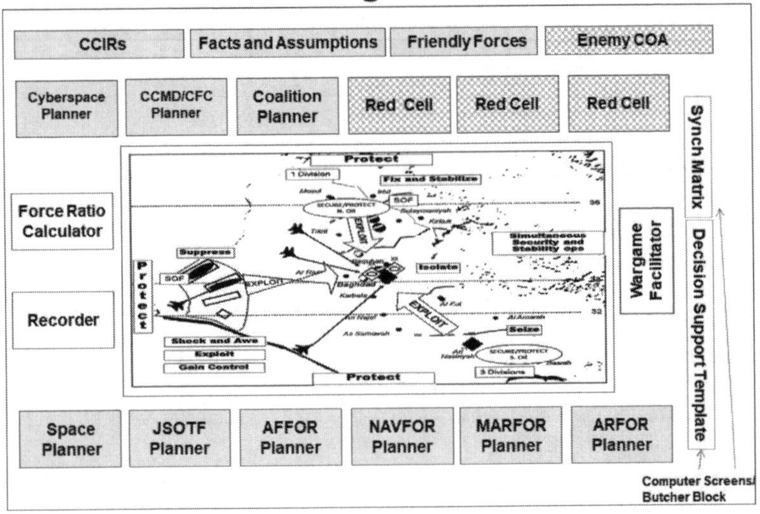

Figure C. Example Wargame Room Organization.

(2) <u>Briefing Sequence</u>. The following briefing sequence is an example of how the Blue Team/facilitator choreographs their wargame (other planners/staff as required):

- Blue Team/Facilitator: reviews conditions and timing for particular COA (includes current *Friendly* Intel picture)
- Red Cell Orientation
- CCMD / CFC Planner
- CJSOTF Planner
- AFFOR Planner
- NAVFOR Planner
- MARFOR Planner

- ARFOR Planner
- Cyberspace Planner
- Space Planner
- Coalition Planner
- Force Ratio Assessment
- Issues from Recorder
- Facilitator: highlights issues and concludes game turn

(3) <u>Rules of the Road</u>.
- Facilitator has final say!!
- Facilitator is the only one who brings problems to arbitrators.
- Don't argue with Red Cell.
- Come prepared…try not to ad lib.
- Keep pace and pay attention.
- Take notes.
  - Submit appropriate information to Recorder.

Name: *Major Payne*
Component: *ARFOR*
COA #: *1*
Critical Event: *Commence CATK*
Topic Item: *ARFOR commence CATK from lodgment in Tunis with 85% strength in personnel Due to planned force flow…need 40 sorties of strike aircraft…*

Decision Point: *CFC-T will launch ARFOR CATK simultaneously with MARFOR once ARFOR Has at least 85% strength in combat forces and sustainment capability for 30 DOS in Classes I, III & IV.*

Assumption: *Once CTAK has been launched, there will be no interruption of planned force flow From N+30 – N+45.*

Submission Cards to Recorder

f. <u>Wargame Process</u>.
- Begins with time or event.
  - *Could be an enemy attack.*
- Next is friendly action, then threat reaction, then friendly counteraction.
  - *If necessary, may continue beyond three moves.*
- Continually assess feasibility.

(1) The facilitator and the red cell CDR get together to agree on the rules of the wargame. The wargame begins with an event designated by the facilitator. It could be an enemy offensive/defensive action or it could be a friendly offensive/defensive action. They decide where (in the AO) and when (H-Hour or L-Hour) it will begin. They review the initial array of forces. Of note, they must come to an agreement on the effectiveness of *intelligence collection* and *shaping actions* by both sides prior to the wargame. The facilitator must ensure that all members of the wargame know what critical events will be wargamed and what techniques will be used. This coordination within the friendly cell and between the friendly and the red cell must be done well in advance.

(a) Again, within each wargaming method the wargame normally has three total moves. If necessary, that portion of the wargame may be extended beyond the three moves. The facilitator decides how many moves are made in the wargame.

(b) During the wargame the players must continually assess the COA's feasibility. Can it be supported? Can we do this? Do we need more combat power, more intelligence assets, more battlespace, more time, do we have the necessary logistics and communications? Has the threat successfully countered a certain phase or stage of a friendly COA? If we find that many of the aforementioned questions were answered "yes," then we may have to make major revisions to our friendly COA. We don't make major revisions to a COA in the midst of a wargame. Instead, we stop the wargame, make the revisions and start over at the beginning.

(2) The wargame is for comparing and contrasting friendly COAs with the threat COAs. We compare and contrast friendly COAs with each other in the fifth step of the JPP, comparison/decision. Avoid becoming emotionally attached to a friendly COA; this will lead to the overlooking of the COA shortcomings and weaknesses.

- *Avoid comparing* one friendly COA with another friendly COA during the wargame.
- Must remain unbiased.
- Do not violate timelines.

(a) A wargame for one COA at the JTF level may take six to eight hours. The facilitator must allocate enough time to ensure the wargame will thoroughly test a COA.

(b) The wargame considers friendly dispositions, strengths, and weaknesses; threat assets, probable COAs and characteristics of the area of operations. Through a logical sequence it focuses the players on essential tasks to be accomplished.

(c) When the wargame is complete and the worksheet and sync matrix (see para. 6.L) are filled out, there should be enough detail to "flesh out" the bones of the COA and begin orders development (once the COA has been selected by the CDR in comparison/decision).

(d) Additionally, the wargame will produce a refined event template and the initial decision support template which we call "decision support tools" (see Figures D & E). These are similar to a football coach's game plan. The tools can help predict what the threat will do and also provide the CDR options for employing the forces to counter a threat action. The tools will prepare the CDR (coach) and the staff (team) for a wide spectrum of possibilities and a choice of immediate solutions.

(e) The wargame stimulates new ideas and provides insights that might have been otherwise overlooked, relying heavily on doctrinal foundation, tactical judgment and experience of the wargamers. The dynamics of the wargame require the red cell CDR and the red cell members to be aggressive, but realistic, in the execution of threat activities.

- Records advantages and disadvantages of each COA as they become evident.
- Creates *decision support tools* (a game plan).
- Focuses OPT on the threat *and* CDR's evaluation criteria.

g. Wargaming Products. We seek certain things to come out of the wargame in addition to wargamed COAs. We enter the wargame with a "rough" event template and must complete the wargame with a "refined," more accurate event template. Remember, the event template with its named areas of interest (NAIs) and time-phase lines (TPLs) will help the J-2 focus the intelligence collection effort. An event matrix can be used as a "script" for the intelligence rep during the wargame. It can also tell us if we are relying too much on one or two collection platforms and if we have overextended these assets.

(1) A first draft of a decision support template (DST) and decision support matrix (DSM) should also come out of the COA wargame. The first time we see decision points and target areas of interest (TAI) should be on the first draft DST as developed in the wargame. As more information about friendly forces and threat forces becomes available, the DST and DSM may change.

Sample Event Matrix

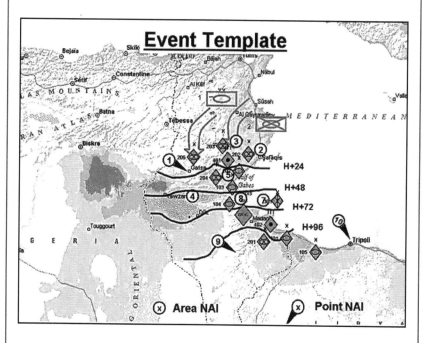

Event Matrix

NAI	Event	NET/ NLT	DP & TAI	Tasked Collection Assets
3,2	Orange WOG first echelon vic Mezzouna displaces from defensive positions	H + 6 H + 12	DP 3	RC-136, TR-1, F/A-18D, Predator Track 3
19, 20	Red SEC moving east across border toward Kasserine Pass	H + 12 H + 24	DP 1	National IMINT, National SIGINT, U-2R, RC-135
1	Orange 205th mech BDE moves north out of Gafsa to link up with lead BDE	H + 6 H + 48	DP 1	National SIGINT, RC-135, Predator Track 6
8	LY 2d echelon forces move north to establish strong point defense vic Gabes	H Hour H +60	DP 2	National SIGINT, Theater IMINT, RC-135, VMU Track 2
2,5	Orange first echelon attacks Sfax	21 July/ 26 July	DP 4	FA-18/FAC/FLIR, National IMINT and SIGINT
10	Orange fire SCUDs at Blue ports and airfields	Phase 1 & 2	DP 5	National / Theater IMINT National /Theater SIGINT
7	Orange establishes CSSC-3 Seersucker Battery vic Jerba Island	Phase 1 & 2	DP 6	JSTARS, Predator Track 1

Figure D. Sample Event Template and Matrix.

(2) Below the products in italics are produced by the intelligence section. Products in underlined are produced by the friendly or the operations section.

- Refines existing *NAI* and *TPL* and identifies new *NAIs*. Identifies likely TAI and decision points.
- Refines the J-2's *situation and event templates* and creates first draft decision support template and matrix.
- *High value targets (HVT)* from J-2 source the OPT's high-payoff targets (HPT).

h. Wargaming Outputs. There are inputs to the COA wargame. Similarly, there are outputs from the COA wargame. We must have the modified COAs, both sketch and narrative, ready to present to the CDR at the COA wargame back brief.

(1) The critical events are associated with the essential tasks which we identified in mission analysis. The DPs are tied to points in time and space when and where the CDR must make a critical decision. Decision points will be tied to the CCIR. Remember, CCIRs generate three types of information requirements: PIRs, FFIRs, and EEFIs. The CDR approves CCIR's. From a threat perspective, PIRs tied to a DP will require an intelligence collection asset to gather information about the threat. In IPB we tie PIRs to NAIs which are linked to threat COAs. The sync matrix is a tool which will help us determine if we have adequate resources.

- Wargamed COAs with graphic and narrative. Branches and sequels identified.
- Information on CDR's evaluation criteria.
- Initial task organization.
- Critical events and DPs.
- *New* resource shortfalls.
- *Refined/new* CCIRs and event template/matrix.
- *Initial* decision support template/matrix.
- "Fleshed out" synchronization matrix.
- *Refined* staff estimates.

(2) The outputs of the COA wargame will be used in the comparison/decision step, orders development, and transition. The outputs on the slide are "products." The results of the wargame are the strengths and weaknesses of each friendly COA, the core of the back brief to the CDR.

i. Red Team. Once the enemy COA has been selected, it's time to build your Red Team.

(1) One of the most critical elements of your wargaming will be the synchronization of the Red Team. Without an aggressive and forward-leaning Red Team, the plan as a whole will be lesser for it. The most important element of wargaming is not the tool used, but the people who participate. Staff members who participate in wargaming should be the individuals who were deeply involved in the development of COAs. A robust cell that can aggressively pursue the adversary's point of view when considering adversary counteraction is essential. This *"red cell"* role-plays the adversary CDR and staff. The red cell develops critical DPs relative to the friendly COAs, projects adversary reactions to friendly actions, and estimates adversary losses for each friendly COA. By trying to accurately portray the capabilities of the enemy, the red cell helps the staff fully address friendly responses for each adversary COA.

(a) The primary purpose of the Red Team is to provide additional operational analysis on the adversary. The Red Team process normally is a peer review of the selected COA against adversary MD or ML COA. The Red Team will fight the MD and/or ML COA, but the goal is not to see if the Red Team can come up with a plan to defeat Blue's COA. The most important piece here is for the Red team to have enough time to develop a Red CONOPS prior to the wargame itself. The Red Team leader and team should be extremely

Sample Decision Support Template

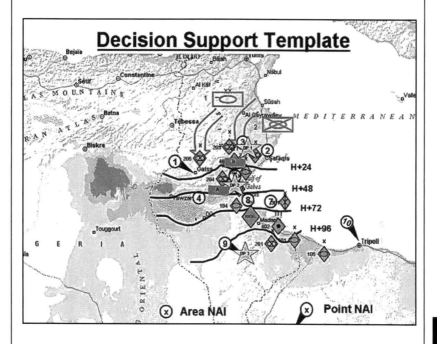

Decision Support Template

Decision Support Matrix

DP	Events & Indicators	NET/NLT	CDR's Options
1	Whether or not enemy 1st echelon units are fixed; (NAI 1-3)	H+24/H+36	1 AD continues turning movement or executes branch plan for envelopment of 1st echelon.
2	1st echelon enemy forces withdrawing into/ through Gabes; 103d Armor Bde covering withdrawal and as possible counterattack force, 204th, 201st, and 104th preparing for BHO vic Gabes; refugees being forced North; (NAI 4-5)	H+96/H+120	Bypass, isolate or clear Gabes; force options: 2d MarDiv of LF6F
3	Enemy delays 2dMarDiv and reorients on mountain passes IOT hold 1 AD and allow forces to withdraw to border; (NAI 9)	H+144/H+168	Options: defeat 2d echelon via encirclement; defeat 2d echelon by annihilation (1AD/2MAW) or allow enemy to withdraw.

Figure E. Sample Decision Support Template and Matrix.

knowledgeable of the Blue plan and should have been in on the development of that plan. The Red Team will, among other duties:

- Project adversary reactions to friendly actions.
- Attempt to accurately portray the capabilities of the adversary.
- Develop critical decision points relative to the friendly COAs.
- Analyze friendly critical vulnerabilities.
- ID adversary HVTs' key terrain, avenues of approach, doctrine and tactics, unconventional threats, reconnaissance and surveillance plans, etc.
- Depict adversary NAIs, TAIs, and DPs and in refining situation templates (threat COAs).

(b) If subordinate functional and Service components establish similar cells that mirror their adversary counterparts, this Red Cell network can collaborate to effectively wargame the adversary's full range of capabilities against the joint force. In addition to supporting the wargaming effort during planning, the Red Cell can continue to view friendly joint operations from the adversary's perspective during execution. The Red Cell process can be applied to noncombat operations to help determine unforeseen or most likely obstacles to, as well as the potential results of planned operations.

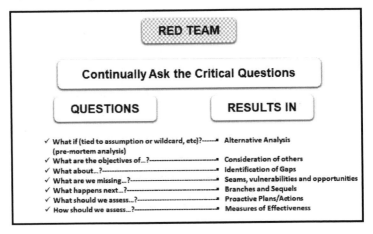

Red Team Critical Questions

j. <u>White Team</u>. A small team of arbitrators normally composed of senior individuals familiar with the plan is a smart investment to ensure that the wargame does not get bogged down in unnecessary disagreement or arguing. The White Team will provide overall oversight to the wargame and any adjudication required between participants. The White Team may also include the facilitator and/or senior mentors as required.

> *"To be practical, any plan must take account of the enemy's power to frustrate it; the best chance of overcoming such obstruction is to have a plan that can be easily varied to fit the circumstances met; to keep such adaptability, while still keeping the initiative, the best way to operate is along a line which offers alternative objectives."*
>
> B.H. Liddell Hart

k. <u>What a Wargame Provides</u>.

- COA Wargame helps the CDR determine how to *best apply friendly strengths against threat critical vulnerabilities (CV)* while *protecting* friendly CVs.
- COA Wargame provides *decision-support tools* which facilitate the *transition to current operations* and subsequent execution.
- The red cell and the wargame test our COAs against a thinking adversary. The wargame ensures we bring viable COAs to the CDR for their decision. The results of COA analysis are better COAs.
- The wargame confirms:
  - Paths to threat CVs
  - Mission analysis

l. <u>A Synchronization Matrix</u> is a decision-making tool and a method of recording the results of wargaming. Key results that should be recorded include DPs, potential evaluation criteria, CCIR, COA adjustments, branches, and sequels. Using a synchronization matrix helps the staff visually synchronize the COA across time and space in relation to the adversary's possible COAs. The wargame and synchronization matrix efforts will be particularly useful in identifying cross-component support resource requirements. Figures F through I are examples of Synchronization Matrix Frameworks and Figures J & K are examples of a Timeline Matrix and Matrix by Phase.

Sequence Number	Critical Event / Phase / Time:								
	Action	Reaction	Counter-Action	Assets	Time	Decision Point	PIR	Procedural & Positive Controls	Remarks

Figure F. Example of Synchronization Matrix Framework

These synchronization matrices might be combined into one that, for example, reflects the contributions that each component would provide, within each joint functional area, over time. The staff can adapt the synchronization matrix to fit the needs of the analysis. It should incorporate other operations, functions, and units that it wants to highlight.

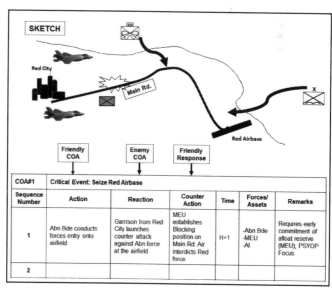

Figure G. Example Wargame Synchronization Matrix.

Example Syncronization Matrixes

Friendly COA #1 Short Name: _____

Enemy COA – (Most Likely / Most Dangerous)

Time / Phase / Critical Event : _____

Situation:

Start:
End:

	Areas	Action	Enemy Reaction	Counteraction
COMPONENTS	ARFOR/Land Component			
	MARFOR/ Land Component			
	NAVFOR/Maritime Component			
	ARFOR			
	JFACC			
	JSOTF			
	Others (e.g. USCG & rest of Interagency)			
JOINT FUNCTIONAL AREAS	Movement & Maneuver			
	Intelligence			
	Firepower			
	Sustainment			
	Command & Control			
	Protection			
	Decision Points			
OTHERS	CCIR			
	Branches			
	Risks			
	Issues			

Figure H. Example of Synchronization Matrix Framework

Continued on next page —

Friendly COA #1 Short Name: <u>Simultaneous NEO</u>

Enemy COA – (<u>Most Likely</u> /Most Dangerous)

Time / Phase / Critical Event: <u>Evacuations</u>

Situation: The host government cannot guarantee the safety of American and third-country nationals (TCN). The environment for a NEO is "uncertain." Rebel factions are becoming more violent.

Start: JTF 780 positioned for NEO
End: Successful NEO and forces return to safe havens.

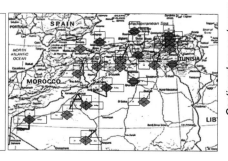

	Areas	Action	Enemy Reaction	Counteraction
C O M P O N E N T S	ARFOR/Land Component	-Preparing to secure evac. site at #2 -Located with NAVFOR afloat	Rebel forces declares all AMCITS "enemies of the revolution" & US	-Launches and secures evac. Site #2 -Evac. AMCITs/TCN onto C-130
	MARFOR/Land Component	-Preparing to secure evac. site at #3 -Located with NAVFOR (afloat)	military presence as evidence of plans to invade their country Several TCN's	-Launches and secures evac. Site #3 -Evac. AMCITs/TUN onto NAVFOR ships
	NAVFOR/Maritime Component	-Supporting MARFOR -Supporting JFACC -Supporting ARFOR (ISB TR)	killed in grenade attack near Site #2	-Supporting MARFOR -Supporting JFACC -Supporting ARFOR (ISB TR)
	ARFOR	-Supporting JFACC -Preparing evac. Control center at site A		-Supporting JFACC -Operates evac. Control center at site A
	JFACC	-Preparing to gain air superiority -Preparing to provide CAS/Coord. airspace		-Gains/maintains air superiority -Provide CAS /Coordinate airspace
	JSOTF	-Preparing to secure evac. site at #1 -Located at ISB B		-Launches and secures evac. Site #1 -Evac. AMCITs/TUN onto to ISB B
	Others (e.g. USCG & rest of Interagency	-DOS – Preparing to implement emergency evac. plans -Pres/SECDEF prepare EXORD		-DOS – Implement emergency evac plan -Pres/SECDEF – Releases EXORD
J O I N T **F U N C T I O N A L** **A R E A S**	Movement & Maneuver	-JTF must be in position NLT___ -JTF be prepared to execute EXORD +4hr		-NAVFOR & JFACC insures freedom of sea and air movements
	Intelligence	-NAIs – Sites 1, 2 & 3 _PIR – Threat against AMCITs, rebel troop movements, rebel attacks, etc.		-Collects info at NAIs -Collects and processes PIR
	Firepower	-Targeting effects continue		-AC-130 spt ARFOR/MARFOR/JSOTF (OPCON to JSOTF, TACON others)
	Sustainment	-HN spt being provided by countries X&Y		-HN spt not available
	Command & Control	-IO/IW -PSYOP spt to ARFOR/MARFOR & JSOTF -JTF HQ located on USS MTW		-NAVFOR provides direct spt to MARFOR & ARFOR -Continues IW/C2W spt -ROE Changes
	Protection	-Weapons status___ -Deception theme___		-Weapons status___ -Deception theme___
O T H E R S	Decision Points	-Recommendation for evac. EXORD		-When to declare "endstates" and mission accomplishment
	CCIR	-Taking of any AMCITs as hostages -Violence directed at AMCITs or TCNs		-# of AMCITs/TCNs evacuated -Any interference with the evacuation
	Branches	-Site #1, 2, or 3 is unusable -Hostages taken		-Implement branch plans
	Risks	-Military presence will provoke rebels		-Significant casualties -Usable to locate all AMCITs
	Issues	-Army helos on carriers Approval for use of evac site ISB in Cty Y		-Policy for evac. of unidentified personnel requesting help

Figure I. Example of Synchronization Matrix Framework

— Continued on next page

Joint Planning Process

Example Syncrhonization Matrixes (cont.)

Continued from previous page

Continued from previous page

Friendly COA #1 Short Name: Simultaneous NEO

Enemy COA – (Most Likely /Most Dangerous)

Time / Phase / Critical Event: Evacuations

Situation: The host government cannot guarantee the safety of American and third-country nationals (TCN). The environment for a NEO is "uncertain." Rebel factions are becoming more violent.

Start: JTF 780 positioned for NEO
End: Successful NEO and forces return to safe havens.

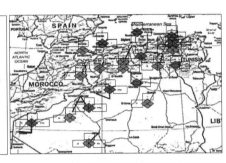

	Areas	Day D-2	D-Day
C O M P O N E N T S	ARFOR/Land Component	-Preparing to secure evac. site at #2 -Located with NAVFOR afloat	-Launches and secures evac. Site #2 -Evac. AMCITs/TCN onto C-130
	MARFOR/ Land Component	-Preparing to secure evac. site at #3 -Located with NAVFOR (afloat)	-Launches and secures evac. Site #3 -Evac. AMCITs/TUN onto NAVFOR ships
	NAVFOR/Maritime Component	-Supporting MARFOR -Supporting JFACC -Supporting ARFOR (ISB TR)	-Supporting MARFOR -Supporting JFACC -Supporting ARFOR (ISB TR)
	ARFOR	-Supporting JFACC -Preparing evac. Control center at site A	-Supporting JFACC -Operates evac. Control center at site A
	JFACC	-Preparing to gain air superiority -Preparing to provide CAS/Coord. airspace	-Gains/maintains air superiority -Provide CAS /Coordinate airspace
	JSOTF	-Preparing to secure evac. site at #1 -Located at ISB B	-Launches and secures evac. Site #1 -Evac. AMCITs/TUN onto to ISB B
	Others (e.g. USCG & rest of Interagency	-DOS – Preparing to implement emergency evac. plans. -Pres/SECDEF prepare EXORD	-DOS – Implement emergency evac plan -Pres/SECDEF – Releases EXORD
J O I N T — **F U N C T I O N A L A R E A S**	Movement & Maneuver	-JTF must be in position NLT___ -JTF be prepared to execute EXORD +4hr	-NAVFOR & JFACC insures freedom of sea and air movements
	Intelligence	-NAIs – Sites 1, 2 & 3 _PIR – Threat against AMCITs, rebel troop movements, rebel attacks, etc.	-Collects info at NAIs -Collects and processes PIR
	Firepower	-Targeting effects continue	-AC-130 spt ARFOR/MARFOR/JSOTF (OPCON to JSOTF, TACON others)
	Sustainment	-HN spt being provided by countries X&Y	-HN spt not available
	Command & Control	-IO/IW -PSYOP spt to ARFOR/MARFOR & JSOTF -JTF HQ located on USS MTW	-NAVFOR provides direct spt to MARFOR & ARFOR -Continues IW/C2W spt -ROE Changes
	Protection	-Weapons status___ -Deception theme___	-Weapons status___ -Deception theme___
O T H E R S	Decision Points	-Recommendation for evac. EXORD	-When to declare "endstates" and mission accomplishment
	CCIR	-Taking of any AMCITs as hostages -Violence directed at AMCITs or TCNs	-# of AMCITs/TCNs evacuated -Any interference with the evacuation
	Branches	-Site #1, 2, or 3 is unusable -Hostages taken	-Implement branch plans
	Risks	-Military presence will provoke rebels	-Significant casualties -Usable to locate all AMCITs
	Issues	-Army helos on carriers Approval for use of evac site ISB in Cty Y	-Policy for evac. of unidentified personnel requesting help

Figure J. Sample Time Line Matrix for a JTF Operation

Time/Phase	Shaping	Deterrence	Seize Initiative	Dominance	Stabilization	Enable CA
Threat **Enemy** Action/Reaction	Component Tasks					
Operational Intel/Surv. Recon Op Env. Aware	AR NAV MAR AF SOF					
Operational **Movement &** **Maneuver** Force Application	AR NAV MAR AF SOF					
Operational **Firepower**	AR NAV MAR AF SOF					
Operational **Protection** Protection AT/NBC Def./ CBRNE/etc.	AR NAV MAR AF SOF					
Operational **Log & SPT** Focused Logistics	AR NAV MAR AF SOF Host Nation SPT					
Operational **CMD & CTL** ROE	Staff C4I Multinational Interagency					
INFO OPS JSOTF/etc.						
Interagency						
Multinational						
ROE						
Decision Pts. **Link to CCIR**						
Risk						

Functional UJTL Operational Level Tasks / *Other Enablers*

Figure K. Sample Matrix by Phase

7. Interpreting the Results of Analysis and Wargaming

Comparisons of advantages and disadvantages of each COA will be conducted during the next step of COA Determination. However, if the suitability, feasibility, or acceptability of any COA becomes questionable during the analysis, the CDR should modify or discard it and concentrate on other COAs. Figure L below is an example of Analysis and Wargaming results. The need to create additional combinations of COAs may also be required.

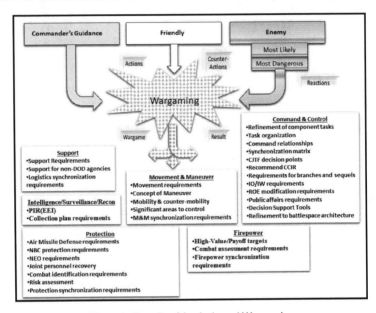

Figure L. Results of Analysis and Wargaming.

8. Flexible Plans

We have several techniques to help us develop adaptability. One of these is to make flexible plans. Flexible plans can enhance adaptability by establishing a COA that provides for multiple options. We can increase our flexibility by providing *branches* to deal with changing conditions on the battlefield that may affect the plan (e.g., changing dispositions, orientation, strength, movement) or by providing sequels for current and future operations. Sequels are COA to follow probable battle or engagement outcomes; victory, defeat, or stalemate.

> *"There were no decisions reached about how to exploit a victory in Sicily. It was an egregious error to leave the future unresolved."*
>
> *-General Omar Bradley on Operation Husky*

a. <u>Branches and Sequels</u>. Many operation plans require adjustment beyond the initial stages of the operation. Consequently, CDRs build flexibility into their plans by developing branches and sequels to preserve freedom of action in rapidly changing conditions. An effective plan places a premium on flexibility because operations never seem to proceed exactly as planned.

b. During the wargaming sequence we will identify operational branches and sequels giving our plans that flexibility and help focus our future planning efforts. Visualizing and planning branches and sequels are important because they involve transitions— changes in mission, type of operations, and often forces required for execution. Unless planned, prepared for, and executed efficiently, transitions can reduce the tempo of the operation, slow its momentum, and surrender the initiative to the adversary. Both branches and sequels should have execution criteria, carefully reviewed before their implementation and updated based on assessment of current operations. Both branches and sequels directly relate to the concept of phasing.

(1) *Branches are options built into the basic plan* and provide a framework for flexibility in execution (Figure M). Branches may include shifting priorities, changing mission, unit organization and command relationships, or changing the very nature of the joint operation itself. Branches add flexibility to plans by anticipating situations that could alter the basic plan. Such situations could be a result of lingering planning assumptions not proven to be facts, anticipated events, opportunities, or disruptions caused by enemy actions, availability of friendly capabilities or resources, or even a change in the weather or season within the operational area. CDRs anticipate and devise counters to enemy actions to mitigate risk. Although anticipating every possible threat action is impossible, branches anticipate the most likely ones. CDRs execute branches to rapidly respond to changing conditions.

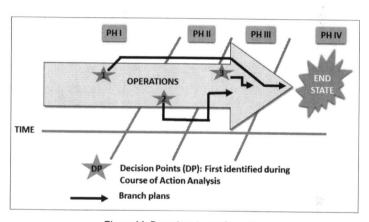

Figure M. Branches to an Operation

(2) *Sequels are subsequent operations* based on the possible outcomes of the current operation — victory, defeat, or stalemate. Sequel planning allows the CDR and staff to keep pace with a constantly evolving situation while staying focused on mission accomplishment (Figure N). They are future operations that anticipate the possible outcomes — success, failure, or stalemate — of the current operations. A counteroffensive, for example, is a logical sequel to a defense; exploitation and pursuit follow successful attacks. Executing a sequel normally begins another phase of an operation, if not a new operation. In joint operations, phases can be viewed as the sequels to the basic plan. CDRs consider sequels early and revisit them throughout an operation. Without such planning, current operations leave forces poorly positioned for future opportunities, and leaders are unprepared to retain the initiative.

> "True genius resides in the capacity for evaluation of uncertain, hazardous, and conflicting information."
>
> Winston Churchill

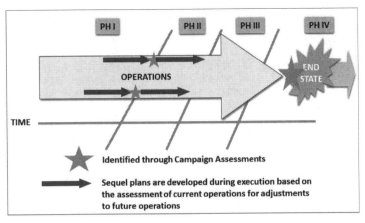

Figure N. Sequels to an Operation

c. The value of *branches and sequels* is that they prepare us for several different actions. *We should keep the number of branches and sequels to a relative few.* We should not try to develop so many branches and sequels that we cannot adequately plan, train, or prepare for any of them. The skillful, well thought-out use of branches and sequels becomes an important means of anticipating future COA. This anticipation helps accelerate the decision cycle and therefore increases tempo and flexibility.

d. Flexible plans avoid unnecessary detail that not only consumes time in their development, but has a tendency to restrict subordinates' latitude. Instead, flexible plans lay out what needs to be accomplished, but leave the manner of accomplishment to subordinates. This allows the subordinates the flexibility to deal with a broader range of circumstances.

e. Flexible plans are plans that can be easily changed. Plans that require coordination are said to be "coupled." If all the parts of a plan are too tightly coupled, the plan is harder to change because changing any one part of the plan means changing all the other parts. Instead, we should try to develop modular, loosely coupled plans. Then if we change or modify any one part of the plan, it does not directly affect all the other parts.

f. Finally, flexible plans should be simple plans. Simple plans are easier to adapt to the rapidly changing, complex, and fluid situations that we experience in combat.

> "The Joint Force is also improving how it frames decisions for the Secretary of Defense in an all-domain, transregional fight. This begins by developing a common intelligence picture and a shared understanding of global force posture, which then serves as a baseline to test operational plans and concepts through realistic and demanding exercises and wargames. By testing our assumptions and concepts, exercises and wargames provide senior leaders with the "reps-and-sets" necessary to build the implicit communication required to facilitate rapid decision-making in times of crisis."
>
> *Gen. Dunford: The Character of War & Strategic Landscape Have Changed*
>
> *Former Chairman of the Joint Chiefs of Staff Gen. Joe Dunford*

V. Course of Action (COA) Comparison

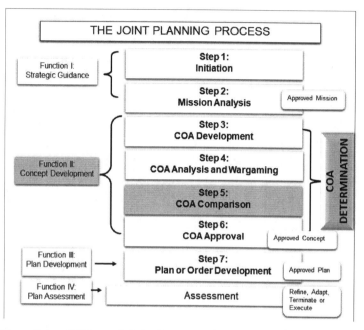

THE JOINT PLANNING PROCESS

Function I: Strategic Guidance

Step 1: Initiation

Step 2: Mission Analysis — Approved Mission

Function II: Concept Development

Step 3: COA Development

Step 4: COA Analysis and Wargaming

Step 5: COA Comparison

Step 6: COA Approval — Approved Concept

COA DETERMINATION

Function III: Plan Development

Step 7: Plan or Order Development — Approved Plan

Function IV: Plan Assessment

Assessment — Refine, Adapt, Terminate or Execute

1. Step 5 to JPP — COA Comparison

COA comparison is a *subjective* process whereby COAs are considered independently of each other and evaluated/compared against a set of criteria that are established by the staff and CDR. The goal is to identify and recommend the COA that has the highest probability of success against the ECOA that is of the most concern to the CDR, so take some time and energy with this step. COA comparison facilitates the CDR's decision-making process by balancing the *ends, means, ways* and *risk* of each COA. The end product of this task is a briefing to the CDR on a COA recommendation and a decision by the CDR.

- Determine comparison criteria
- Define and determine the standard for each criteria
- Assign weight or priority to comparison criteria
- Construct comparison method and record
- Conduct and record the comparison recommend COA

a. In COA comparison the CDR and staff evaluate all friendly COAs – against established evaluation criteria (discussed in previous chapter), and select the COA which best accomplishes the mission. The CDR reviews the criteria list and adds or deletes as they see fit. The number of evaluation criteria will vary, but there should be enough to differentiate COAs. Consequently, COAs are not compared to each other, but rather they are individually evaluated against the criteria that are established by the CDR and staff.

-COA comparison helps the CDR answer the following questions:
  • What are the differences between each COA?
  • What are the advantages and disadvantages?
  • What are the risks?

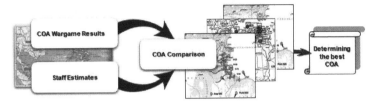

COA Comparison.

b. Staff officers may each use their own matrix to compare COAs with respect to their functional areas as depicted in Table 1 below. Matrices use the evaluation criteria developed before the wargame. Decision matrices alone cannot provide decision solutions. Their greatest value is providing a method to compare COAs against criteria that, when met, produce mission success. They are analytical tools that staff officers use to prepare recommendations. CDRs provide the solution by applying their judgment to staff recommendations and making a decision.

Example: Staff Estimate Matrix (Intel Estimate)

Evaluation Criteria	Frontal COA 1	Envelopment COA 2
Effects of Terrain		√
Effects of Weather	√	
Utilize Surprise		√
Attacks CV		√
Collection Support		√
Counterintelligence	√	
Totals	**2**	**4**

Table 1. Identify and select the COA that best accomplishes the mission.

c. The staff helps the CDR identify and select the COA that best accomplishes the mission. The staff supports the CDR's decision-making process by clearly portraying the CDR's options and recording the results of the process. The staff compares feasible COAs to identify the one with the highest probability of success against the most likely enemy COA (MLCOA) and the most dangerous enemy COA (MDCOA).

2. Prepare for COA Comparison

The CDR and staff developed evaluation criteria prior to wargaming and using this refined evaluation criteria the staff outlined each COA, highlighting advantages and disadvantages. By doing this the staff compared the strengths and weaknesses of the COA and identified their advantages and disadvantages relative to each other. The staff assists the CDR in identifying and selecting the COA that best accomplishes the mission and clearly portraying the CDR's options and recording the results of this process. The staff evaluates feasible COA to identify the one that performs best within the evaluation criteria against the enemy's most likely and most dangerous COAs.

a. Determine/Define Evaluation Criteria. As discussed in Chapter 5-IV, *COA Wargaming and Analysis* criteria are based on the particular circumstances and should be relative to the situation. There is no standard list of criteria, although the CDR may prescribe several core criteria that all staff directors will use. Individual staff sections, based on their estimate process, select the remainder of the criteria. Criteria are based on the particular circumstances and should be relative to the situation.

- Review CDR's guidance for relevant criteria.
- Identify implicit significant factors relating to the operation.
- Each staff identifies criteria relating to that staff function.
- Examples of other criteria might include:
  - Political, social, and safety constraints; requirements for coordination with Embassy/Interagency personnel.
  - Mission accomplishment.
  - Risks.
  - Costs.

b. Define and Determine the Standard for each Criterion.

(1) Establish standard definitions for each criterion. Define the criteria in precise terms to reduce subjectivity and ensure the interpretation of each remains constant between the various COAs.

(2) Establish definition prior to commencing COA comparison to avoid compromising the outcome.

(3) Apply standard for each criterion to each COA.

c. The staff evaluates feasible COAs using those criteria most important to the CDR to identify the one COA with the highest probability of success. The selected COA should also:

(1) Mitigate risk to the force and mission to an acceptable level.

(2) Place the force in the best posture for future operations.

(3) Provide maximum latitude for initiative by subordinates.

(4) Provide the most flexibility to meet unexpected threats and opportunities.

3. Determine the Comparison Method and Record

Actual comparison of COAs is critical. The staff may use any technique that facilitates reaching the best recommendation and the CDR making the best decision. There are a number of techniques for comparing COAs. The most common technique is the decision matrix, which uses evaluation criteria to assess the effectiveness and efficiency of each COA. Here are examples of several decision matrices:

a. <u>Weighted Numerical Comparison Technique</u>. The example below provides a numerical aid for differentiating COAs. Values reflect the relative advantages or disadvantages of each COA for each of the criterion selected. Certain criteria have been weighted to reflect greater value (Figure A and Table 2 are examples).

b. <u>Determine the Weight of each Criterion Based on its Relative Importance and the CDR's Guidance</u>. The CDR may give guidance that result in weighting certain criteria. The staff member responsible for a functional area scores each COA using those criteria. Multiplying the score by the weight yields the criterion's value. The staff member then totals all values. However, the staff must be careful not to portray subjective conclusions as the results of quantifiable analysis. Comparing COAs by category is more accurate than comparing total scores.

Each staff section does this separately, perhaps using different criteria on which to base the COA comparison. The staff then assembles and arrives at a consensus for the criterion and weights. The Chief of Staff/DCJTF should approve the staff's recommendations concerning the criteria and weights to ensure completeness and consistency throughout the staff sections.

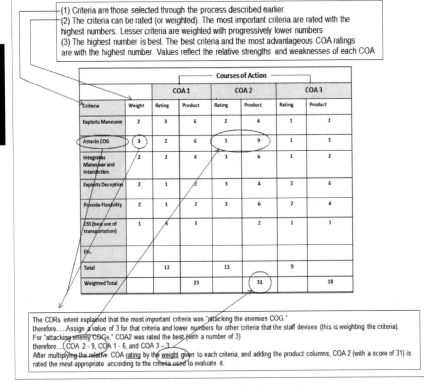

(1) Criteria are those selected through the process described earlier.
(2) The criteria can be rated (or weighted). The most important criteria are rated with the highest numbers. Lesser criteria are weighted with progressively lower numbers
(3) The highest number is best. The best criteria and the most advantageous COA ratings are with the highest number. Values reflect the relative strengths and weaknesses of each COA

Criteria	Weight	COA 1		COA 2		COA 3	
		Rating	Product	Rating	Product	Rating	Product
Exploits Maneuver	2	3	6	2	4	1	2
Attacks COG	3	2	6	3	9	1	3
Integrates Maneuver and Interdiction	2	2	4	3	6	1	2
Exploits Deception	2	1	2	3	4	3	6
Provide Flexibility	2	1	2	2	6	2	4
CSS (best use of transportation)	1	3	3	2	2	1	1
Etc.							
Total		12		13		9	
Weighted Total			23		31		18

The CDRs intent explained that the most important criteria was "attacking the enemies COG."
therefore.....Assign a value of 3 for that criteria and lower numbers for other criteria that the staff devises (this is weighting the criteria).
For "attacking enemy COGs," COA2 was rated the best (with a number of 3)
therefore...(COA 2 - 9, COA 1 - 6, and COA 3 - 3.
After multiplying the relative COA rating by the weight given to each criteria, and adding the product columns, COA 2 (with a score of 31) is rated the most appropriate according to the criteria used to evaluate it.

Figure A. Example Numerical Comparison.

Example #2 COA Comparison Matrix Format							
Evaluation Criteria	Weight	COA #1		COA #2		COA #3	
		Score	Weighted	Score	Weighted	Score	Weighted
Surprise	2	3	6	2	4	2	4
Risk	2	3	6	1	2	2	4
Flexibility	1	2	2	1	1	1	1
Retaliation	1	1	1	2	2	1	1
Damage to Alliance	1	2	2	1	1	1	1
Legal Basis	1	1	1	1	1	1	1
External Support	1	3	3	2	2	1	1
Force Protection	1	3	3	3	3	1	1
OPSEC	1	3	3	2	2	2	2
Total		27 - Green		18 – Yellow		16 - Red	

Table 2. Comparison Matrix Format.

c. <u>Non-Weighted Numerical Comparison Technique</u>. The same as the previous method except the criteria are not weighted. Again, the highest number is best for each of the criteria.

d. <u>Narrative or Bulleted Descriptive Comparison of Strengths and Weaknesses</u>. Review criteria and describe each COA's strengths and weaknesses. See Table 3 below:

Course of Action	Strengths	Weaknesses
COA 1	Narrative or bulletized discussion of strengths using the criteria	Narrative or bulletized discussion of weaknesses using the same criteria
COA 2	Same	Same
COA 3	Same	Same

Table 3. Criteria for Strengths and Weaknesses.

e. Plus/Minus/Neutral Comparison. Base this comparison on the broad degree to which selected criteria support or are reflected in the COA. This is typically organized as a table showing (+) for a positive influence, (0) for a neutral influence, and (-) for a negative influence (Table 4).

Criteria	COA 1	COA 2
Casualty estimate	(–)	(–)
Casualty evacuation routes	(+)	(–)
Suitable medical facilities	0	0
Flexibility	(+)	(–)

Table 4. Plus/Minus/Neutral Comparison.

f. Stop Light Comparison. Criteria are judged to be acceptable or unacceptable with varying levels in between. Ensure you define each color in the stop light on a key along with corrective methods to elevate mid colors to green (Table 5).

COA Criteria	COA 1	COA 2	COA 3
Speed	Red	Green	Yellow
Surprise	Red	Orange	Green
Risk	Orange	Yellow	Green
Total	Red	Orange	Green

Table 5. Stop Light Comparison.

g. Descriptive Comparison. Simply a description of advantages and disadvantages of each COA (Table 6).

COA	ADVANTAGES	DISADVANTAGES
COA 1	•Rapid Delivery •Meets critical needs	•Rough integration of forces •Rough transition •Complex organization •Not flexible at all •Adequate force protection
COA 2	•Rapid delivery •Meets critical needs •Smooth integration •Smooth transition	•Complex organization •Less flexible •Adequate force protection
COA 3	•Smooth integration •Smooth transition •Simplest organization •Adequate force protection •Best force protection	•Less rapid delivery •Does not meet all critical needs

Table 6. Decriptive Comparison.

h. <u>Staff Estimate Comparison</u>. COA comparison remains a subjective process and should not be turned into a mathematical equation. Using √ ,+,- 0 or 1, 2, 3 are as appropriate as any other methods. The key element in this process is the ability to articulate to the CDR why one COA is preferred over another. Table 7 below is an example staff estimate comparison.

-Reviews the mission and situation from its own staff functional perspective.
-Examines the criteria and assumptions for which it is the responsible staff.
-Analyzes each COA from its staff functional perspective.
-Concludes whether the mission can be supported and which COA can best be supported from its particular staff functional perspective

Staff	Decisive Air COA 1	Sequential Build COA 2
J2	✓	
J3		✓
J4	✓	
MARFOR	✓	
ARFOR		✓
AFFOR	✓	
NAVFOR	✓	
JFACC		✓
JFLCC		✓
JFMCC	✓	
TOTALS	6	4

The staff and components recommend COA 2

Table 7. Staff Estimate Comparison Input.

i. <u>Course-of-Action Comparison</u>. Figure B depicts inputs and outputs for course-of-action comparison. Other products not graphically shown in the chart include updated JIPOE products, updated CCIR's, staff estimates, CDR's identification of branches for further planning and a Warning Order as appropriate.

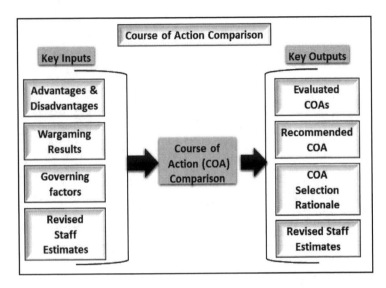

Figure B. Course of Action Comparison.

j. The staff ensures the selected COA is faithfully captured as the concept of operations (CONOPS). The CONOPS – along with the joint functions (movement and maneuver, intelligence, firepower, support, C2, protection) – forms the basis for the operation plan (OPLAN) or order (OPORD).

> "Adventure is just bad planning."
>
> Roald Amundsen (1872 – 1928) Norwegian Arctic & Antarctic explorer

VI. Course of Action (COA) Selection & Approval

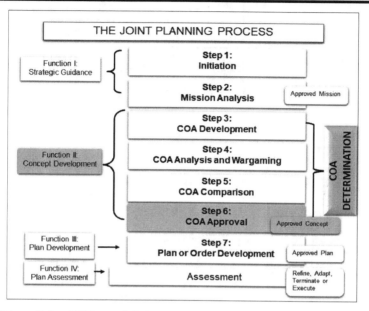

THE JOINT PLANNING PROCESS

Function I: Strategic Guidance	Step 1: Initiation
	Step 2: Mission Analysis — Approved Mission
Function II: Concept Development	Step 3: COA Development
	Step 4: COA Analysis and Wargaming
	Step 5: COA Comparison
	Step 6: COA Approval — Approved Concept
Function III: Plan Development →	Step 7: Plan or Order Development — Approved Plan
Function IV: Plan Assessment →	Assessment — Refine, Adapt, Terminate or Execute

COA DETERMINATION

1. Step 6 to JPP — COA Approval

COA Recommendation. Throughout the COA development process, the CCDR conducts an independent analysis of the mission, possible COA, and relative merits and risks associated with each COA. The CDR, upon receiving the staff's recommendation, combines his/her analysis with the staff recommendation resulting in a selected COA (Figure A). The forum for presenting the results of COA comparison is the *CDR's Decision Brief.* Typically, this briefing provides the CCDR with an update of the current situation, an overview of the COAs considered, and a discussion of the results of COA comparison.

2. Prepare the COA Decision Briefing

Prepare the COA Decision Briefing. This briefing often takes the form of a CDR's Estimate so take good notes of the CDR's comments. This information could include the current status of the joint force; the current JIPOE; and assumptions used in COA development. The CDR selects a COA based upon the staff recommendations and the CDR's personal estimate, experience, and judgment.

> "Our flag does not fly because the wind moves it. It flies with the last breath of each soldier who died protecting it."
>
> - Unknown

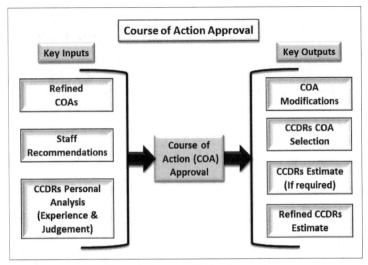

Figure A. Course of Action Approval.

3. Present the COA Decision Briefing

The staff briefs the CDR on the COA comparison and the analysis and wargaming results, including a review of important supporting information (Figure B & C). All principal staff directors and the component CDR's should attend this briefing (physically present or linked by VTC).

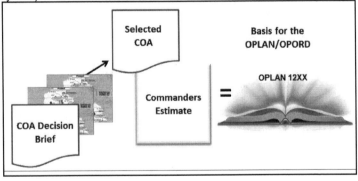

Figure B. COA Approval Inputs.

4. COA Selection/Modification by CDR

COA approval/selection is the end result of the COA comparison process. It gives the staff a concise statement of how the CDR intends to accomplish the mission, and provides the necessary focus for execution planning and OPLAN/OPORD development.

a. Review Staff Recommendations.

b. Apply Results of own COA Analysis and Comparison.

COA Decision Briefing Format

The JPG should prepare a briefing to provide the following to the CCDR:

a. <u>The purpose of the briefing</u>.

b. <u>Enemy situation</u>.

  (1) Strength. A review of enemy forces, both committed/available for reinforcement.

  (2) Composition. Order of battle, major weapons systems, and operational characteristics.

  (3) Location and disposition. Ground combat and fire support forces, air, naval, missile forces, logistic forces and nodes, command and control (C2) facilities, and other combat power.

  (4) Reinforcements. Land; air; naval; missile; nuclear, biological, and chemical (NBC), other advanced weapons systems; capacity for movement of these forces.

  (5) Logistics. A summary of the enemy's ability to support combat operations.

  (6) Time and space factors. The capability to move to and reinforce initial positions.

  (7) Combat efficiency. The state of training, readiness, battle experience, physical condition, morale, leadership, motivation, tactical doctrine, discipline, and significant strengths and weaknesses.

c. <u>Friendly Situation</u>.

d. <u>Mission Statements</u>.

e. <u>CDR's Intent Statement</u>.

f. <u>Operational Concepts and COAs Developed</u>.

  (1) Any changes from the mission analysis briefing in the following areas:

    (a) Assumptions.

    (b) Limitations.

    (c) Enemy and friendly centers of gravity.

    (d) Phasing of the operation (if phased).

  (2) Present COA. As a minimum, discuss:

    (a) COA # ___. (Short name, e.g., "Simultaneous Assault").

     <u>1</u> COA statement (brief concept of operations).

     <u>2</u> COA sketch.

     <u>3</u> COA architecture:

      <u>a</u> Task organization.

      <u>b</u> Command relationships.

      <u>c</u> Organization of the operational area.

    (b) Major differences between each COA.

    (c) Summaries of COAs.

g. <u>COA Analysis</u>.

  (1) Review of JPG's wargaming efforts.

  (2) Add considerations from own experience.

h. <u>COA Comparisons</u>.

  (1) Description of comparison criteria (e.g., evaluation criteria) and comparison methodology.

  (2) Weigh strengths/weaknesses with respect to comparison criteria.

i. <u>COA Recommendations</u>.

  (1) Staff.

  (2) Components.

Figure C: COA Decision Briefing Format

c. Consider Any Separate Recommendations from Supporting and Subordinate CDRs.

d. Review Guidance from the Higher Headquarters/Strategic Guidance.

e. The CDR May:

(1) Concur with staff/component recommendations, as presented.

(2) Concur with staff/component recommended COAs, but with modifications.

(3) Select a different COA from the staff/component recommendations.

(4) Direct the use of a COA not formally considered.

(5) Defer the decision and consult with selected staff/CDRs prior to making a final decision.

5. Refine Selected COA

Once the CDR selects a COA, the staff will begin the refinement process of that COA into a clear decision statement to be used in the CDR's Estimate. At the same time, the staff will apply a final "acceptability" check.

a. <u>Staff Refines CDR's COA Selection into Clear Decision Statement</u>.

(1) Develop a brief statement that clearly and concisely sets forth the COA selected and provides only whatever information is necessary to develop a plan for the operation (no defined format).

(2) Describe what the force is to do as a whole, and as much of the elements of when, where, and how as may be appropriate.

(3) Express decision in terms of what is to be accomplished, if possible.

(4) Use simple language so the meaning is unmistakable.

(5) Include statement of what is acceptable risk.

b. <u>Apply Final "Acceptability" Check</u>.

(1) Apply experience and an understanding of situation.

(2) Consider factors of acceptable risk versus desired outcome consistent with higher CDR's intent and concept. Determine if gains are worth expenditures.

> **Note:** The nature of a potential contingency could make it difficult to determine a specific endstate until the crisis actually occurs. In these cases, the CCDR may choose to present two or more valid COAs for approval by higher authority. A single COA can then be approved when the crisis occurs and specific circumstances become clear.

6. Prepare the Commander's Estimate

a. Once the CCDR has made a decision on a selected COA, provides guidance, and updates his intent, the staff completes the CDR's Estimate. The CDR's Estimate provides a *concise narrative statement* of how the CCDR intends to accomplish the mission, and provides the necessary focus for campaign planning and OPLAN/OPORD development. Further, it responds to the establishing authority's requirement to develop a plan for execution. The CDR's Estimate[1] provides a continuously updated source of information from the perspective of the CCDR. CDRs at various levels use estimates during JPP to support all

---

[1] *Annex J of JOPES Volume I (CJCSI 3122.01) provides the format for a CDR's Estimate. See also Appendix C of this document.*

aspects of COA determination and plan or order development. Outside of formal reporting requirements, a CDR may or may not use a CDR's Estimate as the situation dictates. The CDR's initial intent statement and planning guidance to the staff can provide sufficient information to guide the planning process. Although the CCDR will tailor the content of the CDR's Estimate based on the situation, a typical format for an estimate that a CCDR submits is shown in Figure D below.

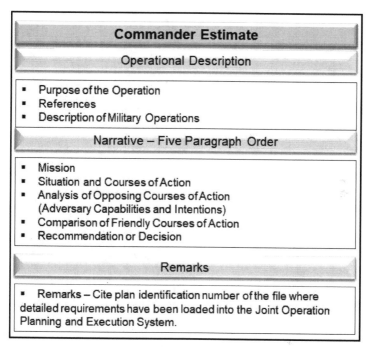

Figure D. Commander's Estimate.

(1) Precise contents may vary widely, depending on the nature of the crisis, time available to respond, and the applicability of prior planning. In a rapidly developing situation, the formal CDR's Estimate may be initially impractical, and the entire estimate process may be reduced to a CDRs' conference.

(2) In practice, with appropriate horizontal and vertical coordination, the CCDR's COA selection could already have been briefed to and approved by the CJCS and SecDef. In the current global environment, where major military operations are both politically and strategically significant, even a CCDR's selected COA is normally briefed to and approved by the President or SecDef. The CDR's Estimate then becomes a matter of formal record-keeping and guidance for component and supporting forces.

b. The supported CDR may use simulation and analysis tools in the collaborative environment to assess a variety of options, and may also choose to convene a concept development conference involving representatives of subordinate and supporting commands, the Services, JS, and other interested parties. Review of the resulting CCDR's Estimate (also referred to as the strategic concept) requires maximum collaboration and coordination among all planning participants. The supported CDR may highlight issues for future interagency consultation, review or resolution to be presented to SecDef during the IPR.[2]

---

[2] *CJCSI 3141.01, Management and Review of Campaign and Contingency Plans*

c. <u>CJCS Estimate Review</u>. The Estimate Review determines whether the scope and concept of planned operations satisfy the tasking and will accomplish the mission; whether the assigned tasks can be accomplished using available resources in the time frames contemplated by the plan; and ensures the plan is proportional and worth the expected costs. Once approved for further planning by the SecDef during an IPR discussion, the CCDR's Estimate becomes the CONOPS for the plan. A detailed description of the CONOPS will be included in Annex C of the plan.[3]

7. Concept Development Summary

The transition to concept development is marked by a decision to develop military options. During the concept development step, CCDRs develop, analyze, and compare viable COAs and develop staff estimates that are coordinated with the Military Departments when applicable. Analysis includes wargaming, operational modeling, and initial feasibility assessments. Concept discussions with the Joint Staff and senior leaders are important and will focus largely on issues the CCDR needs resolved with planning to this point.

a. The CCDR's estimate broadly outlines to the SecDef, senior leaders, planners and subordinates how forces will conduct integrated, joint operations to accomplish the mission. Among other elements and as appropriate, the estimate communicates:

- Review of strategic guidance, assumptions, risk, termination criteria, and mission statement as well as any changes or modifications.
- Recommended COAs and supporting rationale.
- Descriptions and assessments of alternate COAs and the rationale for not recommending them.
- Review of the OE. Feasible ECOAs and comparison of enemy and friendly COAs.
- Commander's intent and desired endstate.
- Assessed strategic and operational centers of gravity, as appropriate.
- Estimated level and duration of the operation.
- Nature, purpose, time-phasing, and interrelationship of operations, including specific relationships to strategic communication.
- Branches, sequels, or other options, including warning and response times, that involve scenarios likely to confront the command.
- Logistics feasibility.
- Gross transportation feasibility.
- Interagency and/or multinational involvement accomplished to date and future potential interagency and/or multinational involvement, any ally/partner nation support.
- The concept for sequencing the operation.

b. **Transition to plan development is marked by approval of a COA and/or plan concept.**

> "[Postwar strategy enabling civil authority] must have an acceptable and locally viable philosophic base; and it must be a strategy suited to, rather than imposed upon, the actual scene."
>
> J.C. Wylie

[3] *JOPES Volume I*

I. Plan or Order Development (Step 7)

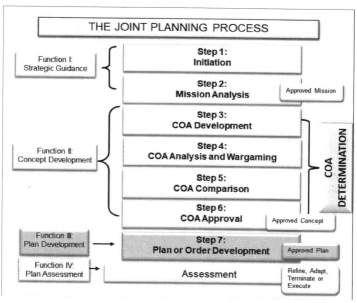

THE JOINT PLANNING PROCESS

Function I: Strategic Guidance	**Step 1:** **Initiation**	
	Step 2: **Mission Analysis**	Approved Mission
Function II: Concept Development	**Step 3:** **COA Development**	
	Step 4: **COA Analysis and Wargaming**	COA DETERMINATION
	Step 5: **COA Comparison**	
	Step 6: **COA Approval**	Approved Concept
Function III: Plan Development	**Step 7:** **Plan or Order Development**	Approved Plan
Function IV: Plan Assessment	Assessment	Refine, Adapt, Terminate or Execute

1. Function III — Step 7: Plan or Order Development

After completing Functions, I and II and Steps 1-6 of the JPP, we now have a concept that's been well staffed which will aid us in developing our plan. Function III and Step 7 to the JPP is Plan or Order Development.

During plan or order development, the CDR and staff, in collaboration with subordinate and supporting components and organizations, expand the approved COA into a detailed plan or OPORD by refining the initial CONOPS associated with the approved COA. The CONOPS is the centerpiece of the plan or order.

2. Format of Military Plans and Orders

Plans and orders can come in many varieties from the very detailed Campaign Plans and OPLANs to simple verbal orders, these orders include; OPORDs, WARNORD, PLANORD, ALERTORD, EXORD, DEPORD, PTDO and FRAGORD. The more complex directives will contain much of the amplifying information in appropriate annexes and appendices. However, the directive should always contain the essential information in the main body. The information may depend on the time available, the complexity of the operation, and the levels of command involved. However, in most cases, the directive will be standardized in the five-paragraph format. Following is a brief description of each of these paragraphs.[1]

---

[1] *JP 5-0, Joint Planning*

• *Paragraph 1 – Situation.* The CDR's summary of the general situation that ensures subordinates understand the background of the planned operations. Paragraph 1 will often contain sub paragraphs describing the higher CDR's Intent, friendly forces, and enemy forces.

• *Paragraph 2 – Mission.* The CDR inserts his/her restated mission (containing essential tasks) developed during the mission analysis.

• *Paragraph 3 – Execution.* This paragraph contains CDR's Intent, which will enable CDRs two levels down to exercise initiative while keeping their actions aligned with the overall purpose of the mission. It also specifies objectives, tasks, and assignments for subordinates (by phase, as applicable—with clear criteria denoting phase completion).

• *Paragraph 4 – Administration and Logistics.* This paragraph describes the concept of support, logistics, personnel, public affairs, civil affairs, and medical services.

• *Paragraph 5 – Command and Control.* This paragraph specifies the command relationships, succession of command, and overall plan for communications.

Note: Specific policy and doctrine for the development of campaign plans and contingency plans can be found in CJCSM 3130.03, Adaptive Planning Formats and Guidance.

3. Development Overview

a. The purpose of plan development is to take the outputs of concept development and produce a feasible plan or an order that is ready to transition into execution. The plan development function focuses on detailing the **way** and refines the required **means** to achieve the **end** at an acceptable risk. The plan development function is scalable based on the time available or complexity of the planning task. At the strategic level, during plan development supported CCMD planners sustain dialogue with OSD, JS, supporting CCMDs, DOD Agencies, subordinate commands and others as required. This enables a plan to responsively adapt to changes in the strategic environment, allowing supporting and subordinate commands to conduct aligned parallel planning.

b. For most plans and orders the CJCS, in coordination with the supported and supporting CCDRs and other members of the JCS, monitors planning activities, resolves shortfalls when required, and reviews the supported CCDR's plan for adequacy, feasibility, acceptability, completeness, and compliance with policy, guidance and doctrine. As required the supported CCDR will have in-progress review (IPR) discussions (see Chapter 6-II, *Plan Review Process*) to confirm the plan's strategic guidance and receive approval of assumptions, limitations, the mission statement, the concept, the plan, and any further guidance required for plan refinement and risk. Normally at this time the Chairman and the Under Secretary of Defense for Policy (USD(P)) will include issues arising from, or resolved during plan review (e.g., key risks, DPs). The intended result of these discussions is SecDef endorsement of the planning to date or acknowledgement of friction points and guidance to shape continued planning. If the President or SecDef decides to execute the plan, all four operational activities — situational awareness, planning, execution and assessment — continue in a complementary and iterative process.[2]

4. CONOPS Refinement

a. The first step of plan development is to refine the initial CONOPS developed during concept development and associated with the selected COA. The planning team refines the CONOPS to clearly and concisely express what the CDR intends to accomplish and how it will be done using available resources. It describes how the actions of the joint force components and supporting organizations will be integrated, synchronized and phased to accomplish the mission, including potential branches and sequels. The refined CONOPS should provide sufficient detail of mission and tasks for supporting and subordinate CDRs to conduct planning aligned with the supported CDR's intent. The refined CONOPS pro-

---

[2] *CJCSI 3141.01 Management and Review of Campaign and Contingency Plans*

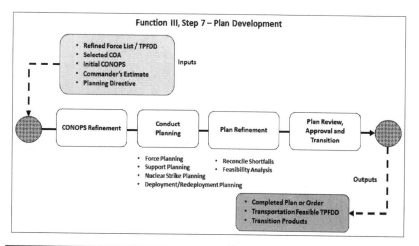

Function III, Step 7 – Plan Development

Inputs
- Refined Force List / TPFDD
- Selected COA
- Initial CONOPS
- Commander's Estimate
- Planning Directive

CONOPS Refinement → Conduct Planning → Plan Refinement → Plan Review, Approval and Transition

- Force Planning
- Support Planning
- Nuclear Strike Planning
- Deployment/Redeployment Planning

- Reconcile Shortfalls
- Feasibility Analysis

Outputs
- Completed Plan or Order
- Transportation Feasible TPFDD
- Transition Products

Use this diagram to follow activities discribed within Paragraphs 4-9.

vides the basis for the development of the plan or order and its Annex C (Operations).[3] The CONOPS:

- States the CDR's intent
- Describes the central approach the JFC intends to take to accomplish the mission.
- Provides for the application, sequencing, synchronization, and integration of forces and capabilities in time, space, and purpose (including those of multinational and interagency organizations, as appropriate).
- Describes when, where, and under what conditions the supported CDR intends to conduct operations and give or refuse battle, if required.
- Focuses on friendly, allied, partner, and enemy COGs and their associated critical vulnerabilities.
- Provides for controlling the tempo of the operation.
- Visualizes the campaign in terms of the forces and functions involved.
- Relates the joint force's objectives and desired effects to those of the next higher command and other organizations as necessary. This enables assignment of tasks to subordinate and supporting CDRs.[4]

b. The staff writes (or graphically portrays) the CONOPS in sufficient detail so subordinate and supporting CDRs understand their mission, tasks, and other requirements and can develop their supporting plans. During CONOPS development, the CDR determines the best arrangement of simultaneous and sequential actions (see Chapter 1-III, *Sequencing Actions*) and activities to accomplish the assigned mission consistent with the approved COA and resources and authorities available. This arrangement of actions dictates the sequencing of activities or forces into the OA, providing the link between the CONOPS and force planning which is preserved and perpetuated via a TPFDD. The structure must ensure unit integrity, force mobility, and force visibility, as well as the ability to rapidly transition to branches or sequels as operational conditions dictate. Supported CCDRs determine the sequencing of forces in concert with supporting commands and capture this sequencing

[3] CJCSM 3130.03 Series. *Planning Formats and Guidance*

[4] JP5-0, *Joint Planning*

in the TPFDD as the CDR's required delivery dates. Planners ensure the CONOPS, force plan, deployment plans, and supporting plans provide the flexibility to adapt to changing conditions and are consistent with the CDR's intent.

c. If the scope, complexity, and duration of the military action contemplated to accomplish the assigned mission warrants execution via a series of related operations, then the staff outlines the CONOPS as a campaign. They develop the preliminary part of the operation in sufficient detail to impart a clear understanding of the CDR's concept of how the assigned mission will be accomplished.

d. During CONOPS development, the CDR assimilated many variables under conditions of uncertainty to determine the essential military conditions, sequence of actions and application of capabilities and associated forces to create effects and achieve objectives. CDRs and their staffs must be continually aware of the higher-level objectives and associated desired and undesired effects that influence planning at every juncture. If operational objectives are not linked to strategic objectives, the inherent linkage or "nesting" is broken and eventually tactical considerations can begin to drive the overall strategy at cross-purposes.[5]

e. This planning results in a plan that is documented in the format of a plan or an order. If execution is imminent or in progress, the plan is typically documented in the format of an order. During plan or order development, the CDR and staff, in collaboration with subordinate and supporting components and organizations, expand the approved COA into a detailed plan or OPORD by refining the initial CONOPS associated with the approved COA. As mentioned earlier the CONOPS is the centerpiece of the plan or OPORD.

5. Conduct Planning: Plan Development Activities

a. The CDR guides plan development by issuing a PLANORD or similar planning directive to coordinate the activities of the commands and agencies involved. A number of activities are associated with plan development and typically will be accomplished in a concurrent, collaborative, and iterative fashion rather than sequentially, depending largely on the planning time available. The same flexibility displayed in COA development is seen here again, as planners discover and eliminate shortfalls and conflicts within their command and with the other CCMDs. The following figure depicts the activities associated with the plan development function.[6]

Plan Development Activities	
➤ Force Planning	➤ Feasibility Analysis
➤ Support Planning	➤ Refinement
➤ Nuclear Strike (No longer in JP 5-0)	➤ Documentation
➤ Deployment/Redeployment Planning	➤ Plan Review and Approval
➤ Shortfall Identification	➤ Supporting Plan Development

Plan Development Activities

a. **Force Planning**.

(1) Force planning is conducted iteratively and in parallel with support planning and deployment/redeployment planning. The primary purposes of force planning are to (1) influence COA development and selection based on force allocations, availability, and

[5] *JP5-0, Joint Planning*

[6] *Ibid*

readiness (2) identify all forces needed to accomplish the supported component CDR's CONOPS with some rigor and (3) effectively phase the forces into the OE. Force planning consists of determining the force requirements by operation phase, mission, mission priority, mission sequence and operating area. It includes force allocation review, major force phasing; integration planning; force list structure development (TPFDD); followed by force list development. Force planning is the responsibility of the CCDR, supported by component CDRs in coordination with the GFM enterprise, including OSD, JS, JS JFC, JFPs, Services, Service components, (FPs, and other supporting commands. Force planning begins early during CONOPS development and focuses on adaptability. The CDR determines force requirements, provides guidance for development of a TPFDD LOI or supplement to the standing Joint TPFDD LOI specific to the OA, and plans force modules to align and time-phase the forces in the notional TPFDD IAW the CONOPS. Major forces and elements are selected from those apportioned or allocated for planning by the CJCS and included in the supported CDR's CONOPS by operation phase, mission and mission priority. Service components then collaboratively make tentative assessments of the specific sustainment capabilities required in accordance with the CONOPS. After the actual forces are identified (sourced), the CCDR refines the force plan to ensure it supports the CONOPS, provides force visibility, and enables flexibility. The CDR identifies and resolves or reports shortfalls with a risk assessment.[7]

(2) In a crisis situation, force planning focuses on the actual units designated to participate in the planned operation and their readiness for deployment. The supported CDR identifies force requirements as operational capabilities in the form of force packages to facilitate sourcing by the Services, JS JFC/JFPs and other FPs' supporting commands. A force package is a list (group of force capabilities) of the various forces (force requirements) that the supported CDR requires to conduct the operation described in the CONOPS. The supported CDR typically describes required force requirements in the form of capability descriptions or unit type codes, depending on the circumstances. The supported CDR submits the required force requests through the JS to the JS JFC/JFPs to develop sourcing recommendations. FPs review the readiness and deployability posture of their available units before recommending which units to allocate to the supported CDR's force requirements. Services and their component commands also determine mobilization requirements and plan for the provision of non-unit sustainment. The supported CDR will review the sourcing recommendations through the GFM process to ensure compatibility with capability requirements and CONOPS. If the sourcing recommendations do not fully support the CONOP, the supported CCDR must provide the operational risk to support the SecDef's decision.[8]

(a) <u>Force Planning Considerations</u>. The total forces identified for supporting and supported plans are initially constrained by the quantity of forces identified in the Force Apportionment Tables. This is a directed constraint for integrated planning for problem sets in the CPG and JSCP. As planning progresses, preferred force identification may provide better fidelity of force availability assumptions. For planning under crisis conditions, execution sourcing occurs in force planning, and force planning assumptions are replaced by assigned or allocated units.

(b) <u>TPFDD Development</u>. The purpose of a TPFDD during planning is to sequence and phase all the forces, support, and sustainment required to execute the plan. Planners may develop a notional TPFDD (based upon planning assumptions) or for crisis conditions a TPFDD based upon execution sourcing. Priority plans are typically directed to have a notional TPFDD while others may truncate this requirement to a simplified list of forces. The JSCP prescribes that level 3T and 4 contingency plans must have a notional TPFDD. If a notional TPFDD is developed as part of the planning process, the force list will be entered into JOPES IT and time phased IAW the CONOPS. Units identified as sourcing solutions

[7] *JP 5-0, Joint Planning*

[8] *JP 3-35, Deployment and Redeployment Operations*

to force requirements within the notional TPFDD are planning assumptions only (preferred forces or contingency sourced forces) and include notional or modified type unit characteristics file (TUCHA) data as a basis for subsequent support planning and transportation feasibility analysis. Under crisis conditions, a TFPDD may be developed directly during force planning with force requirements documented to facilitate execution sourcing and unit deployment.

(c) Preferred Forces Identification. Initially developed as planning assumptions during concept development, CCMD and component planners continue identification/refinement of force planning assumptions as necessary. CCMD and component planners should coordinate with JS JFC, JFPs, Service HQ, and Service FPs to make the best-informed assumptions when identifying preferred forces. These identified units are not sourced and the use of them in this planning function does not indicate they will be contingency or execution sourced. Instead, they provide assumptions critical for support and deployment/redeployment planning. Better informed identification of preferred forces can improve a plan's force sourcing and transportation feasibility.[9]

(d) Mobilization Planning. Initial requirements for mobilization of RC forces to include the scope and authorities should have been identified during concept development. As preferred forces are refined and additional RC forces are identified, CCDR Service components, ICW their Service HQs, further develop mobilization requirements as a part of the force planning process (see DODI 1235.12, *Accessing the Reserve Component (RC)*). Planners must identify the scope and timeline of mobilization required to support the plan and inform OSD leadership during planning dialog or IPRs. During crisis, mobilization decisions points may occur during planning and supported commands should identify such requirements and present mobilization requests to the appropriate level of mobilization authority.[10]

1 Mobilization planning may include consideration of large-scale mobilization for a state of war or national emergency. It may also consider surging and mobilizing the industrial base and training bases.

2 The results of transportation planning may also identify a requirement to plan for the activation of strategic mobility forces and programs. Based upon an assessment of transportation requirements, USTRANSCOM may propose planning for a Secretary decision to activate such programs in conjunction with the Secretary of Transportation per Title 46, U.S.C. (Section 1242) and Title 50, U.S.C. (Section 196).

(e) Non-DOD Capabilities. Planners should document and refine non-DOD capabilities that are part of a plan CONOPS.

1 Interagency Capabilities. The JS J-5 and USD(P) work with the CCMDs with issues involving interagency resources and plan development. At the CDR's discretion, planners will incorporate interagency concerns and inputs into plans to ensure unity of effort towards policy or strategic objectives. Promote Cooperation events provide a forum for the CCMDs and interagency partners to collaborate on planning. Interagency capabilities should be documented in a plan force list or TPFDD. These capabilities are not ordered via the GFMAP but must be coordinated via OSD with the providing department or agency. Interagency capabilities should be documented in a TPFDD so they can be tracked for the purpose of force deployment accounting upon execution.

2 Non-Governmental Capabilities. Like interagency capabilities, non-governmental capabilities (to include NGOs and contractors) that are part of a plan CONOPS may be documented in the force list or TPFDD. Non-government capabilities should be documented so they can be effectively tracked during execution. In-place non-governmental capabilities should also be identified in the TPFDD to reflect existing agreements, contracts, or task orders.

3 Multinational/Coalition Capabilities. Partner military forces or host government support are coordinated via national-level agreements. These capabilities should be documented in the force list or TPFDD for the purpose of force deployment accounting upon

[9] CJCSM 3130.06 Series, *Global Force Management Allocation Policies and Procedures*
[10] Ibid

execution. Host nation support capabilities should also be documented in the TPFDD to reflect negotiated support agreements.

(f) <u>Force Rotations</u>. Service rotations, or internal rotations, for forces that are Military Department-managed deployments can be combined to meet Secretary-ordered deployment lengths. When planning for operations that may be lengthy, consideration should be given to force rotations with unit rotations planned so as to limit the impact on operations. Rotational planning should consider joint reception, staging, onward movement, and integration (JRSOI), turnover time, relief-in-place and transfer of authority and time for the outbound unit to redeploy.[11]

(g) <u>Force Planning during Crisis</u>. Given the time constraints of an emergent crisis, force planning may occur in parallel to the execution functions (allocation and mobilization). Force planning during crisis may then be done based upon execution-sourced forces vice planning assumptions. CCMDs may execution source assigned forces and JS JFC and JFPs execution source in parallel to plan development. Recommendations for force allocation are presented to the Secretary for decision during planning or concurrent a decision to implement an OPORD. Force requirements identified during crisis are documented in a TPFDD in order to facilitate execution sourcing and unit deployment (See Chapter 3, *GFM* for further discussion of execution sourcing.)

b. **Support Planning**. The purpose of support planning is to determine the sequence of the personnel, logistic, and other support required to provide distribution; maintenance; civil engineering, medical, and sustainment in accordance with the concept of operation. Support planning is conducted in parallel with other planning, and encompasses such essential factors as executive agent identification; assignment of responsibility for base operating support; airfield operations; management of non-unit replacements; health service support; personnel management; financial management; handling of prisoners of war and detainees; theater civil engineering policy; logistic-related environmental considerations; support of noncombatant evacuation operations and other retrograde operations; and nation assistance. Support planning is primarily the responsibility of the Service component CDRs and begins during CONOPS development. Service component CDRs identify and update support requirements in coordination with the Services, the Defense Logistics Agency (DLA) and USTRANSCOM. They initiate the procurement of critical and low-density inventory items; determine host nation support (HNS) availability; develop plans for total asset visibility; and establish phased delivery plans for sustainment in line with the phases and priorities of the CONOPS. They develop and train for battle damage repair; develop reparable retrograde plans; develop container management plans; develop force and line of communications protection plans; develop supporting-phased transportation and support plans aligned to the CONOPS and report movement support requirements. Service component CDRs continue to refine their sustainment and transportation requirements as the FPs identify and source force requirements. During distribution planning, the supported CCDR and USTRANSCOM resolve gross distribution feasibility questions impacting inter-theater and intra-theater movement and sustainment delivery. USTRANSCOM and other transportation providers identify air, land and sea transportation resources to support the approved CONOPS. These resources may include apportioned inter-theater transportation, GCC-controlled theater transportation, and transportation organic to the subordinate commands. USTRANSCOM and other transportation providers develop transportation schedules for movement requirements identified by the supported CDR. A transportation schedule does not necessarily mean that the supported CDR's CONOPS is transportation feasible; rather, the schedules provide the most effective and realistic use of available transportation resources in relation to the phased CONOPS.[12]

(1) <u>Support Refinement</u>. During the plan development function, the CCDR's staff creates a detailed OPLAN, OPORD, or CONPLAN, with required annexes. The supported CCDR, subordinate CDRs, supporting CDRs, CSAs, and staff conduct a number of different planning activities, to include force planning, support planning, deployment planning,

[11] *CJCSM 3130.06 Series, Global Force Management Allocation Policies and Procedure*

[12] *CJCSM 3130.02, Adaptive Planning and Execution Policies and Procedures*

redeployment or unit rotation planning, shortfall identification, feasibility analysis, refinement, documentation, plan review and approval and supporting plan development. Planning activities culminate in training and wargaming exercises to provide feedback on the planned concept of support. The joint logistics concept of support specifies how capabilities will be delivered over time, identifies who is responsible for delivering a capability, and defines the critical logistical tasks necessary to achieve objectives during all phases of the operation. Annex W (*Operational Contract Support*) is closely tied to the Concept of Logistics Support (COLS) since contracted support may fill critical operational and logistics capability gaps. The COLS encompasses joint capabilities of all force capabilities, to include multinational, HN, interagency partners, international organizations, NGOs, DOD OCS, plus Active Component and Reserve Component forces.

(a) <u>Logistics Supportability Analysis (LSA)</u>. LSA is conducted by supporting organizations to determine the logistics support they must provide, in accordance with resource-informed planning guidance and to determine the adequacy of resources needed to support mission execution. LSAs ensure logistics is phased to support the CONOPS; establishes logistics C2 authorities; and integrates support plans across the supporting commands, Service components and agencies. LSAs are conducted by each supporting organization to the lowest level of detail needed to quantify the logistics requirements (national stock number level). These LSAs are then integrated by supporting organizations to coordinate roles and responsibilities, capabilities and ensure all understand the sourcing of the support. A joint LSA is created and presented to the CCDR who confirms this support will provide the sustainment needed to successfully execute and complete his/her mission. If there are gaps and shortfalls or high levels of risk that cannot be mitigated internally by supporting organization, the LSA provides the process for presenting issues to senior leaders for resolution.[13]

(b) <u>Logistics Integration</u>. CDRs and staffs apply basic principles, control resources, and manage capabilities to provide sustained joint logistics. Logisticians use the logistic principles of responsiveness, simplicity, flexibility, economy, attainability, sustainability and survivability as a guideline to assess how effective logistics are integrated into plans and execution. To achieve full integration, CDRs and their logisticians coordinate, synchronize, plan, execute, and assess logistic support to joint forces during all phases of the operation.

(c) <u>Core Logistic Functions</u>. Core logistic functions provide a framework to facilitate integrated decision-making, enable effective synchronization and allocation of resources, and optimize joint logistic processes. The challenges associated with support cut across all core logistic functions, especially when multiple JTFs or multinational partners are involved. The core logistics functions listed below provide a common standard of support planning across the joint force.

<u>1</u> <u>Deployment and Distribution</u>. Encompasses the movement and sustainment of the joint force; coordinated and synchronized across the Joint Deployment and Distribution Enterprise (JDDE) by the CDR, USTRANSCOM, in the capacity as the Distribution Process Owner (DPO).[14]

<u>2</u> <u>Supply</u>. Includes management of the supply chain for all classes of supply to include a global network of suppliers. DLA has the primary responsibility for DOD supply chain management.

<u>3</u> <u>Maintenance</u>. Considerations range from field maintenance to depot maintenance and equipment reset. Planners may also address contracted maintenance service.

<u>4</u> <u>Logistic Services</u>. Includes several capabilities to include food service, water and ice service, contingency base services, hygiene series, and mortuary affairs.[15]

<u>5</u> <u>Operational Contract Support (OCS)</u>. Provides the CDR the tools and processes to conduct, plan for, and obtain supplies, services, and construction from commercial sourc-

---

[13] *CJCSI 3110.03 Series, Logistics Supplement to the JSCP and JP 4-0 Joint Logistics*

[14] *JP 4-09, Distribution Operations*

[15] *JP 4-0, Joint Logistics*

es in support of joint operations. OCS includes the associated contract support integration, contracting support, and contractor management functions.[16]

6   Health Services (HS). Improves the health readiness of forces by management of force health protection, health service delivery, and health system support.[17]

7   Engineering. Provides recommendations for the CDR's integration of the engineering capabilities of the joint force to include: general engineering, combat engineering, and geospatial engineering. Further guidance can be found in JP 3-34, *Joint Engineer Operations*.

(2) Transportation Refinement. Transportation refinement simulates the planned movement of resources that require lift support to ensure that the plan is transportation feasible. Contingency sourcing using current and projected unit readiness and mobilization timelines supports TPFDD refinement, enabling a more accurate sourcing and transportation feasibility analysis. If required, the supported CDR evaluates and adjusts the CONOPS to achieve end-to-end transportation feasibility, if possible given resource-informed constraints, or requests additional resources if the level of risk is unacceptable, recognizing additional transportation resources may not be available. Transportation plans must be consistent and reconciled with plans and timelines required by providers of Service-unique combat and support aircraft to the supported CCDR. Planning also must consider requirements of international law; commonly understood customs and practices; and agreements or arrangements with foreign nations with which the U.S. requires permission for overflight, access, and diplomatic clearance. If significant changes are made to the CONOPS, it should be assessed for feasibility and refined to ensure it is acceptable.

(3) Responsibilities for Support Planning.[18]

(a) Directive Authority for Logistics (DAFL). DAFL is statutory authority contained in Title 10, USC, Section 164. The statute specifies that, included among the various authorities that comprise the command authority of CCDRs, "giving authoritative direction to subordinate commands and forces necessary to carry out missions assigned to the command, including authoritative direction over all aspects of military operations, joint training, and logistics" are integral elements of that command authority. DAFL cannot be delegated or transferred. However, the CCDR may delegate the responsibility for the planning, execution, and/or management of common support capabilities to a subordinate CDR or Service component CDR to accomplish the subordinate CDR's or Service component CDR's mission. The CCDR must formally delineate this delegated authority by function and scope to the subordinate CDR or Service component CDR.

(b) Supported CCDR. The supported CCDRs lead integrated logistics planning for their problem sets, inclusive of all associated plans related to the logistics problem both inter-theater and intra-theater. As such, supported CCDRs have coordinating authority for logistics planning. They lead the logistics planning process with all supporting CCMDs to develop a common understanding of logistics requirements, synchronize logistics planning activities, identify problem set logistics resource requirements, and provide logistics supportability analyses (quantitative and qualitative), as well as risk and supportability assessments associated with the plans. The supported CDR designates and prioritizes objectives, timing and duration of the supporting action. The supported CDR ensures supporting CDRs understand the operational approach and the support requirements of the plan. If required, SecDef will adjudicate competing demands for resources when there are simultaneous requirements amongst multiple supported CCDRs. Support planning provides the basis for development of Annexes D (*Logistics*), P (*Host Nation Support*), Q (*Health Services*), and W (Operational Contract Support) for a plan or order.

---

[16] *JP 4-10, Operational Contract Support*

[17] *JP 4-02, Health Service Support*

[18] *JP 4-0, Joint Logistics*

(c) <u>Supporting CDR</u>. Supporting CDRs will ensure their logistics planning is sufficiently integrated and synchronized across the problem set. They assist the supported CCMDs' efforts to develop a unified view of the logistics environment and synchronize resources, timelines, logistics C2, decision points, and authorities. The supporting CDR determines the forces, tactics, methods, procedures, and communications to be employed in providing support. The supporting CDR advises and coordinates with the supported CDR on matters concerning the employment and limitations (e.g., logistics) of required support, assists in planning for the integration of support into the supported CDR's effort, and ensures support requirements are appropriately communicated throughout the supporting CDR's organization.

(d) <u>Component Commands</u>. The Supported CCDR's components develop and continue to refine their mission support, movement infrastructure, sustainment, and distribution requirements as the force planning further identifies force requirements. The Service components develop their support plans in conjunction with their corresponding Service HQs and DOD Agencies (for non-Service controlled materiel). Guidance and consideration for conducting the specific functional steps of support planning are provided in detail in JP 4-0, *Joint Logistics* and its associated series of subordinate logistics publications.

(e) <u>Service HQs, Supporting Commands, and DOD Agencies</u>. In collaboration with the supported CCMD, document logistics requirements, sourcing results, identify capacity or capability shortfalls. Support DOD Agency and Service component development of Logistics Supportability Analysis (LSA), including surge, sustainment, and forces/materiel sourcing (origin and destination) and time-phased sustainment movement requirements.

c. **Deployment/Redeployment Planning.** Deployment and redeployment planning is conducted on a continuous basis for all approved plans/orders and crises. Planning for redeployment should be considered throughout the operation and is best accomplished in the same time-phased process in which deployment is accomplished. In all cases, mission requirements of a specific operation determine the scope, duration, and scale of both deployment and redeployment planning. Unity of effort is critical, since both deployment and redeployment operations involve numerous commands, agencies, and functional processes. Because the ability to adapt to unforeseen conditions is essential, supported CCDRs' deployment plans must be able to support global force visibility requirements. When planning for operations that may be enduring, consideration is given to force rotations. Units must rotate without interrupting operations. Planning should consider JRSOI, turnover time, relief-in-place and transfer of authority, and time it takes for the outbound unit to redeploy. This information is vital for the FPs, JFCs, and JFPs to develop force rotations in the GF-MAP annex schedule if the operation is executed. The joint deployment and redeployment processes consist of four phases: planning; predeployment/pre-redeployment activities; movement; and JRSOI. Both processes are similar; however, each has unique characteristics. These phases are iterative and may occur simultaneously throughout an operation.[19]

(1) <u>Deployment Planning</u>. Deployment planning is initiated in the plan development function and continues to be developed and refined during the execution function. The purpose of deployment planning is to develop a distribution network that supports the full range of activities in supporting the movement of forces and materiel during deployment, sustainment, and redeployment/retrograde phases of an operation. It is conducted iteratively with force and support planning and may identify additional forces necessary to execute deployment functions. It is conducted at all levels of command by both supported and supporting CDRs. Deployment planning activities include all actions required for the deployment of forces up to the point of their employment. It involves planning to move, receive, and integrate forces from origin to final destination. Developing a concept of deployment synchronizes the efforts of the supported, supporting, and subordinate commands, and Service HQs that contribute to this planning. The supported CCDR develops a deployment concept and identifies specific pre-deployment standards necessary to meet

mission requirements. The Service HQs and supporting CDRs provide trained and mission-ready forces as ordered to support the CCDRs deployment concept. Doctrinal discussion of the methods for conducting deployment planning can be found in JP 3-35, *Deployment and Redeployment Operations*.

(2) <u>Movement Planning</u>. Movement planning is the collaborative integration of movement activities and requirements for transportation support. Forces may be planned for movement either by self-deploying or the use of organic lift and nonorganic, common-user, strategic lift resources identified for planning. Competing requirements for limited strategic lift resources, support facilities, and intra-theater transportation assets will be considered by the supported CDR in terms of impact on mission accomplishment. If additional resources are required, the supported CDR will identify the requirements and rationale for those resources.[20]

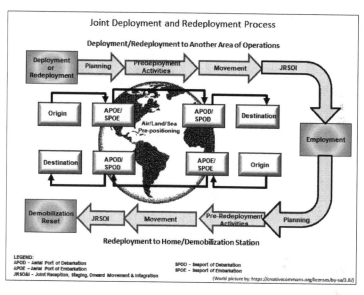

Joint Deployment and Redeployment Process[21]

(a) <u>Supplemental Instruction (SI)/Letter of Instruction (LOI)</u>. CDRs will often publish revised TPFDD development guidance articulating the CDR's deployment and redeployment priorities. Planners then develop a final refinement of the planned TPFDD IAW this revised guidance.

(b) <u>Movement Considerations.</u> As movement plans are developed, planners may need to consider the specific requirements for enroute support to include infrastructure, sustainment, and remain overnight locations. Movement planning may iteratively identify additional force or support requirements necessary to enable movement.[21]

(c) <u>Coordination with USTRANSCOM</u>. The supported CDR and USTRANSCOM coordinate to resolve transportation feasibility issues impacting inter-theater and intra-theater movement and sustainment delivery. USTRANSCOM and other transportation providers identify air, land, and sea transportation resources to support the CONOPS. These resources may include apportioned inter-theater transportation, GCC-controlled theater transportation, and transportation organic to the subordinate commands. USTRANSCOM

---

[20] *JP 3-35, Deployment and Redeployment Operations*
[21] *Ibid*

and other transportation providers develop transportation schedules for movement requirements identified by the supported CDR.

(3) <u>Joint Reception, Staging, Onward Movement, and Integration Planning (JRSOI)</u>. Following the development of movement infrastructure concepts, the supported CCDR's planning team develops the air and sea reception plans, staging plan, and completed JRSOI plan. The requirements to conduct JRSOI may precipitate additional force requirements and cause iterative changes to force planning. JRSOI constraints (e.g., port clearance, intra-theater movement capacity, staging base limitation) imposed on strategic movement must be considered in the TPFDD development.

(4) <u>Plan Integration and Synchronization</u>. For integrated planning involving more than one CCDR, the supported CCDR or Global Synchronizer has coordinating authority for the problem set or global function, respectively. Throughout the planning process, the supported CCDR or Global Synchronizer should integrate and synchronize planning efforts. Resource considerations must include the total requirements of all supported and supporting CCMDs and organizations; all associated plans related to the problem set must be planned within the constraint of apportioned forces.

(5) <u>Redeployment Concept</u>. Effective redeployment operations are essential to ensure Service HQs have sufficient time to fully source and prepare for any subsequent missions or force rotations. Planning for the return of forces, equipment, and material allows for cost effective and streamlined transportation planning with identified roles and responsibilities. A subset of redeployment planning is retrograde planning which allows for the movement of non-unit equipment and material from the point of employment to a reset program or another directed area of operation. It is important for the GCCs to develop a clear overall redeployment and retrograde strategy early on to enable the distribution partners to create a mature theater disposition plan with a comprehensive retrograde estimate. During execution, the supported CCDR's Service components issue redeployment orders (REDEPORDs) IAW the refined redeployment concept. Further discussion of redeployment planning can be found in JP 3-35, *Deployment and Redeployment Operations*.

d. **Nuclear Strike Planning.** *(Although this activity has been removed from JP 5-0, Joint Planning, it is essential that all planning options and considerations are taken into account.)* During some planning efforts, planners may need to conduct nuclear strike planning in addition to force, support, and deployment planning. CDRs must assess the military as well as strategic impact that a nuclear strike would have on conventional operations. Nuclear planning guidance is provided in strategic guidance and the Nuclear Supplement to the JSCP. USSTRATCOM conducts nuclear strike planning and overseas coordination ICW supported GCCs and certain allied CDRs. The planning provided by USSTRATCOM ensures optimal integration of U.S. nuclear and conventional forces prior to, during, and after conflict. Due to the strategic and political consequences associated with nuclear weapons operations and plans, only the President has the authority to authorize use of nuclear weapons.

e. **Shortfall Identification**. Along with hazard and threat analysis, shortfall identification is performed throughout the plan development process. The supported CDR continuously identifies limiting factors and capabilities shortfalls and associated risks as plan development progresses. Where possible, the supported CDR resolves the shortfalls and required controls and countermeasures through planning adjustments and coordination with supporting and subordinate CDR's. If the shortfalls and necessary controls and countermeasures cannot be reconciled or the resources provided are inadequate to perform the assigned task, the supported CDR reports these limiting factors and assessment of the associated risk to the CJCS. The CJCS and the Service Chiefs consider shortfalls and limiting factors reported by the supported CDR and coordinate resolution. However, the completion of assigned plans is not delayed pending the resolution of shortfalls. If shortfalls cannot be resolved within the JSCP timeframe, the completed plan will include a consolidated summary and impact assessment of unresolved shortfalls and associated risks.[22]

[22] *JP 5-0, Joint Operations*

f. **Feasibility Analysis**. This Step is similar to determining the feasibility of a COA, except that it typically does not involve simulation-based wargaming. The focus in this Step is on ensuring the assigned mission can be accomplished using available resources within the time contemplated by the plan. The results of force planning, support planning, deployment planning, and shortfall identification will affect OPLAN or OPORD feasibility. The primary factors considered are whether the apportioned or allocated resources can be deployed to the joint operations area (JOA) when required, sustained throughout the operation, and employed effectively, or whether the scope of the plan exceeds the apportioned resources and supporting capabilities. Measures to enhance feasibility include adjusting the CONOPS, ensuring sufficiency of resources and capabilities, and maintaining options and reserves.[23]

(1) <u>Forces</u>. Supported CCDRs, ICW JS JFC, JFPs, and Military Departments, should determine the feasibility of sourcing the plan's required forces. The Force Apportionment Tables should be referenced for this sourcing feasibility analysis as well as identifying preferred forces. For integrated planning, apportioned forces should constrain the total force requirements of both supported and supporting plans. Consideration should also be given to non-organic capabilities available through contracted support and HNS. During contingency sourcing the JS JFC and JFPs will, collaboratively with FPs, provide contingency sourcing solutions to inform the assessment. Force requirements will be documented in a plan TPFDD in order to analyze transportation feasibility. Planners should also document force requirements in a format that enables the GFM allocation process and execution sourcing should the plan transition to execution (See CJCSM 3130.06, *Global Force Management Allocation Policies and Procedures*).

(2) <u>Sustainment Resources</u>. Supported CCDRs, ICW Military Departments and DOD Agencies, should determine the feasibility of providing the resources required to execute the plan in conjunction with other supporting or concurrent operations. Supporting organizations must provide subject matter experts to identify resource requirements. Resources required will be documented in the TPFDD for subsequent transportation feasibility.

(3) <u>Transportation</u>. For plans with a notional TPFDD, the supported CCDR, ICW CDR, USTRANSCOM will determine the transportation feasibility. The notional TPFDD includes forces and resources previously identified and is the basis for the analysis. In order to evaluate the transportation feasibility of a plan, a TPFDD must be populated with assumed units (identified preferred forces or contingency sourcing) with adequate detail of movement data (TUCHA or refined TUCHA) articulating embarkation requirements. Transportation analysis considers capacity of strategic lift as well as throughput constraints of transit points and JRSOI infrastructure.

6. Plan Refinement

a. Plan refinement is effectively the operational activity of assessment applied during plan development. A*s planning is conducted, planners analyze the plans feasibility and identify shortfalls.* Planners continue to conduct planning in order to reconcile these identified shortfalls. Reconciliation is iterative between force, support, and deployment planning as planners attempt to satisfy or mitigate shortfalls. Shortfall reconciliation may result in a refinement to the area of planning in which the shortfall originated or in another area of planning. For example, a shortfall in forces may be mitigated by contracted support developed through support planning. Plans are refined in one or more areas to satisfy or mitigate all identified shortfalls. Significant shortfalls may lead CDRs to consider changes to include further refinement of the CONOPS. Shortfalls that cannot be mitigated effectively should be identified up the chain of command for the direction of continued planning to leverage additional/other resources or authorities to mitigate. A developed plan is refined in order to resolve or mitigate shortfalls or feasibility limitations. Refinement is conducted by the supported CCDR ICW supporting CCDRs, JS JFC, JFPs, Service HQs, FPs, USTRANSCOM (as both a mobility JFP and DOD Single Manager for transportation), and DLA. The

[23] JP 5-0, *Joint Operations*

intended output is a *feasible plan prepared for JPEC review*. Plan refinement is described further in the following paragraphs:

(1) <u>Forces Refinement</u>. Forces refinement confirms that force requirements are within apportionment or availability constraints and assesses the adequacy of Combat, CS, and CSS force sourcing. It is conducted ICW the supported and supporting commands, JS JFC, JFPs and FPs. If feasibility analysis finds a shortfall of available forces, CDRs may request to increase those quantities or access to specific prepositioned forces/equipment via their chain of command or adjust the plan.

(2) <u>Support Refinement</u>. The purpose of logistic refinement is to confirm the supply and sustainment requirements and assess the adequacy of the resources identified during support planning. Directed by the supported CCDR, logistics refinement is conducted collaboratively by the CCMD Service components, Service HQs, and DLA. During logistics refinement, planners should consider that changes to logistics requirements may iteratively change force requirements for logistics forces.

(3) <u>Deployment Refinement</u>. The purpose of deployment refinement is to adjust the plan to mitigate shortfalls impeding plan feasibility. Deployment refinement includes refinement of the transportation plan, the planned TPFDD, and the JRSOI plan.

(a) <u>Transportation Refinement</u>. The purpose of transportation refinement is to resolve gross distribution feasibility questions impacting inter-theater and intra-theater movement and sustainment delivery. It is conducted by the supported CCDR ICW the JS JFC, USTRANSCOM, and supporting CCDRs. Transportation refinement simulates the planned movement of resources that require lift support to ensure the plan is transportation feasible. The supported CDR evaluates and adjusts the CONOPS to achieve end-to-end transportation feasibility, or requests additional resources if the level of risk is unacceptable. Transportation refinement may include adjusting the movement plan to include changes to points of embarkation (POEs), points of debarkation (PODs), routing, sequence, timing, and mode/source of lift. A transportation feasibility determination requires concurrent analysis and assessment of available strategic and theater lift assets, transportation infrastructure, and competing demands and constraints. Plans are considered transportation feasible when the capability to move forces, equipment, and supplies exists from the point of origin to the destination according to the plan.

(b) <u>TPFDD Refinement</u>. The TPFDD refinement process is conducted concurrent to force planning, sustainment planning, and transportation planning. It may be conducted with a notional TPFDD based upon planning assumptions or an execution sourced TPFDD pending execution. Transportation refinement adjusts the TPFDD flow to ensure transportation feasibility. The supported CDR typically publishes TPFDD refinement guidance during the initial stage of plan development to coordinate this process with the JS JFC and USTRANSCOM, supporting and subordinate commands, other JFPs, Service HQs, and DOD Agencies. Guidance addresses the TPFDD refinement for forces (to include non-unit personnel), logistics (both accompanying supplies and non-unit re-supply), and transportation/JRSOI and may specify the use of collaborative planning tools and procedures.

(c) <u>JRSOI Refinement</u>. The purpose of JRSOI refinement is to use the results of transportation planning and determine the feasibility of intra-theater movement to the destination. JRSOI refinement is conducted in conjunction with transportation refinement by the supported CDR ICW JS JFC, USTRANSCOM, other JFPs, Service HQs, and supporting and subordinate commands. Planning considerations include port clearance, throughput, intra-theater transportation infrastructure, capability to provide sustainment to forces in transit, and build-up at staging bases. Upon completion of the JRSOI refinement, the supported CDR's logistics sustainability analysis assesses the end-to-end transportation viability.

(4) <u>CONOPS Refinement</u>. If shortfalls cannot be reconciled by refinements to forces, support or deployment planning, the CDR may consider changing the CONOPS of the plan. Changes to a CONOPS may include changes to timelines/sequencing of activities, adjustments to priorities, or a change in scope or objectives. If the refined CONOPS deviates significantly from the previously approved COA, the authority initiating the planning should

approve this scope of change. At the strategic level, CCDRs making significant refinements to a CONOPS may need to review those changes via the IPR process (Chapter 6-II, *Plan Review Process*).

(5) <u>Remaining Shortfalls</u>. Along with hazard and threat analysis, shortfall identification is conducted throughout the plan-development process. The supported CDR continuously identifies limiting factors, capability shortfalls, and associated risks as plan development progresses. Where possible, the supported CDR resolves the shortfalls and required controls and countermeasures through planning adjustments and coordination with supporting and subordinate CDRs. If the shortfalls and mitigating countermeasures cannot be reconciled through plan refinement, the supported CDR reports the limiting factors, risk assessment, and mitigation plan to the Chairman. The Chairman and JCS consider the shortfalls and limiting factors reported by the supported CDR and coordinate a resolution. If shortfalls cannot be resolved the plan will include a consolidated summary and impact assessment of unresolved shortfalls and associated risks.

b. <u>Documentation</u>. When the TPFDD is complete and end-to-end transportation feasibility has been achieved and is acceptable to the supported CCDR, the supported CCDR completes the documentation of the plan or OPORD and coordinates access with respective JPEC stakeholders to the TPFDD as appropriate.[24]

7. Plan Review, Approval and Transition

a. This step is the culmination of plan development. *The purpose of plan review and approval is to confirm the acceptability of the plan with the authority that initiated the planning effort.* At the strategic level, CCDR plans are reviewed by JS and OSD senior leaders prior to review and approval by the SecDef or designated representative as required. Subsequent to plan review and approval, the planning team considers development of products to facilitate potential plan transition to execution. Under crisis conditions the review and approval of an order is often significantly abbreviated or potentially virtual dialog to more rapidly transition to execution.

(1) <u>Plan Review and Approval</u>. When all required annexes of the plan are complete, the JS J-5 coordinates with the JPEC for the review of the plan and provides the results to the supported CCDR and CJCS. The CJCS reviews and provides recommendations to SecDef as necessary. The JCS provides a copy of the plan to OSD to facilitate their parallel review of the plan and to inform USD(P)'s recommendation of approval/disapproval to SecDef. After the CJCS and USD(P)'s review, the SecDef or President will review, approve, or direct CCDRs to modify their plans. The President, SecDef, or their designee, is the final approval authority for plans, depending upon the subject matter. Specific responsibilities and procedures for review of plans are addressed in CJCSI 3141.01, *Management and Review of Campaign and Contingency Plans*.

(2) <u>Transition</u>. Transition is critical to the overall planning process. It is an orderly turnover of a plan or order as it is passed to those tasked with execution of the operation. It provides information, direction and guidance relative to the plan or order that will help to facilitate situational awareness. Additionally, it provides an understanding of the rationale for key decisions necessary to ensure there is a coherent shift from planning to execution. These factors coupled together are intended to maintain the intent of the CONOPS, promote unity of effort and generate tempo. Successful transition ensures that those charged with executing an order have a full understanding of the plan. Regardless of the level of command, such a transition ensures that those who execute the order understand the CDR's intent and CONOPS. Transition may be internal or external in the form of briefs or drills. Internally, transition occurs between future plans and future/current operations. Externally, transition occurs between the CDR and subordinate commands.[25]

---

[24] *JP 5-0, Joint Operations*
[25] *Ibid*

(a) Transition Products. Subsequent to plan completion, the supported command should consider development of products to facilitate the transition of the plan into execution. In particular, CCMD planners should consider the methods of requesting forces and authorities in the case of contingency plan execution (see Chapter 7, *Execution*).

<u>1</u> During the planning process, force requirements should be documented in a format that can rapidly transition to execution.

<u>2</u> Some plans may require large numbers of forces to deploy in a short time, requiring a flexible and equally responsive allocation process. To support rapid plan transition to execution, CCDRs should define force requirement using existing Service units, to the maximum extent practicable. CJCSM 3130.06 Series, *Global Force Management Allocation Policies and Procedures* details the different processes and requirements for managing force allocation requirements'.

<u>a</u> Execution. Upon execution, operational force requirements for a contingency or military activity in a campaign plan are execution sourced. Force requirements should be documented with the requisite information necessary to support allocation decisions by the Secretary, execution sourcing by the FPs, and effective transition to unit deployment.

<u>b</u> Exercises. Forces required to conduct exercises are documented in Joint Training Information Management System (JTIMS) and execution sourced by the Services. Exercise planners will build force requirements and FPs will source requirements in the JOPES database. Further information regarding the sourcing of exercise requirements can be found in CJCSM 3500.03 Series, *Joint Training Manual for the Armed Forces of the United States.*

<u>c</u> Joint Individual Augmentees (JIA). JIAs required for a joint headquarters are documented on a JMD in the *Fourth Estate Manpower Tracking System (FMTS).* These are execution sourced via the GFM allocation process.

<u>d</u> TPFDD. All planned movement requirements, including forces and sustainment that require USTRANSCOM strategic lift, must be documented in a JOPES TPFDD. As force requirements are execution sourced, unit refined deployment data is documented and verified. Verified requirements are validated by the supported CCDR and sent to USTRANSCOM to schedule lift. During the plan development function, the TPFDD should be built to support the plan assessments (both the operational activity and functions), including contingency sourcing, as well as prepared to be refined during execution. Under crisis conditions, a TFPDD may be developed directly during planning to facilitate execution sourcing and unit deployment. To facilitate execution sourcing, force requirement in the TPFDD should be documented at the unit level. CJCSM 3122.O2 Series, *Joint Operation Planning and Execution System (JOPES) Volume III Time Phased Force and Deployment Data Development and Deployment Execution* and CJCSG 3122, *Time-Phased Force and Deployment Data (TPFDD) Primer,* detail the TPFDD development process.[26]

(b) Authorities. The authorities required to execute an operation based upon a contingency plan are typically beyond those of a campaign plan. Authority requirements identified during planning should have draft request messages prepared in order to expedite their request during an emerging crisis.

(c) Transition Brief. At higher levels of command, transition may include a formal transition brief to subordinate or adjacent CDRs and to the staff supervising execution of the order. At lower levels, it might be less formal. The transition brief provides an overview of the mission, CDR's intent, task organization, and enemy and friendly situation. It is given to ensure all actions necessary to implement the order are known and understood by those executing the order.[27] The brief may include items from the order or plan such as:

- Higher headquarters mission and CDRs' intent
- Mission
- CDR's Intent

[26] The CJCSM 3122.O2 Series and the CJCSG 3122 will be resinded upon publication of 3130.04, Deployment Policies and Procedures

[27] JP 5-0, Joint Operations

- CCIRs
- Task organization
- Situation (friendly and enemy)
- CONOPS
- Execution (including branches and potential sequels)
- Planning support tools (such as synchronization matrix, JIPOE products, etc.)

(d) <u>Confirmation Brief</u>. A confirmation brief is given by a subordinate CDR after receiving the order or plan. Subordinate CDRs brief the higher CDR on their understanding of CDR's intent, their specific tasks and purpose, and the relationship between their unit's missions and the other units in the operation. The confirmation brief allows the higher CDR to identify potential gaps in the plan, as well as discrepancies with subordinate plans. It also gives the CDR insights into how subordinate CDRs intend to accomplish their missions.

(e) <u>Transition Drills</u>. Transition drills increase the situational awareness of subordinate CDRs and the staff and instill confidence and familiarity with the plan. Sand tables, map exercises, rehearsals of concept (ROC) and rehearsals are examples of transition drills.

b. <u>Complete Supporting Plans/Orders</u>. The development of supporting plans occurs in parallel with the supported command's planning efforts. CDRs will normally review supporting plans upon completion of the supported CDR's plan. For integrated planning by multiple CCMDs, the JPEC review and IPR process may also require a combined review of both supported and supporting plans.

8. Plan Implementation

a. Military plans and orders should be prepared to facilitate implementation and transition to execution. For a plan to be implemented, the following products and activities must occur:

(1) <u>Confirm assumptions</u>. Analyze the current OE and establish as fact any assumptions made during plan development or consider a branch plan to cover the possible repercussions of an invalid assumption during execution.

(2) <u>Model the TPFDD</u>. Model the TPFDD to confirm the sourcing and transportation feasibility assessment. Validate that force and mobility resources used during plan development are currently available. Many critical capabilities reside in the RC (e.g., air and sea port opening), so planners need to know the mobilization authorities as they relate to deployment timelines. Additionally, as reserve units deactivate due to force structure changes, staffs have to re-validate TPFDDs.

(3) <u>Establish execution timings</u>. Set timelines to initiate operations to allow synchronization of execution.

(4) <u>Confirm authorities for execution</u>. Request and receive the President or SecDef authority to conduct military operations.

(5) <u>Conduct execution sourcing from assigned and available forces</u>. If force requirements exceed the capability and capacity of assigned and available forces, submit an emergent RFF through the GFM process, which facilitates a risk-informed SecDef decision to allocate/re-allocate forces from other CCMDs or Services. Develop new assumptions, if required.

(6) <u>Issue necessary orders for execution</u>. The CJCS issues orders implementing the directions of the President or SecDef to conduct military operations. CCDRs subsequently issue their own orders directing the activities of subordinate CDRs.

9. Plan Development Outcomes

Plan development planning function produces the following outcomes distinctive to each level of command.

a. <u>National Level</u>. This planning function produces a completed plan or order to be reviewed by the SecDef and USD(P) or their designated representative. The method of this review methods can range from a scheduled IPR to an informal, potentially virtual dialog for planning done under crisis conditions. Review at this level allows for a completed plan to be considered in the context of other elements of national power. For CPG or JSCP tasked plans, this summary review is informed by the preceding feasibility assessments (sourcing, support, and transportation) and JPEC review that allows for civilian-military dialogue to resolve outstanding issues. Should a plan not be acceptable or have unresolved issues, supported or supporting CCDRs, Services, and DOD Agencies may be provided refined guidance and directed to modify or further develop their plans. Under crisis conditions, plan development may produce an OPORD and an execution sourced TPFDD. Discussion at this level may then prompt both an execution and force allocation decision by the President or Secretary.

b. <u>Supported Command</u>. For the supported CDR, this stage of planning produces a completed plan or order including a force sourcing and transportation feasible notional TPFDD (or TPFDD when planning under crisis conditions) as required. The completed plan includes a base plan with required annexes per CJCSM 3130.03 Series, *Adaptive Planning and Execution Planning Formats and Guidance*. The completed plan and associated notional TPFDD provide the basis for subsequent plan assessment through the life cycle of the plan. Under crisis conditions, the force and transportation feasibility analysis and refinement may be done in place of a more formal plan assessment. Upon decision by the President or Secretary, forces are allocated and the operation order is executed by the supported and supporting CDRs.

c. <u>Supporting Commands, Services, and DOD Agencies</u>. For supporting commands and organizations, this stage of planning produces a supported CCDR completed plan with which supporting commands and organizations may align their own supporting plans. If directed, supporting organizations may submit their completed supporting plans along with the supported plan for JPEC review and briefing to the SecDef. Under crisis conditions, supporting commands, Services, and DOD Agencies may implement the SecDef's decisions for execution and ordered force allocation.

d. <u>Subordinate Commands</u>. For subordinate CDRs, this stage of planning produces a supported CCDR completed plan as a basis for subordinate commands to complete their own plans. When completed, subordinate commands may also submit their plans to their HHQ for review. Under crisis conditions, subordinate commands implement the orders of their HHQ.

9. Summary

The Plan Development function is used in developing an OPLAN, CONPLAN or an OPORD with applicable supporting annexes and in refining preliminary feasibility analysis. During this function, the CDR and staff, in collaboration with subordinate and supporting components and organizations, expand the approved COA into a detailed plan or OPORD by refining the initial CONOPS associated with the approved COA. Plan Development fully integrates mobilization, deployment, employment, conflict termination, sustainment, redeployment, and demobilization activities. In actuality, the plan development process is not sequential but collaborative, adaptive to change and extremely flexible. The steps denoted in this Chapter may overlap, be accomplished simultaneously, or be repeated. Successful plan development depends upon recording and sharing planning guidance and information in the collaborative and parallel planning environment.

> "Plan for what it is difficult while it is easy, do what is great while it is small."
>
> — Sun Tzu, The Art of War

II. Plan Review Process

I. PLAN REVIEW

1. General

a. If a strategy is the implementation of policy while balancing available ends, ways and means, a strategic planning dialog is the iterative conversation among civilian and uniformed senior leaders that considers potential contingencies and the application of military power to address them. A shared understanding of the problem, the goal, and the potential ways to apply military power, and the resources required is critical to presenting these leaders with credible choices. Military power may or may not be the last resort in influencing the problem or trying to bring it to resolution; its allocation is dependent upon senior decision-makers having a clear understanding of what can be accomplished, the inherent risks, and how military power complements other elements of a USG response. This is the ultimate purpose of plan reviews.

b. The plan review process is a continuous review based on "planning to date." This is intended to ensure the SecDef and President have the best planning advice available based on the current OE. It maintains the requirement to ensure the Joint Staff and OSD are updated on plans in development. In the review process CCMDs provide a thorough view of demand on the force in the event of a contingency or crisis. It is recognized that a crisis rarely requires the implementation of a single CCMD plan, but the integration of operations across CCMD AORs, functions, and domains that may require the re-allocation or reassignment of forces to mitigate risk globally.[1]

(1) Plan reviews provide a venue for senior military and civilian leaders to develop a shared understanding of the emerging problem (situation), how military power might address that problem to reach USG preferred resolution (strategic options), what resources can be applied across the government (USG unity of effort); what specific military actions might be taken (operational approach) – or COA; and what decisions, resources, and abilities the military needs in order to take specific actions. The plan reviews also ensure that the entire JPEC, as well as other USG agencies, is involved in the plan development and understands the guidance.

(2) The intent of the plan review process is to produce globally integrated plans to advance U.S. interests and achieve U.S. strategic objectives. This review process addresses the full range of plans, e.g., GCPs, CCPs, and ICPs. These plans provide the SecDef and President the best possible information and options to address the complex and uncertain global environment.

2. Plan Review Purposes

a. The review process has four purposes:

(1) The *first* is to ensure that the plans are executable. That means they are:

(a) Feasible. The assigned mission can be accomplished using available resources within the time contemplated by the plan to advance U.S. security interests. The plan is prepared in a global context and accounts for both ongoing (continuing) operations and the rest of the integrated plan (cross CCMD requirements).

[1] *CJCSI 3141.01, Management and Review of Campaign and Contingency Plans*

(b) _Acceptable_. The plan and related operations are consistent with policy and law; and are within the risk tolerance of the President and the SecDef (The contemplated COA is proportional, worth the cost, consistent with the law of war, and militarily and politically supportable.) [2]

(c) _Complete_. The plans incorporate all major operations and tasks necessary to accomplish the designated objective(s). The plan must identify forces required, decision points, deployment concept and requirements, employment concept, sustainment concept and requirements, time estimates for achieving objectives, mission success criteria and mission conflict termination criteria.

(2) _Second_, this process is a mechanism that allows the CJCS to ensure plans are up to date and provide military advice to civilian leadership and guidance to the CCMDs based on a perspective that looks across CCMDs and Services.

(3) _Third_, this process integrates policy guidance from the SecDef and USD(P) to military leadership, providing perspective and guidance to the Joint Staff and CCMDs. Reviews are also a forum for the SecDef, USD(P), or designated representative to refine strategic direction and policy guidance (national-level objectives, assumptions, limitations, restrictions, and risk) to CCDRs.

(4) _Fourth_, it facilitates integration of plans across domains, functions, and regions; enabling integrated planning and a global perspective.

b. The plan review process addresses GCPs and ICPs as the planning baseline for achieving U.S. national objectives. While not all plans address global issues, at a minimum, plans should exist in a federation of plans where they are mutually supporting and informed.

c. This process supports the necessity for integrated planning across CCMDs and the JPEC.

d. Although the Joint Staff is responsible for the conduct of the plan review process, the entire JPEC conducts the review. The JPEC incorporates the CCMDs, the JS, OSD, the Services, NGB, CSA, and other affected defense agencies to create shared understanding, synchronize efforts across the Joint Force, develop integrated products, and establish the optimal confluence of military plans, operations, and strategy to enable the Chairman to provide military advice from a global perspective.[3]

3. Plan Review Dialogue

a. The plans review process is a dialogue among the SecDef, CJCS, USD(P), and CCDRs. The SecDef, with USD(P) assistance, establishes the review requirements and publishes guidance on the timing of directed plans reviews. The plans review process ensures plans align with the NSS, NDS, CPG, and the JSCP. The plans review process also assists the Chairman in providing military advice to the President, the SecDef, and civilian leadership while assisting the CCMDs in incorporating policy guidance from the OSD and integrating planning across domains, functions, and regions.

b. The plan review process has two complementary lines of effort: _first, ensure plans align with policy; and second, ensure plans are militarily executable and provide realistic military options to the SecDef and President._

(1) OSD manages one line of effort on behalf of the SecDef to ensure plans align with policy in the NSS, NDS, and other strategic documents. The SecDef also determines the acceptable level of risk. The SecDef or USD(P) establishes the review requirements and publishes guidance on which problem sets and plans require review and timing of those reviews for all CPG- and JSCP-tasked problem sets and plans.

[2] _JP 5-0 Joint Operations_

[3] _CJCSI 3141.01, Management and Review of Campaign and Contingency Plans_

(2) The Joint Staff, through the J-5, manages the other line of effort with the JPEC. This line supports the Chairman's responsibility to provide military advice to the SecDef and President.

c. As a plan is produced, it may undergo reviews to ensure it remains consistent with policy, strategic guidance, and intent of the Department. Additionally, changes in the environment (strategic and operational) may require changes to previous approved plans and planning assumptions.[4,5]

4. Plan Review Process

a. The review process is a series of interactions between the CCMD planners, the JPEC (led by the Joint Staff), and OSD representatives to support policy-guided and globally-integrated planning.

b. <u>Plans Review Criteria</u>. CCDRs may have their plans reviewed by OSD and the Joint Staff for any of the following:

(1) There exists a military problem that requires a SecDef or CJCS decision due to incurred risk because of available capabilities (e.g., time distance requires posturing more forces forward to alleviate force flow problems);

(2) There are policy gaps creating military problems not resolvable at the CCMD level;

(3) Priority Challenge Integrated Contingency Plans (by exception);

(4) Major revision to plan (e.g., due to changes in strategic environment, threat capabilities, U.S. capabilities);

(5) Directed by SecDef or CJCS.

c. The number of in-process reviews (IPRs) depends on the maturity of the plan, changes in policy, updates in the global campaigns and their assessments, and SecDef requirements.

(1) IPRs are an in-stride process to ensure necessary updates on plans of concern to the SecDef.

(2) The Joint Staff J-5 will publish a calendar of expected plan reviews annually.

d. Forums exist for plans review that could be executed subsequent to a JPEC staffing of the plan. The review process may take place through paper review—in the case of few or no contentious issues of the plan, or face-to-face/secure video teleconference (SVTC). The lowest level of formal review is an 0-7 /8-level Joint Planning Board (JPB) with subsequent reviews, as required, for Deputy Assistant Secretary of Defense (DASD), Operations Deputies (OPSDEPS), USD(P), JCS Tank, and SecDef approval briefs. At any point, if all contentious issues have been resolved, a paper review may suffice. The nature of the plan (e.g., a global threat covered by a GCP vs. a regional threat covered by a RCP) and the extent of the revision will determine the final level of review.

e. The process is also meant to be agile and efficient for CCMD planners. As appropriate, plan reviews and updates can be done by paper, SVTC, or in-person. When an issue arises in an IPR and the SecDef, USD(P), or designated representative identifies the need for a follow-up, the intermediate steps can be compressed to ensure the information is presented to the SecDef in a timely manner.

5. Plan Initiation and Review of Initial Planning Guidance

a. Most plan reviews will be of plans in some state of revision or update, however in the event a new requirement arises or conditions surrounding a current plan cause change to an existing plan, the SecDef or designated representative will publish planning guidance though the CPG or a SGS.

b. The CCDR or CJCS may also identify changes in the OE that require a new plan or significant revisions to an existing plan. In this case, the CCDR will notify the Chairman and SecDef and request updated guidance. If the CJCS is the initiator, the Chairman will provide the CCDR with initial military planning guidance and request policy guidance from the SecDef.

c. The CCDR, as part of the mission analysis, identifies additional guidance requirements and planning assumptions. These are forwarded through the CJCS to the SecDef for adjustment or approval.

d. <u>Detail</u>.

(1) If the SecDef initiates new planning requirements outside of regularly published guidance (i.e., the CPG), OSD will publish an appropriate planning guidance document.

(2) During mission analysis, the CCMD planning staff will identify additional information requests through the Joint Staff to the OSD(P) staff. As required, the Joint Staff will host JPBs to address questions raised. The Joint Staff will provide updates for global integration to the CJCS.

(a) JPBs can be held at the 0-6 and 0-7 levels. They are normally held by secure video conference (SVTC). The Joint Staff J-5 is responsible for hosting the JPB.

(b) Attendees will include supporting CCMDs, appropriate staff sections from across the Joint Staff, Services, and OSD representation as required. This will ensure the CCMD has access to Joint Staff expertise and that OUSD(P) is informed of plan progress and issues.

(c) For planning efforts that do not necessitate in-person or face-to-face discussions, issues may be addressed by electronic communications.

(3) If necessary, CCDRs may request an IPR coincident with completion of their mission analysis to ensure the assumptions are still valid.

(4) Resolution of discrepancies does not require formal IPRs or meetings.

e. <u>Output</u>. The critical output of this process is agreement between the CCDR, SecDef, Chairman and Joint Chiefs of Staff on the mission objectives (or endstate), resources (e.g., forces, partner contributions, time) available for planning, and critical assumptions. Key topics to ensure are mutually understood include:

(1) What is the military mission? Does everyone agree on the problem to be solved?

(2) What are the strategic and operational assumptions to include allied/partner contributions and the threat?

(3) What is the desired endstate, objective, or conflict termination criteria?

(4) Are there any directed decision timelines?

(5) What resources will be made available and what forces or capabilities require cross-CCMD reprioritization or reallocation to enable operations?

(6) What are the operational limitations (constraints and restraints)?

f. As this process is ongoing, planning continues to proceed and adjusts based on changes in the strategic guidance and discussions.

6. In-progress Reviews[6]

a. The majority of plan reviews will be to update existing plans.

b. Timing. IPRs are scheduled to support the SecDef's requirements. To ensure integration across the Joint Force, SecDef IPRs will be preceded by staffing through JPBs, DASD socializations, (USD(P) reviews and Tanks. These lower-level meetings are part of the planning process and should not be in addition to current planning requirements.

c. Updates. As the plan is updated, CCMD planners should use JPBs and electronic means (e.g., email) to confirm planning to date, ensure policy guidance and assumption are current, and identify changes and updates to resourcing.

(1) CCMD planners should use JPBs to inform the Joint Staff of any additional requirements necessary for success of the plan. Required increases in forces, authorities, or changes to assumptions should be identified early through JPBs to ensure they can be adequately answered before plan completion.

(2) The JPB and OPSDEPs seek to identify the military implications of operations on the Joint Force.

(3) As a participant in the JPB, the Deputy Assistant Secretary of Defense for Plans (DASD(P)) can also identify to CCMD planners any policy changes that will affect the plan.

d. As plan development progresses, the CCMD should conduct regular IPRs to ensure the plan remains consistent with SecDef requirements.

(1) The Joint Staff J-5 will ensure CCMDs have the opportunity to address the Joint Staff and OSD leadership throughout plan development.

(2) CCMDs should ensure the plan is reviewed at the appropriate level at key points in plan development, such as COA selection.

(3) OPSDEPs and JCS Tanks will be scheduled to ensure military leadership is informed at critical points in the planning process.

e. The DASD(P) is responsible for hosting DASD socializations. Attendees will include representatives from regional and functional offices within OSD and the Joint Staff J5. DASD socializations should reveal the policy implication of gaps resulting from military operations. The DASD socialization is part of an IPR and should occur after a JPB.

f. IPRs can be focused to cover portions of the proposed plan in order to ensure critical elements are addressed in sufficient detail.

(1) An initial discussion and review should address the overall concept to provide background for later in-depth discussions.

(a) This should include a discussion of the strategic and operational environment to include threats, allies and partners, and U.S. ability to positively affect U.S. national objectives (U.S. willingness to continue/increase campaign activities).

(b) An assessment of ongoing operations, activities, and investments and their impact on national objectives.

(2) Subsequent discussion might provide more in-depth discussion on key elements of the plan. Examples of detailed discussions for the JPBs and IPRs are:

(a) Deployment concept (access, overflight, anti-access and transload considerations and requirements).

(b) Mobilization timeline.

(c) Decision points and timelines.

(d) Key logistics issues (weapons/precision munitions, resupply, fuel).

[6] CJCSI 3141.01, Management and Review of Campaign and Contingency Plans

(e) Intelligence surveillance, and reconnaissance (ISR).

(f) IO, information infrastructure, and cyberspace operations.

(g) Strategic responses and transition to/from STRATCOM operations while deconflicting/continuing GCC operations.

(h) Force feasibility to include results of latest contingency sourcing, if applicable.

(i) Allied, partner nation, host nation support.

(j) The impact of degraded mobility outputs due to a contested environment on campaign objectives.

g. CCMDs should identify tasks (to include posture changes) that should be added/conducted as part of the GCP/CCP.

(1) Operations, activities and Investments (OAIs) that increase deterrence or better enable execution of the plan.

(2) Action necessary to address assumptions (validate/confirm assumptions).

(3) OAIs that build coalition support and/or capabilities to execute the plan.

(4) Recommendations on what activities should be started, increased, continued, decreased, or stopped.

h. As plan development progresses, CCMDs should continue to conduct additional coordination and reviews. These will be coordinated by the JS J-5 at the request of the CCMD. They include:

(1) JPEC Review (see Section II, *JPEC* of this Chapter and CJCSI 3141,01 *Management and Review of Campaign and Contingency Plans*).

(2) Promote Cooperation events to integrate interagency capabilities. OSD is overall responsible for interagency coordination.[7]

i. Readiness Reviews. Readiness Reviews are detailed reviews of global implications for the Joint Force in the event of conflict. Readiness Reviews begin with an examination of threats, threat capabilities, and likely enemy courses of action. Readiness Reviews then analyze the family of plans related to designated priority challenges to identify friendly resource requirements, potential shortfalls in readiness, resources, and capabilities. The output of a Readiness Review is a Globally Integrated Base Plan (GIBP) that identifies priorities across the Joint Force while in conflict, and outlines POTUS and SecDef decisions for execution.[8]

7. Plan Approval

Upon completion of the review process, plans will be approved as follows:

a. Campaign Plans.

(1) GCPs. The CJCS manages GCPs on behalf of the SecDef. The CJCS approves GCPs on receipt of an endorsement letter from the SecDef.

(2) RCPs and FCPs are managed and approved by the appropriate CA.

(3) CCPs are approved by the CCDR.

b. Contingency Plans.

[7] *See CJCSM 3130.01 Series, Campaign Planning Procedures and Responsibilities and Joint Publication 5-0, Joint Planning for more on Promote Cooperation events*

[8] *A GIBP recommends adjustments to the day-to-day priorities for all CCMDs in the event of a crisis or contingency. The GIBP also identifies Presidential or Secretary-level decisions for execution of the plan. These decisions include activation of the plan, reallocation of strategic assets, and retrograde options. See CJCSI 3141,01 Management and Review of Campaign and Contingency Plans*

(1) SecDef reviews and approves the GIBPs that are the primary branch plans of the GCPs. The Joint Staff conducts Readiness Reviews in support of these reviews as required.

(2) SecDef approves capstone operation plans in complete format/OPLANs and concept plans in concept format/CONPLANs within ICPs for the NDS-directed priority challenges. Reviews of these OPLANs will include, at a minimum, discussion of supporting transportation, cyber, space, strategic, and homeland defense plans.

(3) The SecDef approves level 4 or 3T OPLANS/CONPLANS tasked in the CPG or through the JSCP that are not capstone plans with ICPs for the priority challenges outlined in the NDS.

(4) The CCDRs will approve level 1 to level 3 CONPLANS tasked in the CPG or JSCP and submit for review as required, or if one of the review criteria (paragraph 4.b. above) is met.

c. <u>Process for SecDef Approval</u>. Plans being forwarded to the SecDef or USD(P) should use the following procedures:[9]

(1) When the CCDR determines the plan is sufficiently developed and the plan has completed the JPEC review and analysis, the CCDR signs the plan. SecDef may elect to review the plan for approval.

(2) The CCMD will coordinate with the Joint Staff to ensure JPEC review of the plan is complete prior to submitting the plan to the SecDef for approval. A DASD Socialization and Promote Cooperation will be conducted prior to plan approval. In addition, the USD(P) may host additional meetings to address or resolve significant policy concerns.

(3) DASD(P) will coordinate with JS J-5 to schedule IPRs with the SecDef, DepSec-Def (when requested), or USD(P).

(a) JS J-5 is responsible for scheduling OPSDEP Tank, JCS Tank, and Chairman briefings.

(b) DASD(P), ICW JS J-5, is responsible for scheduling IPRs with USD(P) and SecDef.

(4) The approval packet for the SecDef will be similar to the packet informing the SecDef of planning to date, except that the action memo from the Chairman will request the plan be approved as the plan of record. The approval packet will include an OSD recommendation memo, the CJCS military advice memo, Memorandums for the Record (MFRs) from OSD and Tank meetings, and a slide presentation.

(5) On SecDef approval, OSD will prepare a MFR documenting the SecDef approval of the plan.

8. Transition Reviews[10]

a. During a change in administrations, the CJCS, in coordination with the acting USD(P), prioritizes contingency plans for review by the incoming President and SecDef. The Joint Staff will work with the CCMDs and in-place OSD staff to schedule and prepare these presentations.

(1) As appropriate, presentations may be requested for other incoming administration personnel, such as the National Security Advisor, Secretary of State, Director of National Intelligence and heads of the intelligence agencies, Director of Homeland Security, etc. DASD(P) will be responsible for coordinating these presentations with agencies outside DOD.

(2) Joint Staff J-5 will be responsible for coordinating presentations with appropriate offices and the CCDRs.

---

[9] *CJCSI 3141.01, Management and Review of Campaign and Contingency Plans*
[10] *Ibid*

b. The CA will be responsible for preparing the briefing. The presentation should address timelines, key decisions, and risks associated with the plan.

c. Joint Staff J-5 is responsible for briefing the incoming CJCS. CCMDs may be asked to support with updated material.

d. There is no requirement for Tanks or DASD socializations prior to transition reviews.

e. In addition to those plans identified by the CJCS, CCDRs may recommend plans to the CJCS for updates based on current events.

9. Global and Combatant Command Campaign Plans

a. Global Campaign Plans (GCPs).[11]

(1) GCPs will be reviewed by the JPEC, with the appropriate JS cross functional team (CFT) responsible for coordinating the review, ICW the CA.

(2) GCPs do not require the transportation, sourcing, and logistics assessment as part of their JPEC review.

(3) Typically, GCPs will proceed through several Tank reviews prior to presentation to the SecDef.

(4) GCPs will be subject to annual assessments through routine Joint Staff assessment processes (see CJCSI 3100.01 Series, Joint Strategic Planning System). Additionally, GCPs and updates to the GCPs will be reviewed as directed using the JPB-OPSDEP-JCS Tank process.

b. CCPs may be identified for SecDef review as part of annual senior leadership seminar or conference.

c. GCP reviews should include:

(1) A scene setter (J2 lead) to include an assessment of the threat's strategic and operational goals.

(2) Review of the GCP to include:

(a) CA's summary recommendations.

(b) An overall assessment of the GCP and status by NMS mission area and LOE.

(c) CA's risk assessment.

(d) Key limitations and shortfalls in execution.

(e) Strategic opportunities.

(3) Campaign support to contingency preparedness and transition.

(a) Summary of the GCP's support to contingency preparedness to include deterrence, support to indications and warnings, and achievement of national objectives.

(b) Overview of the contingency challenge (priority threat overview).

(c) How current campaign OA is are setting conditions and validating assumptions for contingency plans.

(d) How the GCP would transition to contingency operations, to include a decision support template/timeline.

(4) Way Ahead.

(a) Recommended strategic prioritization based on the current assessment of the threat and operational environment.

(b) Proposed changes to the GCP.

1 Recommended changes to ongoing and planned OAIs and the related resource impacts.

[11] CJCSI 3141.01, Management and Review of Campaign and Contingency Plans

$\underline{2}$ A discussion of current and proposed posture, to include agreements, authorities, and permissions as well as forward-based force structure (permanent and rotational).

d. To support the annual formal SecDef campaign plan and security cooperation plan review, the J-5 will facilitate an informal socialization of CCPs with all of the CCMDs, JPEC, and OSD during the spring Joint Strategy Working Group (JSWG). The intent is to integrate plans by focusing on gaps, seams, and resources across CCMDs.

e. *JS J-3 will assess GCPs and CCPs for sourcing feasibility*. JS J-3 and JS J-8 will use GCPs and CCPs to inform assignment and allocation input to the GFMAP and GFMIG.

10. Review Synchronization

<u>Scheduling and Battle Rhythm</u>. Plans require periodic updates to account for changes in strategic documents (e.g., NSS, NDS, NMS, CPG, etc.). As strategic guidance influences military planning, the Joint Staff will coordinate changes with the JPEC prior to CJCS approval.

a. GCPs, FCPs, and RCPs are assessed and updated through the Tank process with input from the JPEC and OSD. The Chairman approves all GCPs after endorsement by the SecDef. The CAs approve FCPs and RCPs.

b. The Joint Staff maintains a Joint Force strategic battle rhythm (Joint Force Global Integration Calendar (JFGIC), aligning the SecDef required reviews with the JSPS, other DOD calendar requirements, and leadership availability.

(1) The battle rhythm commences with GCP assessment, review, and approval Tank series. Typically, these assessments build through a sequence of events, such as JPB (0-6 and 0-7 /0-8 level), OPSDEPS Tanks (0-9 level), and JCS Tanks (0-10 level). Some campaign plans may require additional OSD level review.

(2) *The Joint Staff will assess the GCPs at least once per year* and execute additional Tank reviews, such as a Strategy Integration Tank, to address crosscutting issues as needed. JS J-5 employ's the JSWG and JWPS to gain alignment on products and processes for Global Integration.

(3) The CAs will begin to assess RCPs, FCPs, and Defense Critical Missions when those plans are approved; JS J-5 will schedule reviews of those plans as required. The review process results in refinement of the strategy, plans, resources, investments, prioritization, risks, and force development when the findings are pursued through the Planning, Programming, Budgeting and Execution System (PPBES) and GFM processes.

c. Coordinating authorities review and submit their refined campaign plans to the Global Integrator at their respective Tank series (GCP Tank or FCP, RCP, and others at JSCP Tank).

11. Commander's "RATE" Recommendation

As part of the planning process, CCDRs regularly review their library of plans and may identify plans that are no longer current, appropriate, or required. The CCDR makes a recommendation through the JS J-5 to OSD to the SecDef to refine, adapt, or terminate the plan. In specific cases (most often with the campaign plan), the CCDR may also recommend execution of the plan or parts of the plan.[12]

a. <u>Refine</u>. Refinement focuses on additional planning to produce more defined estimate for decision-makers. Refinements for contingency plans are often based on decision by senior leaders for greater detail in the plan; changes in force apportionment, structure, or capabilities; updated intelligence; or changes in strategic guidance. As a CDR refines the

[12] CJCSI 3141.01, *Management and Review of Campaign and Contingency Plans*

plan, there may be changes in timing and decisions points and assumptions. Changes in forces and objectives should not occur in refinement. Refinements to a plan to maintain it in a "living" state usually with the same or similar military objectives or endstates.

(1) Plans recommended for refinement usually do not require formal IPRs. If a paper review is approved, the SecDef or USD(P) approves the refinements through normal coordination mechanisms.

(2) If a paper review is not approved, J-5/Joint Strategic Planning (JSP) and DASD(P) will direct that the review be accomplished with an IPR and will schedule the meeting accordingly.

b. <u>Adapt</u>. Plans should be adapted if the requirement for the plan remains still valid, but significant changes are required. Major changes to force structure, timing, or CONOPS (such as using a different operational approach) generally required adaptation. Adaptation is also appropriate if new or additional objectives are needed. Adaptation is often due to changes in the operational environment or strategic guidance and may require a new mission analysis. Successful execution of campaign plans, which change the starting conditions of a contingency, may also lead to an "adapt" recommendation. Plans undergoing adaptation should be reviewed similar to a new plan.

c. <u>Terminate</u>. CDRs should recommend terminating a plan when the plan may be no longer required. There are several conditions when this may occur or be subsumed into another plan.

(1) Plan subsumed into another plan (cancel plan number). When the plan is subsumed into another plan, the CDR sends a memorandum through the JS J-5/JSP for the Chairman and the SecDef, identifying the plan number that will be cancelled and where the plan information is currently located (e.g., "Plan number XXXX-YY has been superseded [or "is now included with"] by plan number NNNN, effective DTG.") The memorandum identifies the effective date, the new plan number, and identifies that the old plan number is cancelled. J-5/JSP will staff the memorandum by the JSAP process through the JPEC for information only. There is no additional action required, since all the plan requirements are maintained.

(2) CPG- or JSCP-tasked plan. If the CCDR determines that a CPG- or JSCP-tasked plan is no longer required, the CCDR should request permission to archive the plan. The request should include the rationale behind terminating the requirement and forwarded through the JS J-5/JSP for staffing. JS J-5/JSP will staff the request through the JPEC to identify if there are any secondary issues associated with the termination. CPG-tasked plans will be forwarded through DASD(P) for SecDef approval with a Chairman recommendation. JSCP-tasked plans will be sent to the Chairman for decision. If approved, JS J-5 will prepare a memorandum notifying the JPEC of the approval to archive the plan.

(3) <u>CCDR Plan</u>. Numbered plans initiated by a CCDR can be terminated by that CCDR. A memorandum stating that the plan has be archived or cancelled will be sent to the JS J-5/JSP, who will notify the JPEC by an "information only" JSAP.

(4) <u>Support Plan</u>. CCDRs recommending termination of a support plan to another CCDR's plan should notify that CCDR of the recommendation and prepare a co-signed memorandum of agreement on the termination. The cosigned memorandum is then forwarded to the JS J-5/JSP for staffing as an "information only" JSAP.

d. <u>Execute</u>. The SecDef, by direction of the President, determines execution of plans. CCDRs make recommendations through normal command channels.

II. JPEC

1. Overview

a. <u>Purpose</u>. The JPEC review process is a CJCS-directed process, managed by the J-5, which assists the Chairman in providing military advice to the SecDef, the President, and civilian leadership while assisting the CCMDs in integrating planning across domains, functions, and regions. It also ensures that plans submitted to the SecDef and President are executable within current policy and resource constraints.

b. The JPEC creates a shared understanding, synchronizes efforts, develops a way ahead for integrated products, and establishes the optimal confluence of military plans, operations, and strategy.

c. Plan reviews will take the larger problem sets into account. JPBs will address identified seams, threats, risks and mitigation, resource requirements, current and potential issues, strategic deconfliction, and areas for collective action.

2. JPEC Process[13]

a. The headquarters, commands, and agencies involved in joint planning or committed to a joint operation are collectively termed the JPEC. The JPEC is not a standing or regularly meeting entity, but consist of the plan stakeholders. During the review process the JPEC will complete a JSAP on all supported and supporting plans. (*JSAP enables the JPEC to coordinate on the review of a plan. This process supplements the JPBs, OpsDeps, Tanks, and informal dialogues and acts as a forum to review and adjudicate the detailed and technical aspects of a plan that might not arise in other review activities and discussions.*) Prior to submitting a contingency plan for approval, the CCMD should ensure a full JPEC review and analysis is conducted.

b. During the plan development and plan analysis, the supported CCDR, in coordination with JPEC stakeholders, conducts deployment, redeployment, employment, logistics, transportation, and sustainment planning; contingency sourcing; comprehensive feasibility assessments (to include impact on operations and global environment due to adversary action or other global disruptions); and other actions pursuant to guidance and direction received.

c. Plans submitted to the JPEC for review should identify all related CCMD level plans (e.g., functional plans, supporting or adjacent CCMD plans) where a reasonable expectation exists that plan execution will occur in coordination with (ICW) or following the plan under review.

(1) As much as practicable, plans expected to be executed ICW each other should be reviewed together to ensure the plans are integrated (e.g., timelines and resources requirements deconflicted).

(2) Coordinating authorities are responsible for ensuring this integration for contingency plans associated with their campaign plan(s).

d. <u>Plan Integration.</u> The Joint Staff uses several methods to aid in global integration in support of the Chairman.

(1) <u>Tanks.</u> (OPSDEP and JCS)

(a) The intent of the Tank Series is to focus on problem sets over numbered plans to inform the CJCS and JCS for future decision-making and for developing military advice pertaining to military policy recommendations and strategic guidance on directing, employing, managing, developing, and assessing. Senior leader input and decisions should focus on these matters rather than the specifics of war plan execution. Tanks will focus on strategic issues and risks that are specific to the problem set; beginning with a brief focused on adversary objectives and capabilities that threaten U.S., Allies interests and policy objectives, an allied and adversary comparisons, short and long-term capability comparison and strategic gaps. They can be held to cover plan information in general or to address specific issues that require resolution at higher levels.

[13] *CJCSI 3141.01, Management and Review of Campaign and Contingency Plans*

(b) OPSDEPs are chaired by the DJS or DJS (for planning OPSDEP) and held at the 3-star level, with attendance from the J-5 or J-3 of CCMDs, Service, NGB, and agency counterparts. For plan OPSDEPs, an OSD representative may also be invited.

(c) JCS Tanks are chaired by the Chairman or DJS with attendance by CCDRs, Service Chiefs, Chief NGB, and agency directors.

(d) CJCSI 5002.01 Meetings in the JCS Conference Room, provides additional information on Tanks.

(2) Joint Planning Boards (JPBs). JPBs are held at the AO to deputy director-level (2-star) to conduct planning coordination and resolve planning issues at the lowest possible level. JPBs can be used to coordinate timelines and resource distribution between CCMDs that may be required to execute plans simultaneously.

(3) Counterpart Meetings. The Director Joint Staff (DJS), Deputy Director for Joint Strategic Planning (DDJSP), and Chief, Joint Operational War Plans Division (JOWPD) meet regularly with their counterparts in OSD to ensure continuous interchange of information and address issues that may require resolution outside of, or prior to, formal meetings.

e. JPEC will review plans using the six criteria below:

(1) Suitable. The scope and concept of planned operations can accomplish the assigned objectives and are within the planning guidance. Planning assumptions must be logical, realistic, and essential for continued planning.

(2) Feasible. The assigned mission can be accomplished using available resources within the time contemplated by the plan.

(3) Acceptable. The possible strategic and operational advantage gained by the operation meet or exceed the estimated cost and risk. It focuses on the level of risk to mission accomplishment, and whether the proposed plan is consistent with domestic and international law, including the law of war, and is supportable.

(4) Complete. The plan incorporates all assigned tasks and to what degree they include forces required, deployment concept, employment concept, sustainment concept, and time estimates for achieving IMOs corresponding to campaign objectives, description of the end state, mission success criteria, and mission termination criteria. Mission objectives, end-state, termination criteria include clear, objective measurable assessment criteria for decision makers.

(5) Distinguishable. The plan must be sufficiently different from the other courses of action to ensure that OSD leadership possesses a range of flexible military options.

(6) Integrated. The plan must integrate across CCMDs: regions, functions, and domains. CCDRs must ensure plans, including support plans, are multi-national, trans-regional, all-domain and multifunctional and synchronized across time, space, and forces.

f. To accomplish the assessment, J-5/JSP collaborates with the supported CCMD, JFPs, USTRANSCOM, CSAs, and Joint Staff directorates to collect and analyze plan assessment data to finalize a briefing product.

3. JPEC Completion

a. Once the CA, supported command, Service, or CSA completes a final adjudication of the plan, they provide a planner level memorandum to J-5/JSP stating the JPEC review is complete and all critical comments are properly adjudicated. Subsequent discussions must highlight issues of nonconcurrence. Following JPEC review, the CJCS may hold an Ops-Deps and/or JCS Tank for the CCDR to present the plan before briefing it to the USD(P), DepSecDef, or SecDef. Upon completion of the Tank (if required), the CCDR presents the plan to the USD(P), DepSecDef, or SecDef, as directed.

Review CJCSI 3141.01, Management and Review of Campaign and Contingency Plans, for a memorandum example and for greater details on the Plan Review Process and the JPEC.

III. Plan Assessment (Function IV)

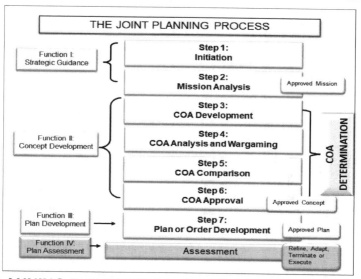

THE JOINT PLANNING PROCESS

Function I: Strategic Guidance
- Step 1: Initiation
- Step 2: Mission Analysis — Approved Mission

Function II: Concept Development
- Step 3: COA Development
- Step 4: COA Analysis and Wargaming
- Step 5: COA Comparison
- Step 6: COA Approval — Approved Concept

COA DETERMINATION

Function III: Plan Development
- Step 7: Plan or Order Development — Approved Plan

Function IV: Plan Assessment
- Assessment — Refine, Adapt, Terminate or Execute

I. PLANNING FUNCTION IV – PLAN ASSESSMENT

1. Overview

Plan assessment is a continuous activity of the operations process and a primary feedback mechanism that enables the command as a whole to learn and adapt. Effective assessment relies on an accurate understanding of the logic used to build the plan. Plans are based on imperfect understanding, assumptions and an operational approach on how the CDR expects a situation to evolve. The reasons or logic as to why the CDR believes the plan will produce the desired results are important considerations when determining how to assess the operation/plan. Continuous assessment helps CDRs recognize shortcomings in the plan and changes in the OE. In those instances when assessment reveals minor variances from the CDR's visualization, CDRs adjust plans as required. In those instances when assessment reveals a significant variance from the CDR's original visualization, CDRs reframe the problem and develop an entirely new plan as required.[1]

a. Plan assessment is part of planning and the plan review process (Chapter 6-II, *Plan Review Process*). Effective plan assessment measures progress toward mission accomplishment (achieving IMOs as applicable and progress toward endstates), identifies changes in the operational and strategic environment, and risk associated with the potential requirement to execute contingency plans. Accordingly, assessment considerations should:

- Be developed in concert with mission success criteria;
- Help guide operational design of campaign and contingency plans;
- Employ common methods that can be developed and applied across all planning and assessment requirements, and be briefed during plan reviews.

[1]FM 5-0 Overview, Combined Arms Doctrine Directorate

b. CCDRs are tasked to develop campaign plans that integrate security cooperation and other foundational activities with operations and contingency plans IAW the strategic policy guidance provided by the CPG and JSCP. Campaign plans also provide for conducting a comprehensive assessment of how the CCMDs activities are contributing to the achievement of IMOs, and how those activities best deter, shape, or mitigate the potential to execute assigned plans. Accordingly, plan assessments should:

- Provide the basis for the Refine, Adapt, Terminate, Execute (RATE) recommendation during IPRs.
- Ensure that assessment of subordinate campaign and contingency plans nest under the assessment of the CCDR's CCP, as well as the FCP they support. This nesting provides the mechanism to synchronize assessment activities across the CCDR's planning requirements and eliminate redundant or contradictory activities.

2. Plan Assessment

The focus of assessment differs during **planning, preparation, and execution**. During *planning*, assessment focuses on gathering information to understand the current situation, the framed problem, and outputs of design methodology to develop an assessment plan. During *preparation*, assessment focuses on monitoring changes in the situation and on evaluating the progress of readiness to execute the operation. Assessment during *execution* involves a deliberate comparison of forecasted outcomes to actual events, using indicators to judge progress toward attaining desired endstate conditions and help CDRs adjust plans based on changes in the situation, when the operation is complete, and when to transition into the next cycle of the operations process.

a. Plan Assessment. Plan Assessment deliberately measures a **completed plan's** effectiveness in accomplishing prescribed objectives. A plan assessment may also identify changes in the OE or strategic direction that may impact the plan, and provide risk-informed recommendations to senior leaders regarding subsequent planning or execution options. As mentioned, plan assessment is also part of the plan review process and is orchestrated by the JS J-5 in support of the Chairman's statutory responsibility to review and assess plans.[2]

(1) Plan Assessment Concurrent to Plan Development. During plan development, a plan or order is analyzed for its feasibility and the plan refined to address identified shortfalls. When planning under crisis conditions, this feasibility analysis may supplant a more deliberate assessment of the completed plan. Under these circumstances, the results of the feasibility analysis are shared by the supported CCDR with the JS and OSD in order to expedite plan approval by the Secretary. If a crisis situation does not prompt immediate execution of a plan or order, a more deliberate plan assessment maybe conducted following plan development.

(2) Plan Assessment Outcomes. The results of a plan assessment may lead to a decision to pursue one of four outcomes for the plan: *refine, adapt, terminate, or execute (RATE)*. All four outcomes can be applied to contingency plan assessments while campaign plan assessments generally do not consider termination as they are in constant execution.

"It is the mark of an educated mind to be able to entertain a thought without accepting it."

— *Aristotle*

---

[2] *CJCSI 3141.01 Management and Review of Campaign and Contingency Plans*

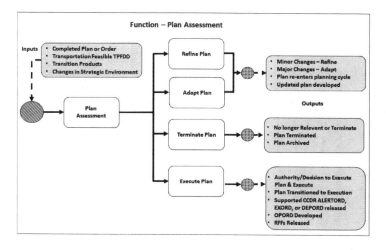

Plan Assessment Inputs and Outputs

(3) Refine, Adapt, Terminate, Execute (RATE). CDRs continually review and evaluate the plan; determine one of four possible outcomes: refine, adapt, terminate, or execute; and then act accordingly. CDRs and the JPEC continue to evaluate the situation for any changes that would require changes in the plan. The CCDR will brief SecDef during routine plan update IPRs of modifications and updates to the plan based on the CCDR's assessment of the situation, changes in resources or guidance, and the plan's ability to achieve the objectives and attain the endstates.[3]

(a) Refine. During all planning efforts, plan refinement typically is an orderly process that follows plan development and is part of the assessment function. Refinement is facilitated by continuous operation assessment to confirm changing OE conditions related to the plan or potential contingency.

(b) Adapt. Planners adapt plans when major modifications are required, which may be driven by one or more changes in the following: strategic direction, OE, or the problem facing the JFC.

(c) Terminate. CDRs may recommend termination of a plan when it is no longer relevant or the threat no longer exists. For CPG- or JSCP-tasked plans, SecDef, with advice from the CJCS, is the approving authority to terminate a planning requirement.

(d) Execution. Execution begins when the President or SecDef authorizes the initiation of a military operation or other activity. An execute order (EXORD), or other authorizing directive, is issued by the CJCS at the direction of the President or SecDef to initiate or conduct the military operations.

> "Even the finest sword plunged into salt water will eventually rust."
> — Sun Tzu

---

[3] JP 5-0, Joint Planning

3. Campaign Plan Assessments

a. The purpose of the campaign assessment is to enable the CCDR and supporting organizations to refine or adapt the campaign plan and supporting plans to achieve the campaign objectives; or, with the Secretary, to adapt the CPG/JSCP-directed objectives to changes in the strategic and operational environments. *The assessment of a campaign plan is largely based upon the operation assessment of its continued execution.* Its purpose is to inform the civilian-military dialogue and IPR process considering changes to current operations or planning. This dialogue is also informed by the results of other campaign plan assessments and the Chairman's Comprehensive Joint Assessment (CJA) which is developed annually and integrates the campaign assessments from the CCDRs and Service Chiefs. The CJA provides the Chairman's directed independent analysis of campaign plan execution and the strategic environment. The Chairman's analysis informs the IPR process and decisions by the Secretary.

b. The campaign assessment is the CCMD's feedback mechanism from campaign execution to campaign planning. It should indicate where the CCMD's ways and means are sufficient to attain their ends, where they are not, and why they are not. For planners:

- How might the command modify its plan's ends, ways, or means in the next cycle of planning?
- Is the plan still suitable to achieve the objectives?
- Are the objectives achievable given environmental changes and emerging political issues?
- Are the assumptions still valid?
- To what degree are the resources employed making a difference in the operational environment?

c. The campaign assessment is also the Department's bridging mechanism from the CCDR's strategy to the strategic, resource and authorities planning processes, informing the Department's strategic direction, assignment of roles and missions, force employment, force posture, force management and force development decision-making.

d. The campaign assessment provides the CCDR's input to the Department on the capabilities needed to accomplish the missions in the contingency plans of their commands over the planning horizon of the CCDR's strategy, taking into account expected changes in threats and the strategic and OE.

e. Overall assessments enable the CCDR to make the case for additional resources or to recommend re-allocating available resources to the highest priorities; and the Secretary and senior leaders to do the same across all CCMDs and to make the case to Congress to add or re-allocate resources through the Future Years Defense Program (FYDP).[4]

Further discussion of campaign plan assessments can be found in CJCSM 3130.01 Series, *Campaign Planning Procedures and Responsibilities*, JP 5-0, *Joint Planning*, and CJCSM 3130.03 Series, *Adaptive Planning and Execution Planning Formats and Guidance*.

4. Assessment Plan Developed in Concert with Campaign Plan

As discussed in Chapter 5-I(b), *Plan Initiation*, planners develop the campaign's *operational assessment plan*, prepared as an annex or appendix of the campaign plan, as they develop the campaign's operational approach. Defining how to measure progress in the

[4] CJCSM 3130.01 Series, *Campaign Planning Procedures and Responsibilities*

campaign helps guide the design of the campaign (and its contingency plans). It also provides a basis, although conditions may have changed or further developed for a *completed plans* assessment during this function of Plan Assessment.

a. Assessment metrics are established concurrently with campaign objectives. Objectives must be achievable and measurable, and stated with a way to measure their achievement. MOPs determine task performance and MOEs evaluate changes in conditions that indicate progress toward objectives. MOPs answer questions such as "was the action taken, were the tasks completed to standard, or how much effort was involved?" MOPs are associated with task accomplishment. MOEs address questions such as, "are our actions producing the desired effects, are we doing the right things to accomplish the objective, or are alternative actions required?"

b. The assessment plan describes how the CCMD will employ manpower, methods, and money consistently over the course of the campaign to document progress toward campaign objectives, the effectiveness of the campaign's ways and means, and changes in the environment. The assessment plan should include data requirements and collection plans; including collection assets tasked and resources requested to collect on the requirements and to analyze the collected data to measure changes. Include assessment tasks in the annual campaign order.

c. Assessment responsibilities should be decentralized to the same extent as the campaign. CCMDs may be conducting security cooperation with multiple security partners, guiding the development of many different military capabilities, and working with many functional experts across the CCMD staff and in its component, subordinate and supporting commands. Assessments utilize data from intelligence collection efforts, Service and special operations components, subordinate and supporting commands, and country teams, all of whom report progress toward campaign objectives to the CCDR throughout the year.

d. CCMDs conduct a comprehensive analysis of their ongoing assessments at least once per year, using their own methodology. The format for this is not prescribed. The following model provides a way to organize some of the key information:

(1) A summary of the previous year's Strategic Environment Estimate/CJA.

(2) A list of campaign objectives prioritized by:

-Strategic (CPG/JSCP) priorities, impact to the National Security Strategy, the prioritized operations, activities, and investments needed to accomplish each objective, resources required per operation, activity, and investment, an estimate of risk incurred if specific operations, activities, and investments are not resourced.

5. Reporting the Assessment

a. CCMDs are required to report their campaign assessments annually via the CJA. The CJA, which includes the Services' assessments of their support to the CCMDs, is used to develop reports such as the Chairman's Risk Assessment (CRA) and the Joint Logistics Estimate, to inform the Program Budget Review (PBR) and Service programmatic planning for the next FYDP, and to enable the Joint Staff and OSD to substantiate policy and programmatic decisions.

b. Each year, the CJA instructs CCMD planners and assessors on the conduct of that year's survey. These questions are representative of the responses solicited in a CJA survey:

- Did the CCMD advance toward accomplishing its military objectives?
- Did those advances also indicate progress toward achieving the objective(s) they were designed to support?
- Did the CCMD reduce the likelihood of crisis and the necessity to execute

its contingency plans?

- How did resource constraints limit progress?
- What is the newly assessed risk to mission the CCMD assumes by falling short of its objectives?
- In the context of the strategic environment, what strategic risks does the USG assume and how are those risks affected by the campaign?
- What will mitigate plan shortfalls and address unacceptable risk?
- Are the planning assumptions still valid? Have some of the assumptions been proven to be facts? Do additional assumptions have to be made to continue planning or operations?

c. Military Departments and Services assess their Campaign Support Plans (CSPs) per the CJA. Upon request, Service component commands will provide their Services with the same data provided to the CCMDs to facilitate the Services' assessments of their CSPs.

See CJCSI 3110.01 Series, *Joint Strategic Capabilities Plan (JSCP) and Supplementals, Title 10 U.S.C.,* sections 113 and 153, *CJCSI 3141.01 Management and Review of Campaign and Contingency Plans,* for additional guidance on campaign assessments and reporting.

6. Assessments and Campaign Plan Reviews

a. CCDRs use assessments to brief progress toward Strategic-directed objectives during annual campaign plan reviews.

b. CCMDs may request a biannual campaign plan review after they have completed an initial assessment. Requests for biannual review will be submitted through the Joint Staff J-5 to the DASD(P), for approval by the (USD(P)).

c. The CCDR's assessment of changes to the environment or emerging political issues may necessitate dialogue with Department senior leaders more frequently than the scheduled review or out-of-cycle.

d. Campaign plan assessments provide a framework for the assessment of a contingency plan. The contingency plan's assessment activities should be synchronized with those of the associated campaign plan. Additionally, the results of a campaign plan assessment inform the review of a contingency plan and shape a potential RATE decision.

7. Global Campaign Plans (GCPs)

The Coordination Authority (CA) is responsible for GCP assessments. These assessments use data and information provided by the CCMDs, Services, NGB, defense agencies, and other Joint Staff assessments to provide the Chairman updates on the GCPs. Joint Staff cross-functional teams (CFTs) support the CA assessments by coordinating JS expertise and inputs, working with OSD, and providing cross GCP insights to the CA to support the analysis. CFTs will use the IPR process (JPBs, Tanks, etc.) to aid the CA in presenting the assessment to the CJCS and SecDef and provide recommendations for the Chairman's advice to the SecDef and President. GCP assessments support the Chairman's role as Global Integrator and other Joint Staff process in the JSPS. Joint Staff CFTs lead GCP assessments.

8. Combatant Command Campaign Plans (CCPs)

a. Annual assessments of CCPs stem from the initial or previous year's assessment to help track overall progress towards achieving IMOs and campaign objectives.

b. Campaign plan assessment should follow a common approach such as that suggested in JP 5-0, *Joint Planning*.

c. The Annual Joint Assessment (AJA) is the primary feedback mechanism for CCPs. CCDRs provide a summary of changes from the current year's AJA responses and recommendations to Joint Staff J-5. These inputs are also used in the GCP assessment.

(1) The Joint Force, through the AJA and Tank series, assess the JSCP plans throughout the fiscal year. CCMDs submit integrated assessments to the CJCS per Joint Staff guidance as part of the AJA.

(2) The integrated assessments evaluate GCP, RCP, and FCP: strategic environment changes, intermediate objectives and progress, risk, priorities, command relationships, resourcing, authorities, EXORDs, posture, and opportunities.

d. CCDRs integrate key recommendations into CCMD AJA responses.

e. When requested, CCP IPRs include campaign plan recommendations as deemed appropriate by CCDRs. Recommendations are narrative text and:

(1) Provide a focused statement clearly defining the responsible DoD or IA office, action required, and timing required for new actions or authorities.

(2) Include a POC from the submitting command to provide more information as required on the recommendation.

(3) Clearly identify findings that support the recommendation.

(4) Include a narrative that identifies the IMOs affected through the recommendation implementation process.

(5) Identify related campaigns or contingency plans affected by the recommendation.

(6) If appropriate, categorize the recommendation within the Doctrine, Organization, Training, Materiel, Leadership and Education, Personnel, Facilities, and Policy (DOTMLPF-P) framework.

9. Contingency Plan Feasibility Analysis

a. <u>Contingency Plan Review</u>. CPG or JSCP-tasked contingency plans are normally reviewed annually or as directed. The JS J-5 ICW USD(P) develops the schedule for plan reviews. Reviews of contingency plans for integrated planning against a problem set may involve supported and supporting CCMDs. Informed by the contingency plan assessment to include the results of the Joint Combat Capability Assessment (JCCA) (a more detailed assessment of the ability to execute a selected contingency plan or group of plans) plan assessment and overarching campaign assessment, the supported CCDR provides recommendations for a RATE decision. The outcomes of contingency plan review shape future planning and may include consideration of developing branch or sequel plans.

b. Reviews will ensure planning assumptions are reasonable, valid, and comply with strategic guidance including the NSS, UCP, CPG, and the NDS and its implementation guidance (NMS and JSCP).

c. Contingency plan reviews measure the Department's ability to successfully execute contingency plans with the highest visibility or having the most severe consequences, as well as those most stressing to ground, maritime, air, space and special operations forces.

d. CCDRs designated as CAs or as supported commands are responsible for conducting the plan assessment for contingency plans prior to submitting them to the SecDef for approval. Level 3T and 4 plans will undergo a full analysis, while other plans will be reviewed by the JPEC through the Joint Staff Action Process (JSAP) staffing process (See CJCSM 3141.01 *Management and Review of Campaign and Contingency Plans*).

10. Summary

During this function, the CCDR refines the complete plan while supporting and subordinate commanders, Services and supporting Agencies complete their supporting plans for his review and approval. All CDRs continue to develop and analyze branches and sequels as required or directed. The CCDR and the Joint Staff continue to evaluate the situation for any changes that would trigger plan revision or refinement. The CCDR may conduct reviews (IPR) as required during plan assessment.

Assessment is the determination of the progress toward accomplishing a task, creating a condition, or achieving an objective. **Assessment is a continuous activity** of the operations process that supports decision-making by ascertaining progress of the operation for the purpose of ***developing and refining plans*** and for making operations more effective. Assessment results enhance the CDR's decision making and help the CDR and the staff to keep pace with constantly changing situations. Assessment involves deliberately comparing intended outcomes with actual events to determine the overall effectiveness of force employment. More specifically, assessment helps the CDR determine progress toward attaining the desired endstate, achieving objectives, and performing tasks.[5] (Figure below displays the continual assessment of the operations process.

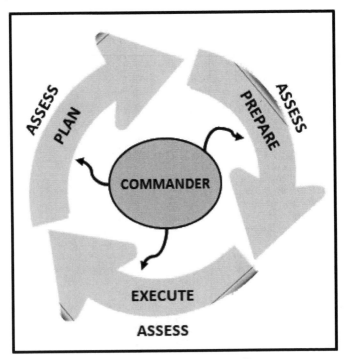

Continual Assessment of Operations Process

> *"Just because you made a good plan, doesn't mean that's what's gonna happen."*
> —Taylor Swift

[5] *FM 5-0, Planning, Orders and Procedures*

I. Execution Functions

1. Function Overview

Execution functions are the required elements for implementing military activities as directed by the President or Secretary. Execution, in this context, applies to the range of military activities including but not limited to operations, exercises, and security cooperation. There are seven functions that comprise Execution:

- allocation
- mobilization
- deployment
- distribution
- employment
- re-deployment
- de-mobilization

While depicted sequentially in Figure A below these functions can be accomplished in parallel and steps can be combined or truncated depending on the time available. During crisis, execution functions may be conducted in parallel with planning functions to rapidly respond to an emergent event. Throughout execution, the operational activities (situational awareness, planning, and assessment) continue and execution functions (Figure B) are adapted to changes in the OE. This Chapter discusses each execution function.

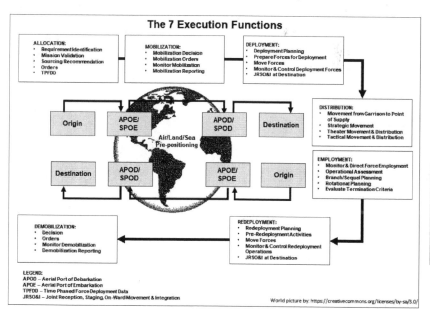

Figure A. The Seven Execution Functions

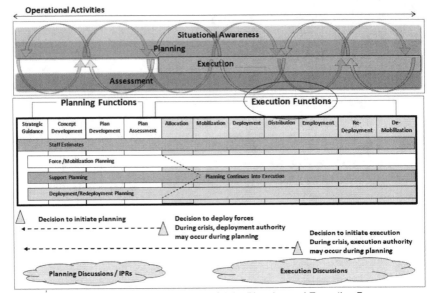

Figure B. Execution Function of Joint Planning and Execution Process

2. Execution Function - Allocation

Allocation, via GFM, is the Secretary's distribution of limited forces or individuals for employment among competing CCMD requirements that cannot be met with assigned or previously allocated forces. Allocation and other sourcing methodologies (e.g., joint exercises, security assistance programs) provide available forces to CCMDs based on stated capability requirements, balanced against risks (operational, future challenges, force management, institutional) and strategic and operational priorities. Figure C below depicts the elements in the allocation execution function.[1] The GFM allocation process is discussed in detail in Chapter 3, *GFM*.

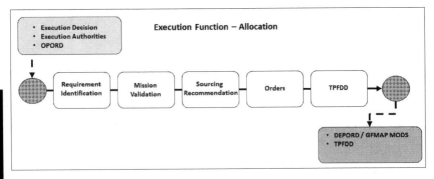

Figure C. Execution Function - Allocation

---

[1] *CJCSM 3130.06 Series, Global Force Management Allocation Policies and Procedures*

a. <u>Requirement Identification</u>. This step comprises the identification of CCMD force requirements for approved military activities to include operations, exercises, and security cooperation. The process for identifying force requirements depends upon the nature of the activity to be conducted.

(1) <u>Assigned Forces</u>. During execution, the supported CCDR may task their assigned forces to fill force requirements in order to perform authorized missions. CCDRs exercise COCOM over assigned forces and employ them for missions, operations, and activities they have authority to execute. These requirements constitute the assigned force demand and are documented in the Joint Capabilities Requirements Manger (JCRM) (the program of record for enabling the GFM allocation process) by the supported CCDR. If additional forces are required, the supported CCDR requests those forces through the GFM allocation process via the annual submission or a RFF. Under crisis conditions assigned forces may be the most responsive to an emergent crisis and may be execution sourced by the supported CCDR to conduct authorized operations.

(2) <u>Operational Requirements</u>. The GFM allocation process is designed to distribute forces to meet CCMDs' force and individual requirements that cannot be met with assigned and previously allocated forces. CCMDs submit annual force submissions or an RFF. The Secretary's decision to allocate forces to operational requirements involves weighing the FPs' risks of sourcing with operational risks to both current and potential future operations.

(a) <u>Security Cooperation Requirements</u>. For security cooperation requirements, CCDRs may employ assigned or allocated forces. Forces conducting approved security cooperation events are either allocated via the GFMAP or documented in the GFMAP as assigned force demand. Separate from the allocation process, Services may provide forces or personnel to support security assistance programs with consideration of Service capacity, Service priority or equity in the security assistance program, and CCMD priorities.[2]

(b) <u>Counter Drug/Counter Narcotics-Terrorism Requirements</u>. CCDRs may employ assigned or allocated forces to conduct approved CD/CNT activities. Forces conducting approved CD/CNT activities are either allocated via the GFMAP or documented in the GFMAP as assigned force demand.[3]

(c) <u>SOF training with foreign forces</u>. Execution sourcing of SOF for JCET events requires specific coordination and approval. Coordination of SOF participation in JCET events is conducted by USSOCOM and execution sourcing of forces is documented in the GFMAP Annex C for allocation.[4]

(d) <u>Joint Individual Augmentation (JIA)</u>. As a part of operational force requirements, planners may identify the need for military or DOD civilians to augment existing staff capability. JIAs are allocated via GFM and the Secretary's decisions for allocation are documented in Annex D of the GFMAP.[5,6]

(e) <u>Defense Support to Civil Authorities (DSCA)</u>. Forces allocated through GFM to support a DSCA mission may include active component (AC) or mobilized RC forces operating under U.S.C., Title 10 authorities. Separately, state or territory governors may employ their own National Guard forces or the National Guard forces of other states or territories ICW their respective governors under U.S.C., Title 32. These National Guard forces operating under U.S.C., Title 32 are not allocated through GFM, but should influence the number and capabilities of U.S.C., Title 10 forces requested via GFM allocation.

(f) <u>GFM Allocation during Crisis</u>. Under crisis conditions, the allocation of forces may occur simultaneous to the decision to initiate execution. A decision from the President or Secretary to execute may be accompanied by SecDef direction to allocate specific forces.

[2] *DoDD 5132.03, DOD Policy and Responsibilities Relating to Security Cooperation*

[3] *CJCSM 3130.06 Series, Global Force Management Allocation Policies and Procedures*

[4] *Ibid*

[5] *Ibid*

[6] *CJCSI 1301.01 Series, Joint Individual Augmentation Procedures*

Changing indications and warnings may require adjustments to force requirements. To support rapid plan execution, CCDRs define force requirements at the unit level so the FP community can quickly identify forces for allocation decisions, with associated risk, by the SecDef while simultaneously supporting the deployment planning and execution process for the supported CCMD.

(3) Exercise Requirements. Exercise requirements are sourced separately from the GFM allocation process. A CCMD conducting an exercise may employ assigned forces or request participation by Service or DOD Agency forces/personnel. The Services provide forces to CCMD exercises considering their capacity to support and their own require-ments to conduct training. FP commitments to provide forces are documented in the Joint Training Information Management System (JTIMS), a tool that automates and supports the global Joint Training System. Service forces participating in CCMD exercises normally do not transfer OPCON, but are under tactical control to the GCC hosting the exercise for the purpose of conducting that exercise and force protection.

b. Validation. After CCMDs submit their operational force requirements, the JS JFC validates each force request against Secretary established criteria in the GFMIG to enforce procedures and verify authorities. The JS JFC will also assign the appropriate priority to each request. The JS J-1 conducts a similar validation process for JIA requirements.[7]

c. Sourcing Recommendations. Once force requirements are validated, the JS JFC and JFPs coordinate with the FPs to evaluate sourcing options and risks, and recommend a sourcing solution for each requirement. The JS JFC staffs the joint sourcing recommenda-tions with the supported CCMD and FPs to ensure risks are accurately presented prior to forwarding the sourcing recommendations in the form of a draft order (also known as GFMAP Modification) to the Secretary via the SDOB.

(1) Delegated Force Transfer Authority. The Secretary may approve limited authority for the short-duration transfer of small units between CCMDs. This authority and the proce-dures to exercise this authority is detailed in the GFMIG. There are also standing EXORDs and OSD memoranda that communicate authorities for CCDRs to temporarily transfer forces in certain instances. These delegations, when exercised, require timely Secretary notification via the Chairman.

(2) Non-organic Support. Non-organic support consists of capabilities that can be provided or supported through HNS or Contracted Support. CCMD/JFC requirements may exceed the apportioned or assigned forces; as such, non-organic support may provide the needed capabilities to mitigate GFM shortfalls. Requirements that are to be supported via non-organic support will not be considered "inherently governmental functions."[8] If contracted support will be utilized, an existing contract or task order must be in place to be considered for execution. Contractors are not captured as part of the global demand on the U.S. military and will not be requested via the allocation process.

d. Orders. The GFMAP is a Secretary-approved deployment order that authorizes the transfer and allocation of forces between CCMDs, Military Departments, and applicable DOD Agen-cies.

(1) Staffing. Consistent with Title 10 Section 151, the CJCS seeks the opinions and advise of other members of the JCS and the CDRs of the unified and specified CCMDs during the staffing process. Subsequently, the GFMAP is staffed with all CCMDs, Services, and DOD Agencies to obtain concurrence and corroborate risks. Successively, the GFMAP is staffed through the JS directorates and OSD staff on to the Chairman, OSD leadership, and, ultimately briefed to the Secretary as one of the orders in the SDOB. During the SDOB briefing, the Secretary reviews the GFMAP and other orders and directs subsequent al-

[7] CJCSM 3130.06 Series, Global Force Management Allocation Policies and Procedures

[8] CJCSM 3130.06 Series, Global Force Management Allocation Policies and Procedures

location or other ordered actions. Once approved by the Secretary, the updated GFMAP is published on the JS J-35 SIPRNET web page.[9]

(2) Deployment Orders. The FPs implement Secretary decisions conveyed in the GFMAP by issuing DEPORDs, through the chain of command, to the units or individuals deploying. Separately, supported CCDRs issue orders to establish command and control relationships for assigned and allocated forces. The JS JFC and JFPs monitor the FPs' progress in meeting the GFMAP orders.

e. Time-Phased Force and Deployment Data (TPFDD). Concurrent with the force allocation process, the TPFDD is updated based upon planning to reflect force requirements and their execution-sourced solutions. These could include DOD civilians, coalition forces, and HNS or contracted support. The TPFDD includes execution data for the time-phased deployment of forces, non-unit cargo, and personnel combined with unit refined movement data. The TPFDD provides a movement and deployment plan for execution-sourced forces.[10]

(1) Requirements. Operational force requirements are usually submitted via JCRM and validated by the JS. When executing under crisis conditions, force requirements may be communicated for Secretary allocation by the most expedient means available in order to respond quickly to emergent events. These expedited communications may be via phone, email, video teleconference (VTC) or other means. The JS documents non-USSOCOM force requirements and USSOCOM documents requirements for their assigned forces. CCMDs should document the requirements in JCRM, a RFF, and Joint Operations Planning and Execution System - Information Technology (JOPES IT) as soon as practicable. To facilitate transitioning from planning to execution, the force requirements for every plan should be documented in the TPFDD at the unit-level to enable execution sourcing by FPs.

(2) Sourcing Solutions. Upon initiating SDOB staffing of the GFMAP, recommended sourcing solutions are documented in the associated TPFDD as execution-sourced solutions by the FP pending the Secretary's approval. The sourcing solutions in JOPES IT should be documented with the JCRM Force Tracking Number (FTN), FTN Line Number or Joint Manning Document (JMD)/Fourth Estate Manpower Tracking System (FMTS) position number as appropriate. As forces are execution sourced, the TPFDD is incrementally updated with unit-refined movement data (usually via Service systems) developed by the specific execution-sourced units.

(3) Exercise Solutions. Services document the unit/personnel sourcing for joint exercises in JTIMS. Sourced exercise requirements in JTIMS are exported for visibility to JCRM. Sourced exercise force requirements are then documented in JOPES IT for exercise deployment/redeployment.

(4) Verification. Execution sourced units develop and document unit-refined movement data in the associated TPFDD, quantifying the unit-specific personnel and cargo embarkation movement requirements. FPs subsequently review and verify the accuracy of the movement data in JOPES IT and notify the supported CCMD (and their Service component) when the TPFDD, or parts of the TPFDD, are ready to be validated.

3. Execution Function - Mobilization

Mobilization is the process by which the U.S. Armed Forces, or part of them, are brought to a state of readiness for war or other national emergency. Mobilization may include activating forces or individuals of the RC (Reserve and National Guard) as well as assembling supplies and material. Requirements for mobilization are identified and developed during CCDR planning in conjunction with Service HQs and the JS. They reflect requirements for

[9] Global Force Management Implementation Guidance

[10] CJCSM 3122.O2 Series, Joint Operation Planning and Execution System (JOPES) Volume III Time Phased Force and Deployment Data Development and Deployment Execution

Execution

force expansion with RC units and individual augmentees (IAs) and for expansion of the CONUS base to sustain the mobilized force for as long as necessary to achieve military and national security objectives.[11,12,13] Figure D depicts the mobilization execution function.

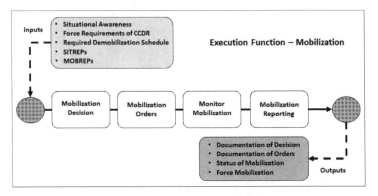

Figure D. Execution Function - Mobilization

a. Mobilization Decision. A decision to mobilize may be prompted by previous mobilization planning, the results of allocation sourcing, or a request by a CCDR with assigned RC forces. The level of decision required to mobilize RC forces is dependent upon the scope and reason for mobilization. The source of authority for mobilization originates with Title 10 U.S.C.

(1) Alert/Notification of Reserve Component Forces. Alerting RC units with adequate lead-time allows for effective implementation of the Reserve train-mobilize-deploy model. In anticipation of a mobilization decision, Military Departments may notify RC units up to 24 months prior to their being considered for activation. Figure E depicts accessing the Reserves for operations.

(2) Limited mobilization authority resides with Military Department Secretaries. Under these authorities, Military Department Secretaries can initiate and direct mobilization. However, short notification mobilizations or those for units and individuals with high deploy-to-dwell ratios may require approval by the Secretary via the SDOB.[14]

(3) For mobilizations requiring President or Secretary approval, the JS in conjunction with the Services prepares and staffs a mobilization decision package for the Chairman, who presents an integrated mobilization plan to the Secretary for review via the SDOB. To the extent possible, normally done with the GFMAP if the force is allocated, the Secretary directs mobilization actions (via SDOB approval) or, if necessary, forwards a recommendation to the President.

There are specific mobilization notification timeline requirements in DoDI 1235.12, *Accessing the Reserve Components (RC)*. Table 1 on the facing page depicts the options of mobilization available. If these timelines are not met, there are requirements for SecDef approval and Congressional notifications. When required, these instances of not meeting the prescribed mobilization alert and mobilization timelines are presented to the Secretary via the SDOB. For rotational forces, this is typically for replacement individuals in units that were mobilized that were later found medically unfit to deploy. Sometimes, it is for entire units – usually when a crisis arises within the approval and notification timelines.

[11] *DoDD 1235.10, Activation, Mobilization, and Demobilization of the Ready Reserve*

[12] *DoDI I235.12, Accessing the Reserve Components (RC)*

[13] *JP 4-05, Joint Mobilization Planning*

[14] *DoDI I235.12, Accessing the Reserve Components (RC)*

Reserve Access Authority

Statute	Utilization Process	Intended Use	Requirements
Involuntary			
Full Mobilization: Section 12301a of Title 10 U.S.C.	Congressional Declaration of War or National Emergency.	Rapid expansion of Military Services to meet an external threat to national security.	-No personnel limitation. -Duration of war or national emergency plus 6 months. -Applicable to all reservists (including inactive and retired).
Partial Mobilization: 12302a of Title 10 U.S.C.	Presidential Declaration of National Emergency.	Manpower required to meet external threat to national security or domestic emergency.	-Max 1,000,000 Ready Reservists on active duty. -Not more than 24 consecutive months.
Presidential Selected Reserve Call-up: 12304) of Title 10 U.S.C.	President determines RC augmentation is required other than during war or national emergency.	Augment the active forces for any named operational mission or to provide assistance for responding to an emergency involving the use or threatened use of weapons of mass destruction, or a terrorist attack or threatened terrorist attack in the U.S. That could result in significant loss of life or property.	-Max 200,000 members of Selected Reserve/Individual Ready Reserve on active duty. -May include up to 30,000 Individual Ready Reserve. -Limited to 365 consecutive days active duty. -Prohibited for support of federal government or a State during a domestic serious natural or man-made disaster, accident or catastrophe. -Prohibited in use of repelling invasions, suppressing insurrections, rebellions, domestic violence, unlawful combinations, or conspiracies, or executing U.S. laws.
Reserve Emergency Call-up: 12304a of Title 10 U.S.C.	SecDef authority in response to Governor's request for federal assistance IAW section 5121 et seq. of Title 42 U.S.C. Presidential determination of major disaster or emergency required.	Manpower required for response to a major disaster or emergency in the U.S. and its territories.	-No personnel limitations -Limited to a continuous period of not more than 120 days. -Does not apply to National Guard or Coast Guard Reserve. -Secretaries of the Military Departments may approve 12304a activations provided the orders are 30 days or less in duration.
Reserve Preplanned Call-up: 12304b of Title 10 U.S.C.	Secretary of Military Department authority to order any unit of the Selected Reserve to active duty for pre-planned and pre-budgeted missions.	Augment AC for any preplanned missions in support of CCMD requirements.	-Max 60,000 on active duty at any one time. -Limited to 365 consecutive days. -Manpower and costs are specifically included and identified in the submitted defense budget for anticipated demand. -Budget information includes description of the mission and the anticipated length of time for involuntary order to active duty. -Secretary invoking section 12304b of Title 10 U.S.C. must submit to Congress a written report detailing circumstances of the call-up.
15 Day Statute: 12301b of Title 10 U.S.C.	Service Secretary authority to order to active duty without consent of persons affected.	Annual training or operational mission.	-15 days active duty once per year. Governor's consent required for National Guard.
Voluntary			
Section 12301d of Title 10 U.S.C.	An authority designated by a Service Secretary may order a RC member to active duty with consent of member.	Active duty in excess of annual training requirements. May be used for training, special work, operational support, etc.	-No set duration. -Consent of the Governor or other appropriate authority of the State concerned required for members of the National Guard.

Table 1. Reserve Access Authority

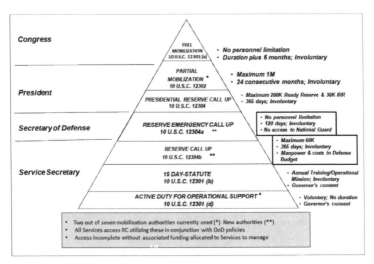

Figure E. Accessing the Reserves for Operations

(4) National Guard forces typically operate per U.S.C., Title 32 under the authority of their state or territory governor and may be early responders to domestic emergencies. Decisions to mobilize National Guard forces under U.S.C., Title 10 authorities are done ICW state or territory governors.

(5) The movement requirements of execution planning potentially trigger a decision to initiate established strategic mobility mobilization programs. Based upon an assessment of transportation requirements, USTRANSCOM may make a recommendation to the Secretary to activate such programs in conjunction with the Secretary of Transportation.[15]

(6) Mobilization during Crisis. Under crisis conditions the decision to allocate and mobilize forces may occur during planning as forces are execution sourced in response to an emergent event. The transition from mobilization to execution in an emergent situation encompasses the activities associated with the time-sensitive development of OPORDs for the deployment, employment, and sustainment of assigned, attached, and allocated forces and capabilities in response to a situation. However, the level of authorization required for mobilization remains consistent with Title 10, U.S.C.

b. Mobilization Orders (MOBORDs). After the decision to initiate mobilization (appropriate to the level of mobilization) the Military Departments publish MOBORDs IAW their respective procedures. OSD may issue implementation instructions and provide additional policy guidance, if required. MOBORD dates should align with the corresponding RC solution ordered in the GFMAP.

c. Monitor Mobilization. The Assistant Secretary of Defense (ASD) Reserve Affairs (RA) is the principal staff assistant to the Secretary for all reserve matters and monitors and enforces mobilization policy and programs.

(1) Sourcing Military Departments and/or NGB monitor the status of mobilization and provide training and readiness oversight. Coordination is made with the JS, OSD, and state/territory governments to ensure adequate resourcing, support, and transition of authorities.

(2) The JS JFC and JFPs monitor mobilization process to ensure timelines support the orders in the GFMAP.

---

[15] Title 46 USC (Section 1242) and Title 50 (Section 196)

d. <u>Mobilization Reporting</u>. Military Departments report unit and individual activation and mobilization information through the *Defense Manpower Data Center and Defense Readiness Reporting System* (DRRS). This reporting, combined with the coordinated advice of the Chairman, informs the President and Secretary of the status of mobilization and enables them to make their statutory reporting to Congress.[16]

4. Execution Function - Deployment

Deployment encompasses preparing forces and individuals for an assigned mission as well as their movement from origin (or home station) to the port of debarkation, through JR-SOI to their point of employment location.[17,18] The process of deployment is depicted below.

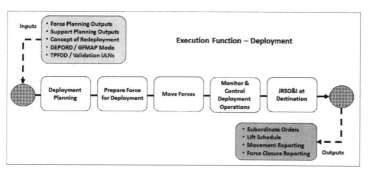

Figure F. Execution Function - Deployment

a. <u>Deployment Planning</u>. As execution is authorized and the forces are identified to support a CCMD's requirements, deployment planning is conducted to execute the deployment of those forces. Requirements for deployment may be based upon CCDR employment of assigned and allocated forces to conduct authorized activities, emergent operational requirements ordered in the GFMAP, joint exercises, or Service training/activities. Deployment planning is based upon the deployment concept and the supported CCMD's CONOPS. Upon establishment of the identified force requirements, the supported CCMD creates a TPFDD reflecting the time-phased movement of those forces required to support the planned CONOPS.

(1) <u>TPFDD Build Process</u>. A TPFDD may be based upon a notional TPFDD with the addition of execution sourced forces and unit-refined movement data. FPs develop unit-refined movement/embarkation data within the corresponding Service systems, creating their portion of the deployment and movement plan. The deployment and movement plan is consolidated as unit level movement data is then exported into JOPES IT. The FP then verifies the movement requirements confirming it is correct and satisfies the GFMAP ordered sourcing.

(2) <u>Supported CCMD Service Components</u>. Movement requirements, documented and verified in the TPFDD by FPs, are then reviewed by the appropriate supported CCMD Service component. The supported CCMD Service component verifies that the execution-sourced forces satisfy mission requirements and the movement plan is properly documented. Verified movement requirements are subsequently forwarded to the supported CCMD.

(3) <u>Supported CCMD</u>. The supported CCMD monitors the TPFDD development as assigned or allocated forces are identified to fill identified requirements.

---

[16] *U.S.C., Title 10*

[17] *JP 4-01, The Defense Transportation System.*

[18] *JP 3-35, Deployment and Redeployment Operations.*

(a) Once TPFDD entries have been verified by the appropriate supported CCMD component, the supported CCMD validates them as movement requirements. Validation by the supported CCMD indicates that the movement requirements are correct and satisfy the supported CCMDs CONOPS with sourcing ordered in the GFMAP. The CCMD validates the specific TPFDD movement requirements via JOPES IT and then forwards the validated lift requirements to USTRANSCOM, authorizing their strategic movement.

(b) The supported CCMD also establishes deployment priorities for their theater that subsequently inform deployment execution by lift providers.

(4) <u>Lift Providers</u>. Lift provider review the validated movement requirements and arrange for appropriate lift. Lift providers may include self-deployment, inter-theater lift assets assigned or allocated to a CCMD, or strategic lift assets assigned to USTRANSCOM. If USTRANSCOM is the lift provider, they review the validated requirements for technical correctness and ensure that movement requirements are within the constraints of allocated lift. USTRANSCOM schedules lift based on the validated movement requirements of all the CCMDs and applying global movement priorities. USTRANSCOM then forwards the movement requirements to its subordinate Service components for scheduling.

(a) USTRANSCOM bases its movement planning and execution upon the planned framework of their *Campaign Plan for Global Distribution*, developed per strategic guidance, with inputs from other CCMDs. The Command's overarching framework of support for plans and operations is contained in USTRANSCOM's *Global Agility Concept Plan* (CONPLAN) 9000.

(b) USTRANSCOM evaluates the force flow and iteratively schedules movements for supported CCMD validated movement requirements in JOPES IT. USTRANSCOM ensures that validated movement requirements are routed and scheduled IAW the supported CCMDs CONOPS. Strategic lift provided by USTRANSCOM supports multiple supported and supporting CCMD requirements. Any lift shortfalls are reconciled and schedules changed in conjunction with the supported CCMD.

(5) <u>Chairman</u>. The Chairman monitors the implementation of the lift priorities in support of execution as directed by the Secretary.

(a) The Chairman assigns movement priorities (usually as part of an EXORD) based upon the Secretaries operational priorities and capabilities reported by USTRANSCOM.

(b) The JS JFC and JFPs monitor deployment operations and the implementation of GFMAP ordered force allocations.

(c) The JS J-4 operates the Joint Logistics Operation Center (JLOC) to maintain Chairman cognizance over deployment operations.

(d) The JS J-4 ICW USTRANSCOM monitors strategic distribution support and infrastructure.

(e) When transportation requirements exceed capacity, the Chairman may adjudicate competing lift requirements ICW CDR, USTRANSCOM and the supported CCMDs.

b. <u>Prepare Forces for Deployment</u>. Supporting CCMDs, Military Departments, and CSAs prepare forces and individuals for deployment. Select forces may deploy within hours or days from receipt of a DEPORD while other units may deploy on a timeline of days to many months. Regardless of the deployment timeline, a myriad of predeployment activities must be accomplished to deploy the force with their required equipment and supplies. These actions range from the strategic to the tactical level. At the strategic and operational levels, TPFDD sourcing/refinement and transportation feasibility may continue well into this phase. At the installation and unit level, activities range from personnel and equipment status confirmed and upgraded to conducting required training. Deployment timelines will dictate available time to conduct prepare-the-force activities, which include: activating deployment and C2 support organizations, conducting movement and support meetings, developing a unit deployment list (UDL) and identifying shipping/handling requirements, and conducting required training. Standards for preparation are as prescribed by the supported CCMD. At the same time, USTRANSCOM postures to direct the movement of those forces.

(1) <u>Readiness Reporting</u>. The Services shall report the readiness of units ordered to deploy. Before being employed, units must be assessed as capable of performing to prescribed standards their assigned mission-essential tasks under conditions expected in the theater of operations in which they would be deployed. Secretary approval is required to reallocate units assessed as not capable of performing to prescribed standards. Readiness is tracked by OSD, Joint Staff, JFPs, Services, and DOD Agencies to ensure continued capability of providing allocated forces ordered in the GFMAP. This reporting informs ongoing situational awareness operational activities monitoring execution activities. It provides a snapshot of readiness to support functional assessments in planning that enable reasonable assumptions for resource informed planning.

(2) USTRANSCOM exercises its UCP responsibilities as mobility JFP, DOD single manager for transportation, DOD single manager for patient movement, Distribution Process Owner (DPO), global distribution operations synchronizer, and for providing mission-tailored Global Standing Joint Force Headquarter capabilities in support of CCMDs and ICW supporting CCMDs, Services, and appropriate USG agencies.

(a) USTRANSCOM analyzes global lift requirements against global strategic lift capacity needed to meet the Chairman's priority for the supported CCMD deployment requirements.

(b) USTRANSCOM performs lateral coordination of recommended resource deliberations with the supported CCMDs. USTRANSCOM prioritizes deployment support globally in support of multiple supported CCMDs.

(c) USTRANSCOM forwards coordinated strategic mobility assets allocations recommendation to the Chairman.

(d) USTRANSCOM operates a Deployment and Distribution Operations Center (DDOC) that directs the global strategic air, land, and sea transportation capabilities ICW supported and supporting commands.

(e) Supported CCMDs establish a Joint Deployment and Distribution Operations Center (JDDOC) that works with the DDOC to balance and regulate the force flow from origin to destination.

(f) Movement control elements confirm diplomatic and ground movement clearances with relevant host nations, state, and USG agencies.

c. <u>Move Forces</u>. Forces deploy via common user lift or may self-deploy as capable.

(1) The USTRANSCOM DDOC manages the strategic common-user transportation assets needed for movement of forces and sustainment, develops lift allocations, and reports the progress and shortfalls associated with deployment to the Chairman and the supported CDR(s). USTRANSCOM supports multiple supported CCMD movement requirements.

(2) The supported CCMD JDDOC manages intra-theater movement of forces to the point of employment ICW the DDOC. Intra-theater movements may be supported by USTRANSCOM lift capabilities.

d. <u>Monitor and Control Deployment Operations</u>.

(1) <u>Supported CCMDs</u>. Balance and regulate the force flow to support the overall CONOPS and CCMD priorities. The supported CCMD may adapt the CONOPS and priorities to changes in the OE and regulates the force flow with consideration for theater distribution and JRSOI capacity limitations.

(2) <u>USTRANSCOM</u>. Coordinates with its Service component commands and commercial lift providers managing the flow of lift assets IAW the supported CCMD's force and sustainment priorities. USTRANSCOM also provides in-transit visibility (ITV) for all force movements supported by their transportation activities and all sustainment movements. Figure G depicts In-Transit Visibility.

Execution

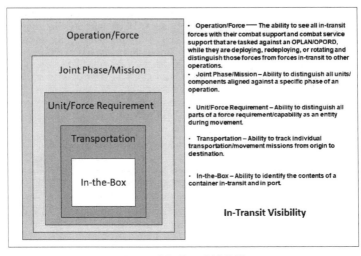

Figure G.In-Transit Visibility

(3) Supporting CCDRs, Service HQs, and CSAs provide visibility of force movement plans by reporting via JOPES IT.

(4) The Joint Logistics Operation Center (JLOC) monitors the deployment of CCDR forces and informs the Chairman of any shortfalls requiring resolution. The Chairman may be engaged to adjudicate identified shortfalls. The Chairman may leverage the analysis and recommendation of the JLOC or Joint Transportation Board (JTB) if established to support his/her adjudication decisions.[19]

e. Joint Reception, Staging, Onward Movement, and Integration at Destination (JR-SOI).[20] JRSOI transitions deploying or redeploying personnel, equipment, and materiel into forces capable of meeting the CCDR's operational requirements or returns them to their parent organization or Service.

- Reception operations include all those functions required to receive and clear personnel, equipment, and materiel through the point of debarkation (POD).

- Staging assembles, temporarily holds, and organizes arriving personnel, equipment, and materiel into forces and capabilities and prepares them for onward movement, tactical operations, or Service reintegration.

- Onward Movement is the process of moving forces, capabilities, and accompanying materiel from reception facilities, marshalling areas, and staging areas (SAs) to tactical assembly areas (TAAs) and/or operational areas (OAs) or onward from the POD or other reception areas to the home/demobilization station.

- Integration is the synchronized transfer of capabilities into an operational CDR's force prior to mission execution or back to the component/Service.

(1) Theater requirements for JRSOI vary significantly based upon the location and nature of the operation. Requirements may include:

(a) Theater-specific training.

(b) Status of Forces Agreement (SOFA) requirements.

(c) Customs or other legal requirements of the host nation.

[19] *JP 4-01, The Defense Transportation System*

[20] *JP 3-35, Deployment and Redeployment Operations*

(2) The supported CCDR is responsible for JRSOI in theater and through the established JDDOC controls force movements to support the CONOPS within the capacity of JRSOI infrastructure.

(3) The supported CCMD Service components, ICW the JDDOC execute JRSOI activities until deploying forces are integrated in theater command and control and capable of tactical mission execution.

(4) Supported commands assuming command and control of arriving forces, plan and conduct integration activities down to the tactical level that complete JRSOI. The reporting up the chain of command of integration activities informs on going situational awareness operational activity supporting execution decision-making.

5. Execution Function - Distribution

As forces deploy, appropriate sustainment must be provided to enable operations. The distribution function includes both the movement of forces for deployment and redeployment as well as sustainment and retrograde.[21,22] The process of distribution is depicted in Figure H below.

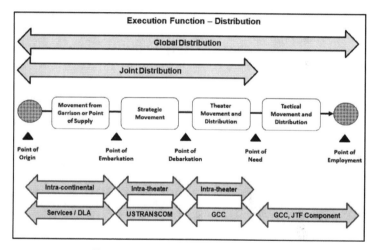

Figure H. Execution Function - Distribution

a. Supported CCDR. The supported CCDR establishes the priorities within their AOR for distribution and directs the intra-theater level of distribution.

(1) Requirements for the distribution of sustainment are identified by the supported CCMD components. Supply requisitions are annotated by the appropriate priority designator and force or activity designator (F/AD) (number used in conjunction with urgency of need designators to establish a matrix of priorities used for supply requisitions) to facilitate prioritization by the Services, Defense Logistics Agency (DLA), and USTRANSCOM.

(2) Any strategic movement required for sustainment is developed through coordination with the Service HQs and documented in the TPFDD. Through refinement of the TPFDD, the supported CCMD sequences overall movement requirements based on the supported CCDR's priorities and balances the distribution flow to remain within the throughput capacity of theater distribution infrastructure. This sequencing and balancing is usually

[21] JP 4-09, Distribution Operations

[22] DoDM 4140.01, Volume I, DOD Supply Chain Material Management Procedures: Operational Requirements

achieved by refining TPFDD required delivery dates and latest arrival date (LAD) for movement requirements prior to validation.

(3) Intra-theater distribution is directed by the supported CCDRs JDDOC ICW the USTRANSCOM DDOC managing inter-theater distribution. The supported CCDR's component forces execute theater distribution operations while the GCC adjudicates allocation of common-user transportation when movement requirements exceed capability.

(4) To more effectively manage distribution, CCDRs exercise Directive Authority for Logistics (DAFL) IAW U.S.C., Title 10 (Section 164) and may delegate responsibility for the management of common support capability to a subordinate command or component. This is normally done ICW the Services and for particular commodities or support services common to two or more Services.[23]

b. <u>Military Departments and Service HQs</u>. Military Departments and Service HQs have U.S.C., Title 10 responsibility for Service logistics and provide sustainment for their Service forces.

(1) Service HQs execute their logistics and sustainment responsibilities through direct communication and coordination with their corresponding CCMD Service component.

(2) The Service HQs respond to sustainment requirements identified via the supported CCMD Service component. The Services prioritize supply distribution based upon the customer's priority designator and F/AD. Any strategic movement required for sustainment is developed through coordination with the CCMD Service component and documented in the TPFDD.

(3) Service HQs, in conjunction with DLA, may coordinate for vendor support for delivery of many commodities as appropriate.

(4) Services can also augment organic logistic capabilities by agreements with USG agencies or allies, or by participating in common, joint, cross-servicing agreements, or existing agreements or task orders.

(5) CCMD Service component commands are responsible for tactical distribution and sustainment of their Service forces within the physical network of the supported GCC's established theater distribution system.

(6) Logistics support for SOF units is the responsibility of the parent Service and usually executed by the Theater Service Component. Given the unique nature of some SOF operations, logistics support can also be provided through other support arrangements.[24]

c. <u>Defense Logistics Agency (DLA)</u>. As part of the distribution process, the DLA is the primary operator of the defense supply and depot system.

(1) DLA is responsible for acquisition, receipt, storage, issuance, and generation of source data for all materiel (other than materiel procured by the individual Services) flowing in the defense distribution pipeline. DLA supports the distribution requirements of multiple supported and supporting CCDRs.

(2) DLA directs the sourcing, packaging, and preparation of sustainment stocks and pre-positioned material to be moved through the distribution pipeline. DLA also coordinates for vendor support for delivery of many commodities as appropriate.

(3) DLA prioritizes supply distribution based upon the customer's priority designator and F/AD.

(4) DLA coordinates as required with USTRANSCOM for the use of the Defense Transportation System (DTS) to meet distribution transportation requirements.

[23] *JP 4-0, Joint Logistics*

[24] *CJCSM 3130.06 Series, Global Force Management Allocation Policies and Procedures*

d. U.S. Transportation Command. USTRANSCOM is the DOD Distribution Process Owner (DPO) responsible for coordinating and synchronizing global distribution operations. CDR, USTRANSCOM coordinates, synchronizes, and executes Joint Deployment and Distribution Enterprise (JDDE) responsibilities to ensure USTRANSCOM simultaneity of support to the supported CCMD, Services, DLA, and other appropriate USG agencies as directed.

(1) USTRANSCOM bases its distribution planning and execution upon the planned framework of their *Campaign Plan for Global Distribution* developed per strategic guidance tasking with inputs from other CCMDs.

(2) As the Single Manager for transportation, USTRANSCOM directs the Defense Transportation System (DTS) and coordinates with the commercial transportation industry to execute movement of forces and sustainment in order to meet national security objectives. The DTS includes contracts with domestic and international commercial carriers for air, ground, inland and coastal waterways, and sea transportation support.

(3) USTRANSCOM assumes responsibility for the movement of materiel for all CCDRs as it enters the DTS linking suppliers (Services, DLA, or Commercial Industry) with customers (supported CCDR or JFC).

(4) USTRANSCOM manages distribution through their established DDOC that directs the global distribution capabilities ICW supported and supporting commands.

(5) To synchronize the distribution process, USTRANSCOM employs the *Integrated Data Environment/Global Transportation Network Convergence* (IGC) as the DOD system of record for in-transit visibility. The IGC enables the integration of supply, cargo, passenger, and unit requirements and movements with airlift, air refueling, and sealift schedules and movements to meet the supported CCDR's requirements.

6. Execution Function - Employment

Employment is the strategic, operational, or tactical use of forces. The supported CDR issues orders and command and controls the employment of assigned and allocated forces to accomplish directed missions. During employment, the outputs of the situational awareness and assessment operational activities inform decision-making at all levels of the chain of command. Changes in the OE or evaluation of progress towards mission accomplishment may prompt the planning or execution of branches/sequels or the termination of operations. The process for employment is depicted in Figure I.

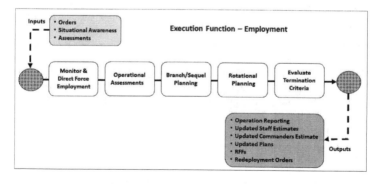

Figure I. Execution Function - Employment

a. <u>Monitor and Direct Force Employment.</u> During execution, the supported CDR monitors and directs employment. The situational awareness and assessment operational activities continue to inform the decision cycle at all levels of the chain of command. As execution of the plan unfolds and the situation develops, the plan is continuously evaluated for validity and changes to current activities which inform the decisions for subsequent planning.

b. <u>Operation Assessment</u>. During the employment of military capabilities, the supported CCMD implements the planned operation assessment to enhance leadership decision-making and make execution and continued planning more effective. The ongoing assessment operational activity is the synthesis of the results of one or more operation assessments and informs CCDR decision making across the theater or functional scope of responsibility.

c. <u>Branch/Sequel Planning</u>. During execution, planning continues as an operational activity in order to address changing conditions in the environment or shifts in political-military direction. Branches or sequels to the current plan may be developed or refined and an updated order issued as circumstances warrant. The supported CCDR and his/her subordinate CDRs direct, monitor, and assess employment activities to align with the updated plan.

d. <u>Rotational Planning</u>. During the employment function, the need for replacement forces may be recognized and those requirements submitted as rotational requirements via the GFM allocation process. Force requirements that will remain past the current fiscal year must be documented as rotational requirements in subsequent fiscal years in order to allow the force sourcing community of interest (JS JFC, JFPs and FPs) to allocate replacement forces. The supported CCMD and FPs must consider planning for the rotation of forces. Through the GFM allocation process, FPs may source multiple units conducting rotational deployments that combine to satisfy a single GFMAP ordered deployment length. Service deployment lengths for rotating forces are vetted during the allocation sourcing process and ordered in the GFMAP. Service HQs (or USSOCOM for SOF) issue DEPORDs for replacement forces implementing the SecDef direction in the GFMAP. The supported CCMD Service components issue REDEPORDs to direct the unit that has completed transfer of authority (TOA) with the replacement unit to redeploy. This ensures Secretary-ordered requirements continue to be met. The Service component REDEPORD does not cancel the CCDR's requirement for the force or the SecDef approved order in the GFMAP to provide the force or capability.

e. <u>Termination</u>. Termination criteria are specified conditions that, when achieved, define an acceptable level of accomplishment of a military objective. Termination is the period in time or set of conditions beyond which the military instrument of national power is no longer the primary means to achieve remaining national objectives.

(1) The operation assessment is designed to evaluate progress toward or achievement of military objectives. This assessment informs the supported CCDRs recommendation to the President and Secretary of when to terminate a directed military activity and its associated authorities.

(2) Termination of a military activity does not simply trigger a return to the steady-state activities of the CCP. Considering all instruments of national power, there may be military efforts above the pre-operational level required to achieve sustainable political/military objectives.

(3) The supported CDR should evaluate and plan for post-termination activities that consolidate and sustain conditions that achieve an enduring acceptable political outcome.

(4) Termination criteria should account for a wide variety of operational tasks that the joint force may need to accomplish, to include disengagement, force protection, transition to post-conflict operations, reconstitution, and redeployment.

(5) Termination of military operations is ultimately a political decision that is directed by the President or Secretary. Established termination criteria is a tool which informs their decision to terminate military operations.

7. Execution Function - Redeployment

Redeployment is the transfer of deployed forces and accompanying materiel from one operational area to support another JFC's operational requirements within a new operational area or home/demobilization station. Redeployment is initiated by a CCDR decision (end-of-mission for a force) or a Service decision for force rotation.[25] The process for redeployment is depicted in Figure J below.

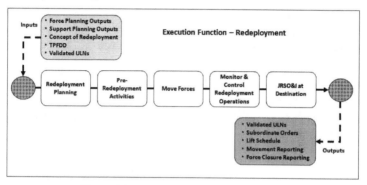

Figure J. Execution Function - Redeployment

a. <u>Redeployment Planning</u>. Redeployment planning is based upon a planned CONOPS and refined during execution at redeployment decision points necessary to meet lead times for effective execution.

(1) <u>Supported CCDR</u>. A CCDR REDEPORD is issued when a force requirement is no longer needed. The order directs the unit to redeploy and informs the GFM stakeholders that the requirement for replacement units is no longer needed. A CCDR REDEPORD is not required to conduct a Service (or USSOCOM for SOF) force rotation or to redeploy a force when it is properly relieved by a replacement force. Movement requirements for redeployment are entered into the TPFDD by the supported CCMD Service component pending validation by the supported CCDR. The CCDR redeployment planning must also consider the drawdown and redeployment of contracted support and retrograde of non-unit equipment and material.

(2) <u>CCMD Components</u>. A CCMD Service component issues a REDEPORD either in response to a CCDR order or on its own authority in compliance with Service (or USSOCOM of SOF) force rotation policy. CCDR Service component REDEPORDs do not cancel the CCDR's force requirement. Ensuring coordination for a replacement rotation force, the supported CCDR Service component issues a REDEPORD. Movement requirements for redeployment are entered into the TPFDD by the supported CCMD Service component while movement requirements for the deployment of replacement force are entered by the Service (or USSOCOM for SOF). Both verify the movement requirements to the CCDR who then validates them in the TPFDD.

(3) USTRANSCOM exercises its UCP responsibilities as mobility JFP, DOD single manager for transportation, DOD single manager for patient movement, DPO, global distribution operations synchronizer, and for providing mission-tailored Global Standing Joint Force Headquarter capabilities in support of CCDRs and ICW supporting CCDRs, Services, and appropriate USG agencies.

[25] *JP 3-35, Deployment and Redeployment Operations*

(a) USTRANSCOM analyzes strategic mobility lift allocations needed to meet the Chairman's priority for all CCDR force redeployment requirements.

(b) USTRANSCOM performs lateral coordination of recommended resource deliberations with the supported CCDRs. USTRANSCOM prioritizes redeployment support globally in support of multiple supported CCDRs.

(c) USTRANSCOM forwards coordinated strategic mobility assets allocation recommendation to the Chairman.

(4) <u>Chairman</u>.

(a) The Secretary determines priority of supported CCDR force requirements and the Chairman executes that priority.

(b) USTRANSCOM keeps the Chairman apprised of progress through the JS.

(c) In cases where redeployment lift requirements exceed USTRANSCOM capacity, the Chairman may need to arbitrate a solution with the supported CCDRs. The Chairman may leverage the analysis and recommendation of the JLOC or JTB, if established, to support his/her adjudication decisions.[26]

(5) <u>Supporting CCMDs</u>. Align their supporting activities with the execution of the supported CCDR's redeployment plan.

(a) Verify Forces in the TPFDD and report redeployment movement requirements within JOPES IT.

(b) Continue to coordinate for the sustainment of forces through the completion of redeployment.

b. <u>Prepare Forces Redeployment Activities</u>. The supported CCMD and its Service components prepare forces and individuals for redeployment. The activities are fundamentally the reverse of the JRSOI activities conducted during the deployment function.

(1) <u>Establish Redeployment Command and Control</u>. To meet its execution responsibilities, USTRANSCOM established a DDOC that directs the global air, land, and sea transportation capabilities ICW supported and supporting commands. Supported CCDRs establish a JDDOC that works with the DDOC to balance and regulate the force flow from origin to destination. Movement control elements confirm diplomatic and ground movement clearances with relevant host nations, state, and USG agencies.

(2) <u>Schedule Force Movements</u>. USTRANSCOM assists the supported CCDR and ensures that validated movement requirements are routed and scheduled IAW the TPFDD.

(3) <u>Force Rotation</u>. As directed, the supported CCMD Service components or subordinate commands establish and coordinate requirements for relief-in-place of rotating forces. Tactical CDRs tasked to redeploy conduct relief-in-place and transfer of authority operations as required prior to starting redeployment movement.

(4) Conduct theater requirements for redeployment based upon the location and nature of the operation. Requirements may include:

(a) Redeployment training or medical screening

(b) Status of Forces Agreement (SOFA) requirements

(c) Customs or other legal requirements of the host nation or destination

c. <u>Move Forces</u>. Forces redeploy via common user lift or may self-deploy as capable.

(1) The supported CCDR is responsible for intra-theater movement of forces to redeployment port of embarkation (POE). The supported CCDR manages and regulates the redeployment flow with consideration of force throughput and staging capacity.

(2) CDR, USTRANSCOM will manage the strategic common-user transportation assets needed for the redeployment of forces, develop lift allocations, and report the progress and shortfalls to the Chairman and the supported CCDRs.

[26] *JP 4-01, The Defense Transportation System*

d. Monitor and Control Redeployment Operations.

(1) Supported CCDR. Balances and regulates the force flow to support the overall CONOPS with consideration for theater distribution, JRSOI capacity, and strategic lift capacity. Force Rotation relief-in-place requirements are considered when managing the redeployment schedule and its alignment with deployment of replacement forces.

(2) USTRANSCOM. Coordinates with its Service components and commercial transportation industry to manage the flow of lift assets IAW the supported CCDR's movement priorities. Provides ITV of redeployment and documents force movements in JOPES IT.

(3) Supporting CCDRs, Military Departments, and CSAs monitor redeployment TPFDD execution and posture to receive redeploying forces.

e. JRSOI at Destination. The responsibility for JRSOI following redeployment will depend upon the subsequent mission of the redeploying force. If redeploying to another AOR for a follow-on mission, the receiving CCDR, in conjunction with its Service components, conducts JRSOI. If redeploying to home station, the originating Service and its subordinate organizations conducts JRSOI. During redeployment to home station, the receiving CCDR or Service assumes responsibility for returning units and personnel when OPCON is relinquished IAW the directing order. The receiving CCDR or Service must have visibility of the redeployment schedule to effectively support its JRSOI activities. That visibility may be through USTRANSCOM provided ITV or via Service coordination of self-redeploying forces.

8. Execution Function - Demobilization

Demobilization activities can begin before the end of the crisis or war as the need for resources diminish and assets for demobilization support become available. Most demobilization actions will commence following the conflict when immediate post-conflict missions have been assigned by the supported CCDR and requirements for military forces and resources decline. Although demobilization, like mobilization, is essentially a Military Department responsibility, the supported and supporting CDRs play coordinating and synchronizing roles. In any event, the CCDRs monitor the status and progress of demobilization and concurrent recovery operations to assess the adequacy of actions to restore readiness of assigned forces to required levels for future conflicts. Following redeployment, the Military Departments deactivate RC units or individuals and return them to a reserve status.[27] The process of demobilization is depicted in Figure K below.

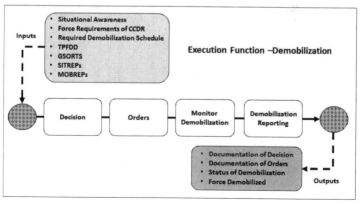

Figure K. Execution Function: Demobilization

[27] JP 3-35, Deployment and Redeployment Operations

a. <u>Demobilization Decision</u>. A decision for demobilization is typically prompted by either the pending expiration in the mobilization authority for a GFMAP allocated RC unit or a reduction in the force requirements of a supported CCDR which impacts RC units. Demobilization timelines should support RC units continuing to meet GFMAP ordered deployment lengths.

b. <u>Demobilization Orders</u>. The original mobilization orders specify a units-ordered demobilization date. To effect demobilization, the Military Departments typically issue additional demobilization instructions to units via the supported CCMD Service component IAW DOD and Service policy. Supported CCDRs are directed to ensure that RC units and individuals are redeployed to home station in sufficient time to accomplish release from active duty processing IAW DOD and individual Service policies.[28],[29]

c. <u>Monitor Demobilization</u>. Sourcing Military Departments and the NGB monitor the status of demobilization. Coordination is made with the JS, OSD, and state/territory governments to ensure adequate resourcing, support, and transition of authorities back to Reserve status.

d. <u>Demobilization Reporting</u>. Military Departments report their respective unit and individual demobilization information through Defense Manpower Data Center and DRRS. This reporting, combined with the coordinated advice of the Chairman, informs the President and Secretary of the status of demobilization and enables them to make their statutory reporting to Congress per U.S.C., Title 10.

9. Summary

The U.S. employs the four instruments of national power (diplomatic, informational, military, and economic) to achieve national strategic objectives. The military instrument's role increases relative to the other instruments as the need to compel an adversary through force increases. Military efforts focus on fielding modular, adaptive forces that can be employed across the range of military operations. Deployment and redeployment operations enable the projection of the military instrument of national power. Joint forces deploy in support of ongoing operations and theater and functional campaign plan activities.

The deployment and redeployment of U.S. forces in support of CCMD requirements are a series of operational events enabled by logistics. These activities are planned and executed by both the supported and supporting commands, Services, NGB, and DOD agencies. The capability to deploy forces to the OA and rapidly integrate them into the joint force, as directed by the JFC, is essential. Mission requirements determine the scope, duration, and scale of deployment and redeployment operations. These operations involve the integrated complementary efforts of numerous commands, Services, agencies, and processes, and as such, ***unity of effort is required for effective and efficient mission accomplishment.***

> *"Ideas are easy. It's the execution of ideas that really sepeartes
> the sheep from the goats."*
> Sue Grafton

---

[28] *DoDD 1235.10, Activation, Mobilization, and Demobilization of the Ready Reserve*
[29] *DoDI I235.12, Accessing the Reserve Components (RC)*

II. Execution

> "In the military, as in any organization, giving the order might be the easiest part. Execution is the real game."
>
> LTG Russel Honore

1. Effective Planning

Planning and preparation accomplish nothing if not executed effectively. Execution is putting a plan into action by applying forces and capabilities to accomplish the mission and using situational awareness to assess progress and make execution and adjustment decisions.[1] Often, the decision to deploy the military will be in conditions significantly different from the original planning guidance or the conditions planned. Assessments and reframing the problem, if required, inform the applicability of, or necessary modifications to the plan in response to changes in the OE. Plans are rarely executed as written regardless of how much time and effort went into the planning process. However, planning provides a significant head start and gives you insight into potential problem areas.

a. CDRs fight the enemy, not the plan. Moltke's dictum of "No plan of operations goes with any degree of certainty to beyond the first contact," rather than condemning the value of planning, reminds CDRs, staffs, and subordinate unit leaders the proper relationship between planning and execution. A plan provides a reasonably forecast of execution. However, it remains a starting point, not an exact script to follow. As General George S. Patton, Jr., cautioned, "...one makes plans to fit circumstances and does not try to create circumstances to fit plans."[2]

b. Effective planning enables transition. Integrated staff effort during planning ensures the plan is a team effort and the knowledge gained across the staff in the planning process is shared and retained. This staff work assists in identifying changes in the OE and guidance, speeding transition to execution.

c. Detailed planning provides the analysis of the threat and the OE. The knowledge and understanding gained enables a well-trained staff to quickly identify what is different between their plan and current conditions and make recommendations based on their prior work.

2. Transition

a. Transition to Execution. As discussed in Chapter 6, *Plan Development,* transition may involve a wide range of briefs, drills, or rehearsals necessary to ensure a successful shift from planning to execution and be subject to the variables of echelon of command, mission complexity, and, most importantly, time.

b. At a minimum, this step includes a CONOPS brief along with the handover and explanation of any execution tools developed during planning, such as a decision support

[1] *FM 3-0, Operations*

[2] *ADP 5-0, The Operations Process*

matrix or an execution checklist. If time and resources allow, the transition step may include ROC drills and confirmation briefs by subordinate units. Successful transition enhances the situational understanding of those who will execute the order, maintains the intent of the CONOPS, promotes unity of effort, and generates tempo.[3]

c. Transition is a continuous process that requires a free flow of information between CDRs and staffs by all available means. At higher echelons where the planners may not be executors, the CDR may designate a representative as a proponent for the order or plan. After orders development, the proponent takes the approved order or plan forward to the staff charged with supervising execution. As a full participant in the development of the plan, the proponent is able to answer questions, aid in the use of the planning support tools, and assist during execution in determining necessary adjustments to the order or plan. Transition occurs at all levels of command. A formal transition normally occurs on staffs with separate planning and execution teams. For transition to occur, an approved order or plan must exist. The approved order or plan and the products of continuing staff actions form the input for transition.[4]

These inputs may include—

- Refined intelligence and IPB products.
- Planning support tools.
- Outlined FRAGOs for branches.
- Information on possible future missions (sequels).
- Any outstanding issues.

d. Regardless of the level of command, a successful transition ensures those who execute the order understand the CDR's intent, the CONOPS, and any planning tools. Transition may be internal or external and in the form of briefs, drills, or the relocation of a planner to the current operations for execution. Internally, transition occurs either between future plans and the future operations center or future operations and current operations centers. Externally, transition occurs between the CDR and his subordinate CDRs.[5] (See Chapter 6, *Plan or Order Development* and JP 5-0, *Joint Planning* for details on the Transition Brief.)

> *"A good plan violently executed now is better than a perfect plan executed next week."*
> *Gen. George S. Patton*

3. Execute

a. In execution the CCDR, staffs, components and supporting CDRs focus their efforts on translating decisions into actions. In the case of a contingency the decision to execute will often be presented as an examination of options in response to a developing crisis or action rather than a specific directive to execute a specific CONPLAN or OPLAN. During execution, the situation may change rapidly and operations the CDR envisioned in the plan may bear little resemblance to actual events in execution

b. The decision to execute will often be presented as an examination of options in response to a developing crisis such as a natural disaster or action by a competitor state or threat (state or non-state) rather than a specific directive to execute a specific CONPLAN or OPLAN.

[3] *JP 5-0, Joint Planning*

[4] *MCWP 5-1, Marine Corps Planning Process*

[5] *Ibid*

c. Initiate Execution. The execution of military activities begins with the direct or delegated authority of the President or Secretary to conduct or execute that military activity. Based upon the character of the activities to be conducted, the distinct vehicles used to convey execution authorities will vary and are discussed briefly in the following paragraphs.

(1) Secretary of Defense Orders Book (SDOB). The SDOB process is the means in which the JS develops and staffs every order that needs approval from the Secretary. The SDOB is detailed further in the *GFMIG* and CJCSI 3130.06, *GFM Allocation Policies and Procedures*. Upon Secretary approval, the Joint Staff, on behalf of the Chairman, issues the appropriate order(s). The Services and CCDRs then pass orders down the chain of command to subordinate CDRs as required. The subsequent orders get increasingly more detailed and transition from the strategic to operational, and eventually tactical. Every CDR in the chain of command issues orders within the authorities he is delegated.[6]

(2) Execution Authorities. Forces may deploy and execute under different authorities dependent upon the activity being conducted and the time available. Depending on the circumstances, authorities to execute and deploy may be communicated separately or together. The global deployment order for all allocated forces is the *GFM Allocation Plan (GFMAP)*. The Joint Staff consolidates all allocation decisions into a single order because it allows the JS, JFPs, JS JFC, CJCS, and Services to keep track of where forces are deploying and facilitates keeping a better understanding of the integrated posture of forces among the CCDRs and the competing priorities (Chapter 3, *GFM, discusses the* allocation process.)

(a) Campaign Plans. CCPs are in constant execution and provide the base plan from which subsequent plans and orders become branch plans. While CCPs are reviewed and endorsed by the Secretary, the authorization to execute a campaign plan does not provide complete authority for the CCDR to execute each of the individual military activities that comprise the plan. Additional CCMD coordination is required to execute the discrete military activities within a campaign plan to include:

1 Posture Plans. As an element of the campaign plan, the posture plan establishes the network of forces, footprints and agreements required to execute campaign and/or contingency operations within an AOR. Posture plans are comprised of posture initiatives, which define a coordinated change to forces, footprint, and agreements that provides a joint capability to meet objectives, reduce gaps, and reduce risk. Posture initiatives are developed and described in the context of closing capability gaps and/or developing required enhancements to existing capabilities in order to meet theater objectives in a region or a sub-region.[7]

2 Force Allocation. In addition to assigned forces, forces allocated via the GFMAP often execute activities in the GCPs and CCPs. The review of a campaign plan by the Secretary does not authorize the transfer of forces to conduct campaign activities.[8]. The *GFMIG* and CJCSM 3130.06, *GFM Allocation Policies and Procedures* detail the GFM allocation process. (The allocation of forces within GFM is discussed in Chapter 3, *GFM.*)

3 Country Team Coordination. Military campaign activities within a foreign country should be coordinated with the U.S. Chief of Mission. CCMD coordination with embassy country teams and strategic guidance, is typically done via the military element of a specific country team. The country team, in turn, coordinates with the host nation for their sovereign approval, when applicable, of proposed U.S. military activities.[9]

4 Country Clearance. When applicable, entry into foreign countries by DoD personnel, aircraft, or vessels require the approval of a foreign clearance request. Requests are submitted via the Aircraft and Personnel Automated Clearance System web-based tool.

---

[6] CJCSM 3130.06, *GFM Allocation Policies and Procedures*

[7] DoDI 3000.12, *Management of U.S. Global Defense Posture (GDP)*

[8] CJCSM 3130.06, *GFM Allocation Policies and Procedures*

[9] Directed by Title 22, U.S.C. (Section 3927)

Requirements vary by each specific country and are governed by the DoD Foreign Clearance Guide.

5 <u>Security Cooperation</u>. Security cooperation includes military activities with international partners to include security assistance, providing defense articles and training, and security force assistance, for building the defense capacity of foreign security forces. Statutory requirements to conduct security cooperation activities is found in Title 10, U.S.C. Security cooperation activities are typically conducted as a part of a CCDR's campaign plan and coordinated with the partner nation through the embassy country team. Security cooperation activities are documented in the *Global Theater Security Cooperation Management Information System (G-TSCMIS).*[10]

6 <u>Theater Security Cooperation (TSC)</u>. TSC support to foreign security forces is contingent meeting the requirements of Title 10, U.S.C. (Section 333)[11] to include concurrence of the Secretary of State, and notification of congress.[12,13]

7 <u>Joint Combined Exchange Training (JCET)</u>. Per Title 10, U.S.C. (Section 322) CDR, USSOCOM is authorized to conduct training with the armed forces or other security forces of a friendly foreign country with the primary purpose of training U.S. SOF units in their mission essential tasks. As an added benefit, JCETs may also support CCMD TCP and U.S. Embassy Chief of Mission country plan objectives. JCETs are coordinated with the partner nation through the country team and submitted for Secretary approval via USD(P) or Assistant Secretary of Defense (ASD) Special Operation/low-intensity Conflict (SO/LIC). Approval requires a completed HRV by the DOS of any partner forces to be trained. Execution of a JCET requires OSD (SO/LIC) notification via the JS J-37 45 days prior to the deployment.[14]

8 <u>Exercises</u>. The authority to plan and execute exercises originates with the U.S.C, Title 10 authorities for Military Departments and CCDRs to conduct training. Both Services and CCMDs are authorized to deploy forces (Service-retained or CCDR assigned/ allocated) to conduct unit level training.[15]

(b) <u>Operations</u>. As operations are another military activity, the authority to execute a military operation originates from a decision by the President and/or Secretary and are typically conveyed via a Chairman issued EXORD. CCDRs and Services subsequently issue orders down the chain of command to subordinate CDRs to implement the direction of the CJCS EXORD.

1 <u>Contingency Plans</u>. When directed, CCDRs may execute contingency plans to respond to emergent threats to national security interests. Contingency plans developed prior to a crisis can provide the basis for a response. However, the actual OE will likely differ from the hypothetical situation anticipated during planning. Planners must review and validate the plan assumptions based upon situational awareness of actual conditions. Deviations from planning assumptions may require re-entering the planning process in order to refine or adapt the plan (see Chapter 6, *Plan Assessment*).

2 <u>Crisis Action</u>. In a time-constrained environment, often called crisis action, planners may expedite, compress or perform in parallel the operational activities and functions of the JPP to meet the time demands of the situation. Under these circumstances, the JPP remains fundamentally unchanged, in order to provide a consistent, logical approach for effectively integrating planning and execution functions with ongoing operational activities.

3 <u>Standing EXORDs</u>. Secretary approved, and Chairman released, standing EXORDs delegate to one or more CCDRs specific execution authorities. Existing EXORDs

[10] *JP 3-20, Security Cooperation*

[11] *Title 10, U.S.C. (Section 333)*

[12] *CJCSI 3710.01 Series, DoD Counterdrug Support*

[13] *CJCSM 3130.06 Series, Global Force Management Allocation Policies and Procedures*

[14] *Ibid*

[15] *CJCSM 3110.011-1 Series, Contingency Planning Supplement to the JSCP*

can be found on the JS J-35 GFM Division web page (SIPRNET). Directed CCMDs develop and issue their own standing EXORDs accordingly prescribing further execution details to their subordinate and supporting commands. Subordinate commands, in turn, issue EXORDs communicating the authorities granted, as appropriate.

4 <u>Combined Authorities</u>. Forces frequently deploy and operate under a combination of the listed authorities. For example, a force may be allocated for an operation, and while conducting the operation also participate in exercise or security force assistance (SFA). When operating under multiple authorities, CDRs should clearly delineate the linkages between authorities and specific activities. Planners should clearly identify the specific authorities when developing force requests for allocation.[16,17]

d. *If an existing plan is appropriate*, the CDR and staff should review and update the plan. When prior execution planning has been accomplished through adaptation of an existing plan or the development of an emergency OPORD, most of the guidance necessary for execution will already have been passed to the implementing commands, either through an existing plan or by a previously issued WARNORD, PLANORD, ALERTORD, PTDO, DEPORD, or REDEPORD. Under these circumstances, the EXORD need only contain the authority to execute the planned operation and any additional essential guidance, such as the date and time for execution. Reference to previous planning documents is sufficient for additional guidance.

e. *If no existing plan meets the guidance*, the CDR and staff conduct crisis planning (planning in reduced timeline). More often than not, the CDR and staff have conducted some previous analysis of the OE that will speed the planning process. In the no-prior-warning response situation where a crisis event or incident requires an immediate response without any prior formal planning, the EXORD must pass all essential guidance that would normally be issued in the WARNORD, PLANORD, and ALERTORD. Under such rapid reaction conditions, the EXORD will generally follow the same paragraph headings as the PLANORD or ALERTORD. If some information may be desirable but is not readily available, it can be provided in a subsequent message because the EXORD will normally be very time-sensitive.

f. During execution, the situation may change rapidly. Operations the CDR envisioned in the plan may bear little resemblance to actual events in execution. Subordinate and supporting CDRs need maximum latitude to take advantage of situations and meet the higher CDR's intent when the original order no longer applies.

> "No war is over until the enemy says it's over. We may think it over, we may declare it over, but in fact, the enemy gets a vote."
> General James Mattis, USMC

4. Planning During Execution

Planning continues during execution, with an initial emphasis on refining the existing plan and producing the OPORD. As the operation progresses, planning generally occurs in three distinct but overlapping time-frames: *future plans, future operations, and current operations* as the figure "*Planning During Execution*" on the following page depicts.

a. <u>The Joint forces J-5 effort focuses on future plans</u>. The time-frame of focus for this effort varies according to the level of command, type of operation, CDR's desires, and other factors. Typically, the emphasis of the future plan's effort is on planning the next phase of operations or sequels to the current operation. In a campaign, this could be planning the next major operation (the next phase of the campaign).

---

[16] DoDD 5100.01, Functions of the Department of Defense and its Major Components
[17] CJCSM 3130.06, *GFM Allocation Policies and Procedures*

> "Determine the thing that can and shall be done, and then we shall find the way." Abraham Lincoln

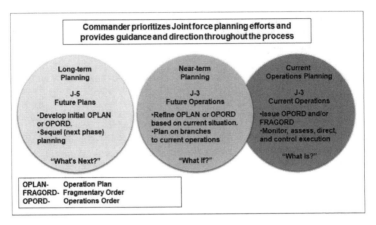

Commander prioritizes Joint force planning efforts and provides guidance and direction throughout the process

Long-term Planning	Near-term Planning	Current Operations Planning
J-5 Future Plans	**J-3 Future Operations**	**J-3 Current Operations**
•Develop initial OPLAN or OPORD. •Sequel (next phase) planning	•Refine OPLAN or OPORD based on current situation. •Plan on branches to current operations	•Issue OPORD and/or FRAGORD •Monitor, assess, direct, and control execution
"What's Next?"	"What If?"	"What Is?"

OPLAN- Operation Plan
FRAGORD- Fragmentary Order
OPORD- Operations Order

Planning During Execution

b. Planning also occurs for branches to current operations (future operations planning). The time-frame of focus for future operations planning varies according to the factors listed for future plans, but the period typically is more near-term than the future plans time-frame. Future planning could occur in the J-5 or JPG, while future operations planning could occur in the Joint Operations Center (JOC) or J-3.

c. Finally, current operations planning addresses the immediate or very near-term planning issues associated with ongoing operations. This occurs in the JOC or J-3.

d. During execution of a plan, accomplishment of the plan's tasks are monitored and measured for how successfully each objective was completed, along with the input of new data and information as it is obtained to allow selection of branches or sequels, if applicable, or the plan to be modified as necessary. During execution, planning will continue for future operations within the plan to include branches and sequels. The ***planning functions*** discussed in previous chapters will continue to be the basis for planning, although planners may reenter the planning process at any of the earlier functions.

(1) <u>Campaign Plan</u>. Activities within campaign plans are in constant execution. During execution, the CDR will likely have reason to consider updating the operational approach. It could be triggered by significant changes to understanding of the OE and/or problem, validation or invalidation of assumptions made during planning, identifying (through continuous assessment process) that the tactical actions are not resulting in the expected effects, changes in the conditions of the OE, or the endstate. The CDR may determine one of three ways ahead:

- The current OPLAN is adequate, with either no change or minor change (such as execution of a branch)—the current operational approach remains feasible.

- The OPLAN's mission and objectives are sound, but the operational approach is no longer feasible or acceptable—a new operational approach is required.

Execution

- The mission and/or objectives are no longer valid, thus a new OPLAN is required—a new operational approach is required to support the further detailed planning.

(2) <u>Contingency Plan</u>. For a contingency plan, members of the planning team may not be the same as those responsible for execution. They may have rotated out or could be in the planning sections of the staff rather than operations. This is the most likely situation where the conditions used in developing the plan will have changed, due to the time lag between plan development and execution. Staff from the planning team (J-5) need to provide as much background information as possible to the operations team (J-3). The planning team should be a key participant, if not the lead, in updating the plan for the current (given) conditions. This enables the command to make effective use of the understanding gained by the staff during the planning process. The operations team should be the co-lead for the plan update to ensure they understand the decision processes and reasoning used in development of the operational approach and COAs. This will speed plan the update, ease transition, and minimize the time required to revisit the issues that arose during the initial plan development.

(3) <u>Crisis Plan</u>. For a crisis plan, the planning team will analyze approved contingency plans with like scenarios to determine if an existing plan or its information may apply. Contingency plans rarely, if ever, align exactly to a crisis scenario, however some details may be useful. If information within a contingency plan is appropriate to the situation, that information may be used in helping to formulate the crisis plan which, when approved, will be executed through an OPORD or FRAGORD. In a crisis, planning usually transitions rapidly to execution, so there is limited deviation between the plan and initial execution. Planners from the command J-5 can assist in the planning process through their planning expertise and knowledge gained of the OE during similar planning efforts.

(4) <u>Assessment</u>. Assessment could cause the JFC to shift the focus of the operation, which the JFC would initiate with a new visualization manifested through new planning guidance for an adjusted operation or campaign plan.

> *"The most successful people are those who are good at plan B."*
> *James Yorke*

5. Plan Adjustment

Events that offer better ways to success are opportunities. CDRs recognize opportunities by continuously monitoring and evaluating the situation.[18]

a. Uncertainty and risk are inherent in all military operations. Recognizing and acting on opportunity means taking risks. Reasonably estimating and intentionally accepting risk is not gambling. Carefully determining the risks, analyzing and minimizing as many hazards as possible, and executing a supervised plan that accounts for those hazards, contributes to successfully applying military force. Gambling, in contrast, is imprudently staking the success of an entire action on a single, improbable event (i.e., assuming away the threat).[19]

b. When CDRs embrace opportunity, they accept risk. It is counterproductive to wait for perfect preparation and synchronization. The time taken to fully synchronize forces and warfighting functions in a detailed order could mean a lost opportunity. It is far better to quickly summarize the essentials, get things moving, and send the details later. Leaders optimize the use of time with WARNORDs, FRAGORDs, and verbal updates.

Execution

[18] *FM 3-0, Operations*
[19] *Ibid*

c. Assessment in execution identifies variances, their magnitude and significance, and the need for decisions and what type—whether execution or adjustment. The CDR and staff assess the probable outcome of the operation to determine whether changes are necessary to accomplish the mission, take advantage of opportunities, or react to unexpected threats.

d. During execution, assessing allows the CDR and staff to determine the existence and significance of variances from the operations as envisioned in the initial plan. The staff makes recommendations to the CDR about what action to take concerning variances they identified.

6. Summary

A CDR's visualization of the situation allows subordinate and supporting CDRs—and in some cases higher headquarters—to adjust their actions rapidly and effectively. Throughout execution, the Chairman through their staff monitors movements, deployment and employment of forces, assesses achievement of tasks, and resolves shortfalls as necessary. The Chairman monitors and directs action needed to ensure successful completion of military operations. Execution continues until the operation is terminated or the mission is accomplished or revised.

Execution

A. Summary

> *"Don't let day to day operations drive out planning"*
> Donald Rumsfeld

Today's operational environment continues to present multiple, diverse and difficult strategic challenges for the United States. A national level of effort involving seamless integration and coordination of multiple elements of national power—diplomacy, information, economics, finance, intelligence, law enforcement, and military (and in some cases, alliance and coalition partners) is required to win current and future fights and to ensure the viability of a government capable of "defending the people" and our vital national interests at home and abroad.

> *"Today's strategic landscape is also extraordinarily volatile, and the nation faces threats from an array of state and nonstate actors. These realities are why some have called today's operating environment the most challenging since World War II."*
> Gen. Joe Dunford, Chairman of the Joint Chiefs of Staff

Although conflict, violence, and war endure, the methods through which political goals are pursued are always evolving. How this change in the character of conflict will play out and what the Joint Force must do to prepare to meet the demands of tomorrow requires our collective attention.[1]

Forecasting the future, particularly the deep future, is a daunting task, but the global trends are rapidly gathering momentum and shaping every facet of society and international discourse, including security policy and warfare.[2]

The National Defense Strategy acknowledges an increasingly complex global security environment, characterized by overt challenges to the free and open international order and the re-emergence of long-term, strategic competition between nations. This increasingly complex security environment is defined by rapid technological change, challenges from adversaries in every operating domain, and the impact on current readiness resulting from the longest continuous stretch of armed conflict in our Nation's history.

The U.S. military finds itself at a historical inflection point, where disparate, yet related elements of the operational environment are converging, creating a situation where fast-moving trends across our multiple elements of national power are rapidly transforming the nature of all aspects of society and human life – including the character of warfare. These trends include significant advances in science and technology, where new discoveries and innovations are occurring at a breakneck pace, a dizzying pace of human interaction and a world:[3]

- that is connected through social media where cognition, ideas, and perceptions, are almost instantaneously available.
- where economic disparities are growing between and within nations and regions.

[1] *Joint Operating Environment 2035*

[2] *The Operational Environment and the Changing Character of Future Warfare Forecasting the Future: Toward a Changing Character of Warfare, U.S. Army Training and Doctrine Command*

[3] *Ibid*

- with competition for natural resources, especially water, becoming more common.

Where geopolitical challenges to the post-Cold War U.S.-led global system in which near-peer competitors, regional hegemons, ideologically-driven non-state actors, and even super empowered-individuals, are competing with the U.S. for leadership and influence in an ever-shrinking world.

These trends must be considered in the military sphere, matched with advances in our adversaries' capabilities and operational concepts, and superimposed over a U.S. military that has been engaged in a non-stop state of all-consuming counter-insurgency warfare for the last 20-plus years. The result is a U.S. military that may find itself with the very real potential of being out-gunned, out-ranged, out-protected, outdated, out of position, and out of balance against our adversaries. These potential foes have had time to refine their approaches to warfare, develop and integrate new capabilities, and in some cases expedite growing changes in the character of warfare.[4]

In this, today's global operating environment, we, as joint planners, must adjust our view of the adversary and the environment writ large. A logical way of gaining that increased understanding is to break the OE into its major parts, examine these parts individually, and then study the relationships and interaction between them to comprehend not only what is occurring, but why, and then plan to it, realizing that as the environment evolves, so will OUR plan!

The CCDRs environment of today is potentially overwhelming with information. Our job as planners is to present that information in a logical flow to the CCDR, using JPP as a tool to assist us plan. Remember, the plan you're working on is probably not the CCDR's only plan, nor concern. The CCDR's time is valuable and your job is to get the point across in a relevant, well-thought-out manner. Give the CCDR solutions, not problems, or if you are lacking the solution, give the CCDR options - well worked, learned and vetted.

"Planning for Planners" has outlined an approach to planning utilizing the construct of the Joint Planning Process. As this document has presented, the JPP is key to making logical, sequential and learned decisions. It is a standardized planning process that is conceptually easy to understand and capable of being applied in campaign, contingency or crisis environments; however, it is only a guide.

The traditional military-centric single center of gravity [focus] planning approach that worked so well in the Cold War doesn't allow us to accurately analyze, describe, and visualize today's networked, adaptable, asymmetric adversary. This adversary has no single identifiable source of all power. Rather, because of globalization, the information revolution, and, in some cases, the non-state characteristic of our adversary, this form of adversary can only be described (and holistically attacked) as a system of systems. This environment requires astute multi-dimensional and adaptive planners. Our resources are limited, but our adversaries are not. Plan well, plan often, don't be married to your plan and never stop asking the most important questions in planning; "WHAT IF," and as General Zinni, USMC is often quoted as saying, "THEN WHAT?"

"And for the support of this declaration, with a firm reliance on the protection of Divine Providence, we mutually pledge to each other our Lives, our Fortunes, and our sacred Honor."
Final lines of the Declaration of Independence, July 4, 1776

---

[4] *The Operational Environment and the Changing Character of Future Warfare Forecasting the Future: Toward a Changing Character of Warfare, U.S. Army Training and Doctrine Command*

B. References

Primary Sources

CJCSI 1301.01, *Joint Individual Augmentation Procedures*

CJCSG 3130, *Adaptive Planning and Execution Overview and Policy Framework*

CJCSG 3122, *Time-Phased Force and Deployment Data (TPFDD) Primer*

CJCSI 1301.01, Series, *Joint Individual Augmentation Procedures*

CJCSI 3100.01, *Joint Strategic Planning System*

CJCSI 3110.01, *Joint Strategic Campaign Plans*

CJCSI 3110.O3 Series, *Logistics Supplement to the JSCP*

CJCSI 3141.01, *Management and Review of Campaign and Contingency Plans*

CJCSI 3710.01, Series, *DoD Counterdrug Support*

CJCSI 3401.01, *Joint Combat Capability Assessment*

CJCSI 5715.01, *Joint Staff Participation in Interagency Affairs*

CJCSI 8501.01, *Chairman of the Joint Chiefs of Staff, Combatant Commanders, Chief, National Guard Bureau, and Joint Staff Participation in the Planning, Programming, and Budgeting System*

CJCSM 3105.01, *Joint Risk Analysis*

CJCSM 3110.011-1, Series, *Contingency Planning Supplement to the JSCP*

CJCSM 3122.01, *Joint Operation Planning and Execution System (JOPES) Volume I*

CJCSM 3122.03, *Joint Operation Planning and Execution System (JOPES) Volume II*

CJCSM 3122.02, Series, *Joint Operation Planning and Execution System (JOPES) Volume III Time Phased Force and Deployment Data Development and Deployment Execution*

CJCSM 3130.00, *Crisis Action TPFDD Development and Deployment Execution*

CJCSM 3130.01, *Campaign Planning Procedures and Responsibilities*

CJCSM 3130.03, *Planning and Execution Formats and Guidance*

CJCSM 3130.04, *Deployment Policies and Procedures.*

CJCSM 3130.06, Series, *Global Force Management Allocation Policies and Procedures*

CJCSM 3314.01, *Intelligence Planning*

CJCSM 3500.04, *Universal Joint Task List*

CJCSM 3500.05A, *JTFHQ Master Training Guide*

DoDD 1235.10, *Activation, Mobilization, and Demobilization of the Ready Reserve*

DODD, 1322.18, *Military Training.*

DODD 1400.31, *DOD Civilian Work Force Contingency and Emergency Planning and Execution.*

DODD 1404.10, *DOD Civilian Expeditionary Workforce*

DODD 3000.05, *Military Support for Stability, Security, Transition and Reconstruction (SSTR)*

DODD 3000.06, *Combat Support Agencies*

DODD 3020.42, *Defense Continuity Plan Development*

DODD 5100.1, *Functions of the Department of Defense and its Major Components*

DODD 5100.3, *Support of the Headquarters of Combatant and Subordinate Joint Commands*

DoDD 5132.03, *DoD Policy and Responsibilities Relating to Security Cooperation*

DODD 5143.01, *Under Secretary of Defense for Intelligence (USD(I))*

DODI 1100.22, *Policy and Procedures for Determining Workforce Mix*

DoDI I235.12, *Accessing the Reserve Components (RC)*

DODI 3000.05, *Stability Operations*

DoDI 3000.12, *Management of U.S. Global Defense Posture (GDP)*

DODI 3020.41, *Contractor Personnel Authorized to Accompany the US Armed Forces*

DODD 5100.1, *Functions of the Department of Defense and its Major Components*

DODD 5100.3, *Support of the Headquarters of Combatant and Subordinate Joint Commands*

DODI 5122.05, *Assistant Secretary of Defense for Public Affairs (ASD(PA))*

DODI 8260.03, *The Global Force Management (GFM) Data Initiative (DI)*

DODM 4140.01, Volume I, *DoD Supply Chain Material Management Procedures: Operational Req's*

DOD Dictionary of Military and Associated Terms (DOD Dictionary)

DTM-17-004, *Department of Defense Expeditionary Civilian Workforce, Change 1*

Budget of the U.S. Government

Defense Strategy Review

Global Force Management, Business Rules

Global Force Management, Initial Capabilities Document

Global Force Management Implementation Guidance. (GFMIG)

Guidance for Development of the Force

Joint Information Operations Planning Handbook

Joint Operating Environment (JOE) 2035

Joint Operating Concept (JOC), Deterrence Operations

National Defense Strategy of the United States of America

National Military Strategy of the United States

National Military Strategic Plan for the War on Terrorism

National Response Framework

National Security Presidential Directive 1 (NSPD-1)

National Strategy for Homeland Security

National Strategy to Secure Cyberspace

National Security Strategy of the United States

Quadrennial Diplomacy and Development Review, DOS

The Goldwater-Nichols Department of Defense Reorganization Act of 1986 (10 USC 161 et. seq. PL 99-433)

Title 10, US Code, as amended

Unified Command Plan

U.S. National Strategy for Public Diplomacy and Strategic Communication

Joint Publication 1, *Doctrine for the Armed Forces of the United States*

JP 2-0, *Joint Intelligence*

JP 2-01, *Joint and National Intelligence Support to Military Operations*

JP 2-01.3, *Joint Intelligence Preparation of the Operational Environment*

JP 2-03, *Geospatial Intelligence Support to Joint Operations*

JP 3-0, *Joint Campaigns and Operations*

JP 3-07, *Stability Operations*

JP 3-08, *Interorganizational Coordination during Joint Operations*

JP 3-13, *Information Operations*

JP 3-20, *Security Cooperation*

JP 3-27, *Homeland Defense*

JP 3-33, *Joint Task force Headquarters*

JP 3-35, *Deployment and Redeployment Operations*

JP 4-O, *Joint Logistics*

JP 4-01, *The Defense Transportation System*

JP 4-05, *Joint Mobilization Planning*

JP 4-09, *Distribution Operations*

JP 5-0, *Joint Planning*

JP 5-00.2, *JTF Planning Guidance and Procedures*

Joint Doctrine Note 2-19, Strategy

Joint Doctrine Publication (JDP) 01, United Kingdom, *Campaigning*

Joint Doctrine Publication (JDP) 3-00, United Kingdom, *Campaign Execution*

Annexes

Joint Doctrine Publication 5-00 (JDP-5-00) United Kingdom, *Campaign Planning*

Books

Clausewitz, Carl von., *On War*. Edited and translated by Michael Howard and Peter Paret. Princeton, NJ: Princeton University Press, 1976.

Franks, Tommy with Malcom McConnell, *American Soldier*. Regan Books, 2004.

Freeman, Douglas Southall, *Lee's Lieutenants, a Study in Command*. Scribner's,1943.

Gordon, Michael R. and Trainor, Bernard E., *Cobra II*. Pantheon Books, New York, 2006.

Koontz, Harold and O'Donnell, Cyril, *Principles of Management: An Analysis of Managerial Functions* 5th ed. New York: McGraw-Hill,1972.

Reynolds, Nicholas E., *U.S. Marines in Iraq, 2003 Basrah, Bagdad and Beyond: U.S. Marines in the Global War on Terrorism*, History Division, United States Marine Corps, Washington D.C. 2007.

Ricks, Thomas E., Fiasco, *The American Military Adventure in Iraq*. The Penguin Press, New York, 2006.

Strange, Dr. Joe, *Centers of Gravity and Critical Vulnerabilities: Building On The Clausewitzian Foundation So That We Can All Speak The Same Language*. Marine Corps University Foundation, 1996.

Strange, Dr. Joe, Marine Corps University, *Perspectives on Warfighting*, Number 4, 1996.

Sun Tzu, *The Art of War*. Oxford, Oxford University Press, 1963.

Vego. Dr. Milan, *Joint Operational Warfare*, 20 September 2007.

Weigley, Russell F., *The American Way of War: A History of United States Military Strategy and Policy*. Bloomington, Indiana University Press, 1977.

William Manchester, *American Caesar,* (New York: Dell Publishing, 1983), 617.

Wright, Donald P and Reese, Timothy R. *On Point II. Transition to the New Campaign: The United States Army in Operation Iraqi Freedom May 2003-January 2005.*

Service Publications/War Colleges

ADRP 3-0, *Operations*

ADP 5-0, *The Operations Process*.

AFDD 2-5, *Information Operations*.

AFDD 2-6, *Air Mobility Operations*.

ATP 5-0.3, MCRP 5-10.1, NTTP 5-01.3, AFTTP 3-2.87, Multi-Service Tactics, Techniques, and Procedures for Operation Assessment.

Air Force War and Mobilization Plan.

Army War College, *Campaign Planning Primer*.

FM 3-0, *Operations*.

FM 3-07, *Stability Operations*.

FM 3-24/MCWP 3-33.5, *Counterinsurgency*.

FM 3-93, *The Army in Theater Operations*.

FM 5-0, *The Operations Process*.

FM 100-5, *Operations*.

Marine Corps Doctrinal Publication (MCDP) 1-2, *Campaigning*.

Marine Corps Doctrinal Publication-5 (MCDP-5), *Planning*.

Marine Corps Warfighting Publication (MCWP) 5-1, *Marine Corps Planning Process*.

National Defense University. *Interagency Management of Complex Crisis Operations Handbook.* January 2003.

Naval War College, *NWP 5-01, Navy Planning*.

The Operational Environment and the Changing Character of Future Warfare Forecasting the Future: Toward a Changing Character of Warfare, U.S. Army Training and Doctrine Command, G-2

Sweeny, Dr. Patrick C., Naval War College, *NWC 4111H, Joint Operations Planning Process Workbook*.

United States Army War College, Department of Military Strategy, Planning, and Operations, Carlisle Barracks. *Campaign Planning Handbook,* COL Mark Haseman.

Warfare Studies Institute, *Joint Air Operations Planning Course, Joint Air Estimate Planning Handbook*.

Articles/Monograph/Thesis/Reports

Burdon, John; Conway, Timothy; Santacroce, Michael A., *The Challenge of Achieving Adaptive Planning and Execution Objectives; Synchronizing with Global Force Management*. 15 September 2011.

Cole, Ronald H. (1999). *Grenada, Panama, and Haiti: Joint Operational Reform*" (PDF). Joint Force Quarterly (20 (Autumn/Winter 1998-99)): 57–74.

Conway, Timothy W., Santacroce, Michael A., *Utilizing the Current Allocation Process for Requesting*

Combat Support Agency (CSA) Support to CCDRs. 03 November 2010.

CRS Report R43838, *A Shift in the International Security Environment: Potential Implications for Defense*—Issues for Congress, by Ronald O'Rourke.

Dickson, Keith. *Operational Design: A Methodology for Planners.*

Dunford, Joe., *The Character of War & Strategic Landscape Have Changed.*

Dunford, Joseph F.Jr., *2018 National Military Strategy Framework*, Joseph Francis Dunford Jr., 19th Chairman of the JCS.

Echevarria, Antulio J., *Clausewitz's Center of Gravity: It's Not What We Thought.* Strategic Studies Institute, U.S. Army War College, Carlisle Barracks, Pa. 2003.

Futrell, Robert F., *Ideas, Concepts, Doctrine*, Vol. 1 (Maxwell AFB, Ala.: Air University Press, 1989).

Gardner, David W., *Clarifying Relationships Between Objectives, Effects and End States with Illustrations and Lessons from the Vietnam War.* Joint Advanced Warfighting School. 5 April 2007.

Goodwin, LCDR Ben B., *War Termination and the Gulf War: Can We Plan Better?* Naval War College. 9 February 2004.

Guthrie, Colonel USMC, John W., U.S. Army War College, *The Theater Commander: Planning for Conflict Termination.* 15 March 2006.

Janiczek, Rudolf M. A., *Concept at the Crossroads: Rethinking Center of Gravity.* Strategic Studies Institute, U.S. Army War College, Carlisle Barracks, Pa. October 2007.

Luck, Gary, Gen (Ret), *Insights on Joint Operations: The Art and Science, Best Practices, The Move Toward Coherently Integrated Joint, Interagency, and Multinational Operations.* September 2006.

Manwaring, Max G., *The Inescapable Global Security Arena*, Carlisle Barracks, PA: Strategic Studies Institute, U.S. Army War College, 2002, p. 2.

Melshen, Dr. Paul., *Mapping Out a Counterinsurgency Campaign Plan: Critical Considerations in Counterinsurgency Campaigning.* 17 March 2006.

Nemfakos, Charles • Blickstein, Irv • Seitz McCarthy, Aine • Sollinger, Jerry M.; *The Perfect Storm: The Goldwater-Nichols Act and Its Effect on Navy Acquisition.*

Poole, Walter S., Joint History Office, Office of the Chairman Joint Chiefs of Staff, *The Effort to Save Somalia August 1992-March 1994.* August 2005.

Reed, James W., Should Deterrence Fail: *War Termination in Campaign Planning*, Parameters. Summer 1993 pp. 41-52.

Richard K., *Engaging Civil Centers of Gravity and Vulnerabilities.*

Santacroce, Michael A.; Conway, Timothy W.; Eremita, Nicholas. *Moving Out With Global Force Management.* 23 March, 2010.

Sele, Richard K., *Engaging Civil Centers of Gravity and Vulnerabilities.* Military Review. September-October 2004.

Smith,Leighton W., *A Commander's Perspective*, as found in, Dennis J. Quinn (ed), *The Goldwater-Nichols DOD Reorganization Act: A Ten-Year Retrospective* (Washington, DC: National Defense University Press, 1999) p. 29. See also Clark A. Murdock, *Beyond Goldwater-Nichols: Defense Reform for a New Strategic Era, Phase 1* Report (Washington, DC: Center for Strategic and International Studies Press, 2004). http://csis.org/files/media/csis/pubs/ bgn_ph1_report.pdf, p. 14.

Stewart, Richard W., ed. (2005). *Chapter 12: Rebuilding the Army Vietnam to Desert Storm.* American Military History, Volume II. United States Army Center of Military History.

U.S. Department of Defense, Department of Defense Press Briefing by Deputy Secretary Work and Gen. Selva on the FY2017 Defense Department Budget Request in the Pentagon Press Briefing Room, February 9, 2016, http://www.defense.gov/News/News-Transcripts/Transcript-View/Article/653524/ department-of-defense-press-briefing-by-deputy-secretary-work-and-gen-selva-on.

Walsh, Mark R. and Harwood, Michael J., *Complex Emergencies: Under New Management*, Parameters. Winter 1998, pp. 39-50.

National Publications/Official Reports/Policy/Web/Other

Adaptive Planning Combatant Commander Organizational Workload to Capacity Study, USJFCOM. July 2009.

Al-Qa'ida (the Base), International Policy Institute for Counter-Terrorism, on the World Wide Web at www.ict.org.il/inter_ter (accessed 3 April 2002).

Conduct of the Persian Gulf War, Final Report to Congress. April 1992.

Katzman, Kenneth. CRS Report for Congress. *Iraq: Post-Saddam Governance and Security.* January 17, 2007.

Millennium Challenge Act of 2003. Title VI.

National Security Presidential Directive 1 (NSPD-1), *Organization of the National Security Council System.* February 13, 2001.

The National Security Policy Process: *The National Security Council and Interagency System.* Alan G. Whittaker, PhD., Frederick C. Smith, and Ambassador Elizabeth McKune. Annual Update: April, 2007.

C. Timelines/Dates & OPLAN Annexes

1. Times

(C-, D-, M-days end at 2400 hours Universal Time (Zulu time) and are assumed to be 24 hours long for planning.) The Chairman of the Joint Chiefs of Staff normally coordinates the proposed date with the CDRs of the appropriate unified and specified commands, as well as any recommended changes to C-day. L-hour will be established per plan, crisis, or theater of operations and will apply to both air and surface movements. Normally, L-hour will be established to allow C-day to be a 24-hour day.

a. C-Day. The unnamed day on which a deployment operation commences or is to commence. The deployment may be movement of troops, cargo, weapon systems, or a combination of these elements using any or all types of transport. The letter "C" will be the only one used to denote the above. The highest command or headquarters responsible for coordinating the planning will specify the exact meaning of C-day within the aforementioned definition. The command or headquarters directly responsible for the execution of the operation, if other than the one coordinating the planning, will do so in light of the meaning specified by the highest command or headquarters coordinating the planning.

b. D-Day. The unnamed day on which a particular operation commences or is to commence.

c. F-Hour. The effective time of announcement by the Secretary of Defense to the Military Departments of a decision to mobilize Reserve units.

d. H-Hour. The specific hour on D-day at which a particular operation commences.

e. H-Hour (amphibious operations). For amphibious operations, the time the first assault elements are scheduled to touch down on the beach, or a landing zone, and in some cases the commencement of countermine breaching operations.

f. I-Day. (CJCSM 3110.01/JSCP). The day on which the Intelligence Community determines that within a potential crisis situation, a development occurs that may signal a heightened threat to U.S. interests. Although the scope and direction of the threat is ambiguous, the Intelligence Community responds by focusing collection and other resources to monitor and report on the situation as it evolves.

g. L-Hour. The specific hour on C-day at which a deployment operation commences or is to commence.

h. L-Hour (Amphibious Operations). In amphibious operations, the time at which the first helicopter of the helicopter-borne assault wave touches down in the landing zone.

i. M-Day. The term used to designate the unnamed day on which full mobilization commences or is due to commence.

j. N-Day. The unnamed day an active duty unit is notified for deployment or redeployment.

k. R-Day. Redeployment day. The day on which redeployment of major combat, combat support, and combat service support forces begins in an operation.

l. S-Day. The day the President authorizes Selective Reserve call-up (not more than 200,000).

m. <u>T-Day</u>. The effective day coincident with Presidential declaration of national emergency and authorization of partial mobilization (not more than 1,000,000 personnel exclusive of the 200,000 call-up).

n. <u>W-Day</u>. Declared by the President, W-day is associated with an adversary decision to prepare for war (unambiguous strategic warning).

2. Operational Plan Annexes

A   Task Organization
B   Intelligence
C   Operations
D   Logistics
E   Personnel
F   Public Affairs
G   Civil-Military Operations
H   Meteorological and Oceanographic (METOC) Operations
J   Command Relationships
K   Command, Control, Communications, and Computer (C4) Systems
L   Environmental Considerations
M   Geospatial Information and Services
N   Assessments
P   Host Nation Support
Q   Health Services
R   Reports
S   Special Technical Operations
T   Chemical, Biological, Radiological, and Nuclear (CBRN) Response (CBRN-R)
U   Notional Counterproliferation Decision Guide
V   Interagency-Interorganizational Coordination
W   Operational Contract Support
X   Execution Checklist
Y   Commander's Communication Strategy
Z   Distribution
AA Religious Support

Annexes A-D, K, and Y are required annexes for a Crisis OPORD per JOPES. All others may either be required by the JSCP or deemed necessary by the supported CCDR.

D. Commander's Estimate

1. Purpose

a. The CDR's estimate, submitted by the supported CDR in response to a CJCS WAR-NORD, provides the CJCS with time-sensitive information for consideration by the NCA in meeting a crisis situation. Essentially, it reflects the supported CDR's analysis of the various COAs that may be used to accomplish the assigned mission and contains recommendations as to the best COA (recommended COAs submitted for President, SecDef approval may be contained in current OPLANs or CONPLANs or may be developed to meet situations not addressed by current plans. Regardless of origin, these COAs will be specifically identified when they involve military operations against a potential enemy). Although the estimative process at the supported CDR's level may involve a complete, detailed estimate by the supported CDR, the estimate submitted to the CJCS will normally be a greatly abbreviated version providing only that information essential to the President, SecDef and the CJCS for arriving at a decision to meet a crisis.

b. Supporting CDRs normally will not submit a CDR's estimate to the CJCS; however, they may be requested to do so by the supported CDR. They may also be requested to provide other information that could assist the supported CDR in formulating and evaluating the various COAs.

2. When Submitted

a. The CDR's Estimate will be submitted as soon as possible after receipt of the CJCS WARNORD, but no later than the deadline established by the CJCS in the WARNORD. Although submission time is normally 72 hours, extremely time-sensitive situations may require that the supported CDR respond in 4 to 8 hours.

b. Follow-on information or revisions to the CDR's Estimate should be submitted as necessary to complete, update, or refine information included in the initial estimate.

c. The supported CDR may submit a CDR's Estimate at the CDR's own discretion, without a CJCS WARNORD, to advise the SecDef and CJCS of the CDR's evaluation of a potential crisis situation within the AOR. This situation may be handled by a SITREP instead of a CDR's Estimate.

3. How Submitted

The CDR's Estimate is submitted by record communication, normally with a precedence of IMMEDIATE or FLASH, as appropriate. GCCS Newsgroup should be used initially to pass the CDR's estimate, but must be followed by immediate record communication to keep all crisis participants informed.

4. Addressees

The message is sent to the CJCS with information copies to the Services, components, supporting commands and combat support agencies, USTRANSCOM, and other appropriate commands and agencies.

5. Contents

a. The CDR's Estimate will follow the major headings of a CDR's Estimate of the Situation as outlined in Appendix A to Enclosure J but will normally be substantially abbreviated in content. As with the WARNORD, the precise contents may vary widely, depending on the nature of the crisis, time available to respond, and the applicability of prior planning. In a rapidly developing situation, a formal CDR's Estimate may be initially impractical, and the entire estimative process may be reduced to a CDR's conference, with corresponding brevity reflected in the estimate when submitted by record communications to the CJCS. Also, the existence of an applicable OPLAN may already reflect most of the necessary analysis.

b. The essential requirement of the CDR's Estimate submitted to the CJCS is to provide the SecDef in a timely manner, with viable military COAs to meet a crisis. Normally, these will center on military capabilities in terms of forces available, response time, and significant logistic considerations. In the estimate, one COA will be recommended. If the supported CDR desires to submit alternative COAs, an order of priority will be established. All COAs in the WARNORD will be addressed.

c. The estimate of the supported CDR will include specific information to the extent applicable. The following estimate format is desirable but not mandatory and may be abbreviated where appropriate.

(1) <u>Mission</u>. State the assigned or deduced mission and purpose. List any intermediate tasks, prescribed or deduced, that the supported CDR considers necessary to accomplish the mission.

(2) <u>Situation and COA</u>. This paragraph is the foundation of the estimate and may encompass considerable detail. Because the CJCS is concerned primarily with the results of the estimate rather than the analysis, for purposes of the estimate submitted, include only the minimum information necessary to support the recommendation.

(a) <u>Considerations Affecting the Possible COA</u>. Include only a brief summary, if applicable, of the major factors pertaining to the characteristics of the area and relative combat power that have a significant impact on the alternative COAs.

(b) <u>Enemy Capability</u>. Highlight, if applicable, the enemy capabilities and psychological vulnerabilities that can seriously affect the accomplishment of the mission, giving information that would be useful to the President, SecDef, and the CJCS in evaluating various COAs.

(c) <u>Terrorist Threat</u>. Describe potential terrorist threat capabilities to include force protection requirements (prior, during, and post mission) that can affect the accomplishment of the mission.

(d) <u>Own COA</u>. List COAs that offer suitable, feasible, and acceptable means of accomplishing the mission. If specific COAs were prescribed in the WARNORD, they must be included. For each COA, the following specific information should be addressed:

<u>1</u>. Combat forces required; e.g., 2 FS, 1 airborne brigade. List actual units if known.

<u>2</u>. FP.

<u>3</u>. Destination.

<u>4</u>. Required delivery dates.

<u>5</u>. Coordinated deployment estimate.

<u>6</u>. Employment estimate.

<u>7</u>. Strategic lift requirements, if appropriate.

(3) <u>Analysis of Opposing COA</u>. Highlight enemy capabilities that may have significant impact on U.S. COAs.

(4) <u>Comparison of Own COA</u>. For the submission to the CJCS, include only the final statement of conclusions and provide a brief rationale for the favored COA. Discuss the advantages and disadvantages of the alternative COAs, if significant, in assisting the President, SecDef, and the CJCS in arriving at a decision.

(5) <u>Recommended COA</u>. State the supported CDR's recommended COA (recommended COA should include any recommended changes to the ROE in effect at that time) (CJCSM 3122.01)).

E. Acronyms & Abbreviations

A

AC	Active Component
ACT	Advanced Civilian Teams
AFD	Assigned Force Demand
ALERTORD	Alert Order
AMC	Air Mobility Command
AO	Area of Operations
AOI	Area of Interest
AOR	Area of Responsibility
APOD	Aerial Port of Debarkation
APOE	Aerial Port of Embarkation
ASD	Assistant Secretary of Defense
ATO	Air Tasking Order

C

C2	Command and Control
C3	Command, Control, and Communications
C3I	Command, Control, Communications, and Intelligence
C4	Command, Control, Communications, and Computers
C4I	Command, Control, Communications, Computers, and Intelligence
CA	Coordination Authority
CATF	Commander Amphibious Task Force
CC	Critical Capabilities
CCDR	Combatant Commander
CCMD	Combatant Command
CCO	Complex Contingency Operation
CCP	Combatant Command Campaign Plan
CCIR	Commander's Critical Information Requirements
CDR	Commander
CF	Critical Factors
CFLCC	Coalition Forces Land Component Command
CJA	Comprehensive Joint Assessment
CJCS	Chairman of the Joint Chiefs of Staff
CJTF	Commander Joint Task Force
CLF	Commander Landing Forces

CMO	Civil-Military Operations
CMOC	Civil-Military Operations Center
COA	Course of Action
COCOM	Combatant Command (Command Authority)
COG	Center of Gravity
COLISEUM	Community On-line Intelligence System for End Users Managers
COMINT	Communications Intelligence
CONOPS	Concept of Operations
CONPLAN	Concept Plan
COS	Chief of Staff
CPG	Contingency Planning Guidance
CR	Critical Requirements
CRF	Critical Response Force
CSA	Combat Support Agencies
CS	Civil Support, Combat Support
CSS	Combat Service Support
CV	Critical Vulnerabilities

D

D2D	Deployment-to-Dwell
DASD(P)	Deputy Assistant Secretary of Defense for Plans
DCM	Deputy Chief of Mission
DCJTF	Deputy Commander JTF
DDOC	Deployment and Distribution Operations Center
DEPORD	Deployment Order
DHS	Department of Homeland Security
DIA	Defense Intelligence Agency
DIME	Diplomatic, Informational, Military, Economic
DISA	Defense Information Systems Agency
DJ-3	Joint Staff Director of Operations
DJ-7	Joint Staff Director for Operational Plans and Joint Force Development
DLA	Defense Logistics Agency
DOD	Department of Defense
DODD	Department of Defense Directive
DODI	Department of Defense Instruction
DOS	Department of State

DOTMLPF-P	Doctrine, Organization, Training, Material, Leadership and Education, Personnel, Facilities, and Policy
DP	Decisive Point
DPO	Distribution Process Owner
DPG	Defense Planning Guidance
DRRS	Defense Readiness Reporting System
DR4A	Directed Readiness, Assignment, Allocation, Apportionment and Assessment
DSCA	Defense Support of Civil Authorities
DSR	Defense Strategy Review
DTA	Dynamic Threat Assessment

E

ECOA	Enemy Course of Action
EEFI	Essential Elements of Friendly Information
EEI	Essential Elements of Information
eJMAPS	Electronic Joint Manpower and Personnel System
EW	Electronic Warfare
EXORD	Execute Order

F

FCC	Functional Component Command
FDO	Flexible Deterrent Option
FFIR	Friendly Force Information Requirements
FRAGO	Fragmentation Order
FTN	Force Tracking Number
FYDP	Future Years Defense Program

G

GCC	Geographic Combatant Commander
GCSS	Global Combat Support System
GCP	Global Campaign Plan
GDP	Global Defense Posture
GEOINT	Geospatial Intelligence
GFM	Global Force Management
GFMAP	Global Force Management Allocation Plan
GFMB	Global Force Management Board
GFMIG	Global Force Management Implementation Guidance
GIBP	Globally Integrated Base Plan
GIF	Global Integrated Framework
GWOT	Global War on Terrorism

H

HA	Humanitarian Assistance
HN	Host Nation
HNG	Host-Nation Government
HNS	Host-Nation Support
HPT	High-Priority/Payoff Target(s)

I

IA	Interagency
IAW	In Accordance With
IGO	Intergovernmental Organization
ILO	In-Lieu-Of
IMO	Intermediate Military Objective
IRR	Individual Ready Reserve
IO	Information Operations
IPB	Intelligence Preparation of the Battlespace
IPG	Initial Planning Guidance
IPIE	Intelligence Preparation of the Information Environment
IPL	Integrated Priority List
IPR	In-Progress Review
IT	Information Technology
ITV	In-Transit visibility

J

J-1	Manpower and Personnel Directorate of a Joint Staff
J-2	Intelligence Directorate of a Joint Staff
J-3	Operations Directorate of a Joint Staff
J-4	Logistics Directorate of a Joint Staff
J-5	Plans Directorate of a Joint Staff
J-6	Command, Control, Communications, and Computer System
JAO	Joint Area of Operations
JAOC	Joint Air Operations Center
JATF	Joint Amphibious Task Force
JC2WC	Joint Command and Control Warfare Center
JCCA	Joint Combat Capability Assessment
JCRM	Joint Capabilities Requirements Manger
JCS	Joint Chiefs of Staff
JDDOC	Joint Deployment and Distribution Operations Center
JFAST	Joint Flow and Analysis System for Transportation
JFC	Joint Force Commander

JFLCC	Joint Force Land Component Commander
JFMCC	Joint Force Maritime Component Commander
JFP	Joint Force Provider
JFSOCC	Joint Force Special Operations Component Commander
JIACG	Joint Interagency Coordination Group
JIA	Joint Individual Augmentee
JIB	Joint Information Bureau
JIC	Joint Intelligence Center
JIOC	Joint Intelligence Operations Center
JIPOE	Joint Intelligence Preparation of the Operational Environment
JIPTL	Joint Integrated Prioritized Target List
JLOC	Joint Logistics Operations Center
JLOTS	Joint Logistics Over-the-Shore
JMC	Joint Movement Center
JMD	Joint Manning Document
JMET	Joint Mission Essential Task
JMETL	Joint Mission Essential Task List
JOA	Joint Operations Area
JOC	Joint Operations Center
JOPES	Joint Operation Planning and Execution System
JOWPD	Joint Operational War Plans Division
JP	Joint Pub
JPP	Joint Planning Process
JPEC	Joint Planning and Execution Community
JPG	Joint Planning Group
JRSOI	Joint Reception, Staging, Onward Movement, and Integration
JS	Joint Staff
JS JFC	Joint Staff Joint Force Coordinator
JSCP	Joint Strategic Campaign Plan
JSPS	Joint Strategic Planning System
JTB	Joint Transportation Board
JTF	Joint Task Force
JTF HQ	Joint Task Force Headquarters

L

LAD	Latest Arrival Date
LOC	Lines of Communications
LOE	Line of Effort
LOO	Line of Operation
LSA	Logistics Supportability Analysis

M

M2D	Mobilization-to-Dwell
MCOO	Modified Combined Operations Overlay
MDCOA	Most Dangerous Course of Action
METT-TC	Mission, Enemy, Terrain, Troops, Time Available and Civilian
MLCOA	Most Likely Course of Action
MILDEC	Military Deception
MOE	Measure of Effectiveness
MOP	Measure of Performance

N

NAI	Named Area of Interest
NDS	National Defense Strategy
NGB	National Guard Bureau
NGO	Nongovernmental Organization
NLT	Not Later Than
NMS	National Military Strategy
NSA	National Security Agency
NSC	National Security Council
NSC/DC	National Security Council/Deputies Committee
NSC/PC	National Security Council/Principals Committee
NSC/IPC	National Security Council/Interagency Policy Committee
NSCS	National Security Council System
NSD	National Security Directive
NSPD	National Security Presidential Directive
NSR	National Security Review
NSSD	National Security Study Directive
NSS	National Security Strategy

O

OA	Operational Area
OCJCS	Office of the Chairman of the Joint Chiefs of Staff
OE	Operating Environment
OEF	Operation Enduring Freedom
OGA	Other Governmental Agency
OIF	Operation Iraqi Freedom
OPCON	Operational Control
OPG	Operations Planning Group
OPLAN	Operation Plan
OPORD	Operation Order
OSD	Office of the Secretary of Defense
OUSD(P)	Office of the Under Secretary of Defense for Policy

Annexes

P

PC	Promote Cooperation
PD	Presidential Directive
PDD	Presidential Decision Directive
PIR	Priority Intelligence Requirements
PLANORD	Planning Order
PMESII	Political, military, economic, social, information, infrastructure
POD	Port of Debarkation
POE	Port of Embarkation
PPBE	Planning, Programing, Budgeting and Execution
PPD	Presidential Policy Document
PRD	Presidential Review Directive
PTDO	Prepare to Deploy Order

R

RATE	Refinement, Adaptation, Termination, or Execution
RC	Reserve Component
RCP	Regional Campaign Plan
RDD	Required Delivery Date (at destination)
REDEPORD	Re-deployment Order
RFF	Request for Forces
RFI	Request for Information
ROC	Rehearsal of Concept

S

SDOB	Secretary of Defense Operations Book
SECDEF	Secretary of Defense
SFP	Service Force Providers
SGS	Strategic Guidance Statements
SOP	Standing Operating Procedures
SPG	Strategic Planning Guidance
SPOD	Seaport of Debarkation
SPOE	Seaport of Embarkation

T

TDP	Theater Logistics and Distribution Plan
TIA	Theater Intelligence Assessment
TPFDD	Time-Phased Force and Deployment Data
TPFDL	Time-Phased Force and Deployment List
TPP	Theater Posture Plan, Tactics, Techniques and Procedures
TUCHA	Type Unit Characteristics File

U

UCP	Unified Command Plan
UJTL	Universal Joint Task List
ULN	Unit Line Number
USAFRICOM	United States African Command
USAID	United States Agency for International Development
USCENTCOM	United States Central Command
USCYBERCOM	United States Cyber Command
USD(P)	Under Secretary of Defense for Policy
USELEMNORAD	United States Element North American Aerospace Defense Command
USEUCOM	United States European Command
USNORTHCOM	United States Northern Command
USPACOM	United States Pacific Command
USSOCOM	United States Special Operations Command
USSOUTHCOM	United States Southern Command
USSF	United States Space Force
USSPACECOM	United States Space Command
USSTRATCOM	United States Strategic Command
USTRANSCOM	United States Transportation Command

V

VCJCS	Vice Chairman of the Joint Chiefs of Staff
VOCO	Voice Orders of the Commander

W

WARNORD	Warning Order
WMD	Weapons of Mass Destruction
WX	Weather

Y

YR	Year

Z

Z	Zulu
ZULU	Time Zone Indicator for Universal Time

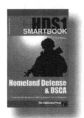

Purchase/Order

SMARTsavings on SMARTbooks! Save big when you order our titles together in a SMARTset bundle. It's the most popular & least expensive way to buy, and a great way to build your professional library. If you need a quote or have special requests, please contact us by one of the methods below!

View, download FREE samples and purchase online:
www.TheLightningPress.com

Order SECURE Online
Web: www.TheLightningPress.com
Email: SMARTbooks@TheLightningPress.com

24-hour Order & Customer Service Line
Place your order (or leave a voicemail)
at 1-800-997-8827

Phone Orders, Customer Service & Quotes
Live customer service and phone orders available
Mon - Fri 0900-1800 EST at (863) 409-8084

Mail, Check & Money Order
2227 Arrowhead Blvd., Lakeland, FL 33813

Government/Unit/Bulk Sales

The Lightning Press is a **service-disabled, veteran-owned small business**, DOD-approved vendor and federally registered—to include the SAM, WAWF, FBO, and FEDPAY.

We accept and process both **Government Purchase Cards** (GCPC/GPC) and **Purchase Orders** (PO/PR&Cs).

Keep your SMARTbook up-to-date with the latest doctrine! In addition to revisions, we publish incremental **"SMARTupdates"** when feasible to update changes in doctrine or new publications. These SMARTupdates are printed/produced in a format that allow the reader to insert the change pages into the original GBC-bound book by simply opening the comb-binding and replacing affected pages. Learn more and sign-up at: **www.thelightningpress.com/smartupdates/**

VISA MasterCard PayPal